Wind Pow

The twin th[...]scalating fossil
fuel prices [...]urces a global
imperative. [...]come ever more
important. Though no panacea, wind power has the potential to make a substantial
contribution to meeting electricity needs in many countries. This concise and accessible
account of the history and future development of wind power technology will appeal to a
wide audience of engineers, scientists, policymakers and students, and anyone interested
in learning more about the potential of wind power.

Wind power has been used increasingly to generate electricity since the early 1990s in
Europe, the United States and elsewhere, and the global installed wind power capacity
has been doubling every three to four years. It is now making the transition from being a
minor contributor to becoming an important source of electricity. However, the level of
understanding of what can be expected from wind systems in the future – and at what
cost – is generally poor. This book requires no prior technical knowledge, and describes
the development of wind power technology from medieval times to the present. It looks
forward to the important role that wind power is expected to play – as a clean,
competitive and abundant energy source – in helping to meet our future energy needs.

- Considers the factors underpinning wind power's exponential growth since 1990, to
  help understand what we can expect from wind power in the future.
- Shows how the cost of wind energy compares with electricity produced by burning
  fossil fuels.
- Explains simply how wind turbines (and windmills) work.
- Explains why wind power's intermittency does not prevent its large-scale use.

PETER MUSGROVE's wind power interests go back to 1974 when he was teaching
engineering at the University of Reading. Though wind power was then the UK's least
favoured renewable energy option he strongly argued the case for large offshore wind
farms and gained recognition for their potential. He was also very much involved in the
UK vertical axis wind turbine programme through the 1970s and 1980s. He was one of
the prime movers in the 1978 formation of the British Wind Energy Association (BWEA)
and was its first Chairman. He left academia for industry in 1988 and as National Wind
Power's Head of Development through the 1990s was involved in all aspects of wind
farm development and wind turbine assessment. He retired in 2003 and was given the
BWEA's first Lifetime Achievement Award.

Wind Power

PETER MUSGROVE

CAMBRIDGE
UNIVERSITY PRESS

CAMBRIDGE UNIVERSITY PRESS

Cambridge, New York, Melbourne, Madrid, Cape Town, Singapore, São Paulo, Delhi,
Dubai, Tokyo

Cambridge University Press
The Edinburgh Building, Cambridge CB2 8RU, UK

Published in the United States of America by Cambridge University Press, New York

www.cambridge.org
Information on this title: www.cambridge.org/9780521762380

First published 2010

Printed in the United Kingdom at the University Press, Cambridge

A catalogue record for this publication is available from the British Library

ISBN 978-0-521-76238-0 Hardback
ISBN 978-0-521-74763-9 Paperback

For my grandchildren
James, Terry, Faith, Jonathan and Euan
and their generation

Contents

Preface

The high standard of living that we in developed countries enjoy is underpinned by the massive consumption of oil, gas and coal, the so-called fossil fuels. However we now recognise that the carbon dioxide emissions caused by the combustion of these fuels is changing the global climate, with potentially devastating consequences for us all. Electricity generation is the largest single source of carbon dioxide emissions, and the need to move to the production of electricity in ways that avoid these emissions has stimulated an upsurge of interest in renewable sources of energy.

Wind power is the renewables technology that has made most progress in recent years. And though modern wind turbines have much in common with traditional windmills they are very much larger and much more efficient. Since 1973, when the first oil crisis encouraged a number of countries to commence their development, wind turbines have progressed to the point where they can generate electricity at a cost that is comparable with that produced by burning fossil fuels, though without the carbon dioxide emissions and price volatility. In 2008 the global installed wind power capacity passed 100 000 MW and wind power provided just over 1% of global electricity; by 2020 the installed capacity is likely to exceed 1 million MW, with wind power providing about 10% of global electricity, a percentage that is expected to grow steadily in the following decades.

This book tells the story of the development of wind power technology from its medieval beginnings through many centuries of success to its nineteenth-century decline and its late-twentieth-century revival, and looks forward to the important role that wind power seems likely to play – as a clean, competitive, reliable and abundant energy source – in helping to meet twenty-first-century energy needs. The core text of Chapters 1 to 9 requires no prior technical knowledge – just a reasonably numerate interest in how we can meet our future energy needs without causing further damage to the climate. The appendices and extensive notes give more detail for those who require it.

Acknowledgements

My interest in wind power began in 1974 when I was a Lecturer in the Engineering Department at the University of Reading and I am grateful to all my colleagues there, and in particular to Peter Dunn who was then Head of the Department, for their encouragement and support.

In 1988 I left academia to go into industry and I would like to thank David Lindley, then Managing Director of the Wind Energy Group, for inviting me to join him at WEG and for his support in facilitating my transition. I moved as Head of Development to National Wind Power (NWP), a related wind farm development company, when it was formed in 1991 and remained there until my retirement in 2003. I am very happy to acknowledge the debt that I owe to my many former NWP colleagues for all that I learnt from them, and though I cannot name them all – the list would be too long – I particularly wish to express my thanks to Alan Moore, Pam McEvoy and John Warren for their friendship, encouragement and support.

I have learnt from, and enjoyed the company of, many who have contributed to the global development of wind power. Again space does not allow me to name them all but I would particularly like to thank Henrik Stiesdal for his friendship over many years, and for the many technical discussions which greatly improved my understanding of wind turbine technology; I am happy also to take this opportunity to express my appreciation to him for reading my draft of Chapters 4, 5 and 6, and for making many constructive suggestions.

Last, but certainly not least, I wish to express my thanks and gratitude to my wife Sonja for her patience and support over many years, and in particular through the past four years when 'the book' has taken priority over a host of other potential retirement activities.

1

Wind power and our energy needs:
an overview

Traditional windmills

Though their origins are uncertain traditional windmills first came into use in the late twelfth century in the south and east of England, in Flanders and in northern France. They subsequently spread quite rapidly into and beyond neighbouring countries, and by the early fourteenth century several thousand had been built in England alone. With a diameter, from sail-tip to sail-tip, of about 13 m they would give on average an output equivalent to the work that could be done by 15 to 20 people, and they were particularly well suited for the laborious but necessary task of grinding grain. Pitstone mill in Buckinghamshire, shown in Figure 1.1, dates from 1627 and is the oldest surviving windmill in England; apart from being slightly larger – it has a diameter of 17 m – it is typical of early windmill designs, and when operating the four sails would have canvas stretched over the wooden lattice framework.

Though watermills had been used since Roman times the locations where they could be sited, by rivers or by fast flowing streams, were limited. Windmills, by contrast, could be sited almost anywhere. Their design evolved and their use proliferated and for 700 years they had an important role in the rural economy, used primarily for grinding grain though other applications such as land drainage became important in some areas. Over the centuries they steadily became larger, more efficient and more complex until by the nineteenth century they were typically about 25 to 30 m diameter with an average output – depending on size and location – equivalent to the work that could be done by 100 to 200 people.

Though windmills were very successful, and were built in their thousands, the introduction of the steam engine by Thomas Savery in 1698 marked the beginning of the end for them. Early steam engines were costly and inefficient, and consumed so much coal that they were mostly used on coal fields for pumping water out of coal mines, or for high-value applications such as pumping water out of Cornish

Figure 1.1 Pitstone windmill, 1627, Buckinghamshire, England.

tin and copper mines; consequently their use at first grew only slowly. However as the eighteenth century progressed steam engines became more efficient,[1] thanks to the inventions of men such as Thomas Newcomen and James Watt, and by the middle of the nineteenth century they were in widespread use, powering Britain's Industrial Revolution. Steam-powered mills could produce flour that was both whiter and cheaper, and the use of windmills steadily declined.

Power and energy in the modern world

The late nineteenth century saw the development of relatively light weight and low-cost internal combustion engines, burning first coal gas and later liquid fuels such as petrol, and paving the way for the motor car and the aeroplane. It also saw the development of the first systems for generating and distributing electricity. Initially these were small scale and localised, and electricity was expensive. However electricity is such a convenient and versatile means for transporting power that electricity generation and distribution systems grew rapidly through the twentieth century. We in the developed world now have the convenience of power available to us on demand, in our homes, our offices and our factories, at a price which by any historic standard is extremely low. A man or woman working physically hard all day long can deliver an energy output equivalent to that required to light a 60 watt bulb for about 8 hours; the same amount of energy would today cost a domestic user of electricity about 7 pence, and an industrial user even less.

We measure the power we use in watts, named after James Watt whose invention of the separate condenser for steam engines in 1769 greatly improved its efficiency and economic viability. For periods of just a few minutes a fit adult can produce a power output of about 300 watts,[2] which is usually abbreviated as 300 W; however for prolonged periods – such as many hours per day for many days in succession – an average output of 60 W is more realistic.[3] For commercial applications the watt is too small for convenience and we use either the kilowatt (kW), which is equal to one thousand watts, or the megawatt (MW), which is equal to one million watts. Energy is calculated as the power we use multiplied by the time we use it; using a kilowatt of power continuously for one hour therefore requires one kilowatt-hour (abbreviated kWh) of energy, as does using a hundred watts for ten hours. The kilowatt-hour is the unit of electricity recorded by the meters in our homes, and an average UK home consumes about 4500 kWh of electricity per year.[4] Since there are 8760 hours in a year the *average* power consumption in a typical UK home is therefore about 510 W, which is equivalent to the power that could be produced by 25 domestic servants each working 8 hours per day!

The electricity we use in our homes amounts to about one-third of the total UK electricity consumption; the other two-thirds is used on our behalf in shops and offices and by industry. Most is generated by burning gas, coal and oil but about 15% comes from nuclear power stations. And though our overall electricity consumption is substantial, the energy used to provide electricity represents only about one-third of the total UK energy consumption; the other two-thirds is used to provide heat in our homes and offices and factories, to power our cars and trucks and trains and planes, and for a wide variety of other tasks. This energy too is provided by the combustion of gas, coal and oil, the so-called fossil fuels.

Global warming and climate change

Thanks to the use we have made of these fossil fuels, which until recently have been low cost and abundant, the standard of living in the developed world has been transformed in the past two centuries. But as a result of our greatly increased energy consumption we have in recent decades been consuming these fossil fuels nearly one million times faster than the rate at which they were formed, beneath the Earth's surface, more than a hundred million years ago: and in burning these fuels we have poured countless million tonnes of the greenhouse gas carbon dioxide into the atmosphere, together with large quantities of other pollutants. As a result we have initiated a period of rapid global warming and climate change; and as the developing world seeks to follow our example by also burning increasing volumes of fossil fuels, so that they too can transform their standard of living, the pace of global warming and climate change is expected to accelerate. Though a small but vociferous minority are still reluctant to accept that global warming is happening, or that it is caused by greenhouse gas emissions, the overwhelming consensus amongst climate scientists is that the phenomenon is real and needs urgent action. As Sir Nicholas Stern noted in 2007, in the summary of his comprehensive report on *The Economics of Climate Change*, 'The scientific evidence is now overwhelming: climate change is a serious global threat, and it demands an urgent global response.'[5]

From the end of the last glaciation, about 10,000 years ago, up to the year 1750 the concentration of carbon dioxide in the atmosphere was approximately constant at about 275 parts per million (ppm); since then, as countries have industrialised, the concentration has risen to about 380 ppm and is now rising at about 5% per decade.[6] Worldwide the 10 warmest years since records began in 1850 have all occurred since 1995, and Figure 1.2 shows the warming trend of

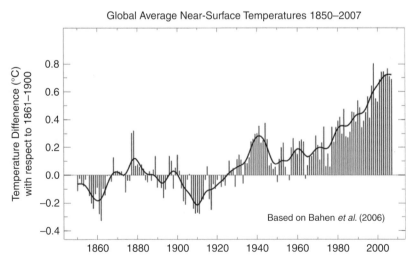

Figure 1.2 Global average temperatures since 1850. *Source*: Image © Crown Copyright 2007, the Met Office.

recent decades very clearly.[7] As can be seen global temperatures have risen by about 0.7 °C over the past century and, as the Stern Review notes, on current trends global average temperatures could rise by two to three degrees within the next fifty years or so.[8]

The consequences of this rapid warming cannot be predicted with certainty but there is the real risk of 'damaging and potentially irreversible impacts on ecosystems, societies and economies'. Changed patterns of rainfall and increased sea levels are likely to affect developing countries most severely, and if we fail to reduce our emissions of carbon dioxide and the other greenhouse gases millions of people are likely to become environmental refugees, with widespread suffering, economic disruption, and consequent social and political instability. As the Royal Commission on Environmental Pollution (RCEP) stated in 2000 in its comprehensive report on energy use, global warming and climate change 'There is a moral imperative to act now. If this generation took no measures to curb rising emissions, it would be condemning our children, grandchildren and generations beyond them to considerable dangers. In the light of where the harshest impacts are likely to fall, this would perpetrate an enormous global injustice.'[9] In order to limit the future damaging effects of global warming the RCEP recommended that action be taken to limit increases in carbon dioxide levels to a maximum of 550 ppm, that is to about double the pre-industrial level, with this ceiling reached by 2050; and for the UK this would require a challenging 60% reduction in carbon dioxide emissions by 2050. This target for reducing UK emissions was accepted by the Government in its 2003 Energy White Paper, and was restated more precisely in the 2007 Energy White Paper as a commitment to reduce carbon dioxide emissions by at least 60% by 2050 (against a 1990 baseline). The Climate Change Act of 2008 then broadened the commitment so that it covered all greenhouse gas emissions, and increased the commitment to an 80% reduction by 2050.[10]

Some of those who have opposed action to curb global warming have stated that the cost of reducing greenhouse gas emissions is unaffordably high. However when the Government first committed to large cuts in greenhouse gas emissions it also endorsed a review undertaken by the Intergovernmental Panel on Climate Change (IPCC) which concluded that stabilising carbon dioxide emissions at no more than 550 ppm would lead to an average GDP loss for developed countries of around 1% in 2050; and the Government pointed out that the nation's wealth, as measured by GDP, would by then be three times larger than at present.[11] The subsequent and comprehensive Stern Review on the economics of climate change estimated that the cost of action to reduce greenhouse gas emissions and avoid the worst impacts of climate change would amount to around 1% of global GDP per year; by contrast the cost of inaction would be many times greater, 'equivalent to losing at least 5% of global GDP each year, now and forever' and with the risk that the cost of inaction might rise to 20% or more of GDP.[12] The Stern Review consequently concluded that 'prompt and strong action is clearly warranted'.

Since UK emissions of carbon dioxide account for only 2% of the global total its actions to reduce emissions will have no material effect on climate change unless they are part of a concerted international effort. However this does not justify inaction and the UK Government has stated that it will seek to persuade other countries to follow its lead. As the RCEP comments 'If the UK does not show that it is serious about doing its part, it cannot expect other nations – least of all those which are much poorer – to do theirs.'[13] Given that the UK led the world into the Industrial Revolution it would seem particularly appropriate that it should now seek to lead the way in demonstrating that the enhanced standard of living which we now enjoy, thanks to our increased energy consumption, is compatible with much reduced emissions of carbon dioxide.

Renewable energy sources

Reducing carbon dioxide emissions by 80% by 2050 is a challenging target, but one that the UK Government believes can be achieved. Increased energy efficiency has the potential to significantly reduce greenhouse gas emissions by reducing the amount of energy we need for a given task, but we will also need to greatly increase the proportion of energy provided by renewable energy sources. Given the present high levels of energy consumption in the UK, as in other developed countries, there are those who doubt that renewables can make a significant contribution. But though the total worldwide rate of energy consumption is high – and averages 12 billion kW – it is dwarfed by the flow of energy that the Earth receives from the Sun, which is more than 7000 times larger and averages 87 000 billion kW.[14]

However although the total incoming flow of solar energy is so very large the power density of direct solar radiation is relatively low; in bright sunlight it amounts to just about one thousand watts per square metre of surface (1000 W/m^2). And at UK latitudes the annual *average* solar energy received is about one tenth of this, that is 100 W/m^2 (though closer to the equator the average rises to about 250 W/m^2). A consequence of these relatively low power densities is that large areas must be covered with power conversion devices in order to give significant electrical output power levels. The most promising direct solar energy conversion technology is photovoltaics, abbreviated PV, and PV arrays first came to prominence with their use to provide power for satellites in the 1960s. Much progress has been made with this technology since then, and the economics have improved to the point where PV systems are now cost effective on land for low-power applications where the cost of connection to an electricity distribution network is high, such as lighting in remote locations. However electricity from PV systems is still much too expensive for large scale grid-connected applications; costs are typically[15] in the range 40 to 70 p/kWh and need to be reduced by at least a factor of five if this technology is to make a substantial contribution to our future energy needs.

Wind energy conversion

The renewables technology that has made the greatest progress in reducing costs in recent decades has undoubtedly been wind energy conversion, using wind turbines such as those shown in Figure 1.3 which are the modern successors to traditional windmills.[16] The turbines in the figure are part of the thirty turbine North Hoyle wind farm, Britain's first offshore wind farm which is located 8 km

Figure 1.3 North Hoyle wind farm, 2003, off the North Wales coast. *Source*: Image courtesy of Vestas.

off the North Wales coast and was completed in 2003.[17] The fibreglass blades are each just under 40 m long, giving a rotor diameter of 80 m, and each turbine has a power rating of 2 MW (= 2000 kW). As a measure of the progress that has been made in recent years with wind turbine technology one may note that it would take about *one thousand* traditional windmills like the one shown in Figure 1.1 to provide the same annual energy output as just *one* of the North Hoyle wind turbines; and each of the North Hoyle wind turbines gives an annual energy output equal to the annual electricity consumption of about 1500 homes.

Wind energy is an indirect form of solar energy, and winds result from the fact that the Earth's equatorial regions receive more solar energy than the polar regions. The resulting temperature differences cause large-scale convection currents in the atmosphere, and this air in motion is what we perceive as winds. The detailed processes are complex,[18] as is evident from the weather maps that can be seen daily in newspapers and on television. The overall result, however, is that about 2% of the incoming solar energy gets converted to wind energy.[19] Though this percentage is small it still means that the rate at which solar energy is converted in the atmosphere to wind energy is more than one hundred times greater than the total worldwide power consumption.

Energy payback and wind energy costs

Average wind speeds vary significantly from place to place but at on-land locations with good wind exposure, such as those that are chosen for wind farms, the average wind speed at the hub height of a modern wind turbine is typically in the range 7 to 8 m/s; this corresponds to an average power density in the wind of about 500 W/m^2.[20] Offshore wind speeds are higher and typically average about 9 m/s, which gives an average power density in the wind of about 860 W/m^2. The blades of a modern wind turbine efficiently intercept the air flowing through the whole area swept by the blades, even though the blades occupy only about four or five per cent of this area.[21] Power densities relative to the blade area are therefore typically in the range 10 to 20 kW/m^2 and these relatively high power densities, coupled with the engineering progress made in wind turbine technology over the past 30 years, lead to energy payback periods of just a few months; *in other words all the energy used in a wind turbine's construction, installation, operation and eventual decommissioning is repaid by its energy output during its first few months operation.*[22]

Short energy payback periods go hand-in-hand with favourable economics and the cost of electricity from on-land wind farms is typically in the range 3 to 6 p/kWh, with electricity from offshore wind farms costing typically in the range 6 to 10 p/kWh.[23] The UK has the best on-land wind resource in Europe and – surrounded as it is by shallow seas – it also has by far the best offshore wind resource.[24] The total resource is many times larger than the total UK electricity consumption and this high resource, coupled with costs that are lower than for any other high-resource renewable energy option,[25] has led to wind power being seen as the technology that will make the largest contribution to meeting UK renewable energy targets up to 2020, and probably well beyond.[26]

Intermittency issues

Though detractors, especially those who prefer other generating options, often declare that wind power's intermittency precludes its use on a large scale – because of the risk that the lights would go out during lengthy periods of calm – this assertion is quite wrong. Wind power's primary role is to reduce our use of fossil fuels, and the extent of our dependence on fossil fuel imports, whilst at the same time delivering substantial reductions in greenhouse gas emissions. UK system integration studies have repeatedly shown that wind power's inter-mittency (or – more accurately – its variability) only reduces fuel savings by about 5% when compared with the savings that correspond to a completely predictable power output.[27] And maintaining system reliability, including having enough capacity provided by low capital cost gas-fired power stations to cover those occasions when demand is high and wind speeds are low, typically adds less than a penny per kilowatt-hour to the cost of wind powered generation.[28]

Wind or nuclear?

Comparisons are inevitably made between wind and nuclear power generation costs, but with conflicting results. The Cabinet Office Strategy Unit in its 2002 *Energy Review* concluded that by 2020 onshore wind and offshore wind would both provide electricity at lower cost than nuclear power stations,[29] but the Department of Trade and Industry in its 2006 *Energy Review* concluded that nuclear was cheaper.[30] As the House of Commons Environmental Audit Committee pointed out – in its 2006 report[31] on nuclear, renewables and climate change – much depends on the underlying assumptions, not least in relation to the perception of the financial risk associated with nuclear power investments. What is clear is that the generation cost from Sizewell B, the last nuclear power station to be built in Britain, was in excess of 6 p/kWh (in the early 1990s) and the forecasts of much lower costs in future are based on the use of third generation nuclear reactor designs, none of which has yet been built; and there are continuing concerns relating to safety, radioactive waste disposal, decommissioning and nuclear proliferation.[32] However both nuclear power stations and wind farms have the potential to make substantial contributions to UK electricity needs with near zero greenhouse gas emissions and the UK Government has decided to encourage and support them both.

Costs compared with conventional generation

Although the cost of wind-generated electricity is at present somewhat higher than the cost of generation from fossil fuels the magnitude of the differential clearly depends on the price of fossil fuels, for which the key marker is the price of crude oil in US dollars per barrel. Through the post-war years and up to 1973 this, when adjusted for inflation, had been fairly stable at about $13/barrel (at 2007 price levels).[33] However the oil crisis in late 1973, which prompted the revival of interest in renewable sources of energy, was followed by 12 years during which oil prices (again at 2007 price levels) averaged about $62/barrel. This period of high oil prices encouraged the evolution of the modern wind turbine, as is described in Chapter 5, but oil prices slumped in 1986 and were to stay low, averaging just $28/barrel at 2007 price levels, for the next 17 years. These lower oil prices led initially to reduced interest in the use of renewable energy sources, however by the early 1990s concerns were mounting that the carbon dioxide emissions from power stations were a principal cause of global warming and climate change; and these concerns were sufficient to persuade several countries to implement programmes to increase their use of renewable energy sources, not least wind energy.

Many energy experts expected oil prices to continue at their relatively low post-1986 levels well into the twenty-first century. The Department of Trade and Industry, for example, published its *Energy Projections for the UK* in 2000 with a high price scenario which assumed that oil prices out to 2020 would be no more than $20/barrel; the corresponding low price scenario was that oil prices out to 2020 would be just $10/barrel.[34] However growing demand, especially from fast-developing countries such as China and India, combined with only limited increases in oil production led to rising oil prices after 2003. These reached a peak of nearly $150/barrel in mid-2008, though the global financial crisis and recession that developed through 2008 led to the price falling to below $50/barrel by the end of the year.

No one knows how the price of oil will vary over the next decade and beyond, but there is the expectation that worldwide demand for energy will continue to grow, as developing countries continue to raise their living standards, and that oil production will grow less rapidly. The underlying trend for the price of oil is therefore almost certainly upwards, but as has been seen in the past a relatively small mismatch between supply and demand can cause a large change in the oil price and we can therefore expect to see considerable price volatility superimposed on this rising trend. The UK Government's 2008 central forecast is that the 2020 oil price will be $70/barrel, and the United States Energy Information Administration's forecast for 2020 is similar.[35] Gas prices are linked to oil prices and with oil at $70/barrel the cost of electricity from the gas-fired power stations that are Britain's main source of electricity would be about 7 p/kWh.[36] This is higher than the cost of electricity from on-land wind farms, and comparable with the cost of power from offshore wind farms, even before wind power is given any financial credit for the fact that it does not produce carbon dioxide emissions.

Though coal is abundant, and was in the past the dominant fuel for electricity generation, its use has declined in recent years and this decline is expected to continue. The greenhouse gas emissions from coal-fired power stations are about twice as high (per unit of electricity generated) as the emissions from gas-fired power stations. The increased use of coal would therefore aggravate global warming problems, unless the coal combustion is accompanied by carbon capture and storage. Though this has been much discussed in recent years no coal-fired power station equipped with carbon capture and storage facilities has yet been built, and the practicality and costs are very uncertain.

Wind power's growth and potential

The world faces a climate change crisis as a result of the global warming caused by our extravagant use of fossil fuels. If we are to sustain the high energy-consuming lifestyles to which we in the developed world have grown accustomed, and to which the developing world aspires, we need to move away from our dependence on fossil fuels as speedily as we can, and make the transition to meeting as high a proportion as possible of our energy needs from renewable sources. Wind power, thanks to the remarkable engineering progress made in the past thirty years, is now recognised as the most promising and cost-effective of the high-resource renewable energy options and wind energy systems are now being deployed worldwide at a rapidly growing pace.

The year ending December 2007 saw nearly 20 000 MW of wind power capacity installed – equivalent to ten thousand turbines like those shown in Figure 1.3 – and the total worldwide installed capacity to the end of 2007 was 94 000 MW.[37] Of this total 57 000 MW was installed in Europe with more than 22 000 MW in Germany and more than 15 000 MW in Spain. And though Denmark's small size limited the installed capacity there to just over 3000 MW this was sufficient to provide 20% of Denmark's electricity. By contrast the installed wind power capacity in the UK was a very modest 2400 MW, sufficient to provide just $1\frac{1}{2}$% of the electricity consumption and disappointingly small for a country that has the best wind resource in Europe. And though the United States had 17 000 MW installed by the end of 2007, which supplied just over 1% of its electricity consumption, this too was disappointingly small for a country that has a good wind resource and is more than twenty times larger than Germany.

Provisional data for 2008 indicates that it was another record year with 27 000 MW installed, taking the global total wind power capacity at the end of 2008 to 121 000 MW.[38] In the United States the capacity installed in the year exceeded 8000 MW, and the total US capacity increased to just over 25 000 MW. This puts it ahead of Germany, where a 1700 MW increase in 2008 gave it a year-end total of just under 24 000 MW. In China the installed capacity more than doubled, for the third consecutive year, and the installation of 6300 MW took its wind power capacity to 12 200 MW, a total which is exceeded only by the US, Germany and Spain.

Wind energy has the potential to make very much larger contributions to electricity needs in the UK, the US and in many other countries at a price which is comparable with the price of electricity from fossil fuels; and utilising this potential would help to provide energy price stability as well as reducing dependence on fuel imports and giving significant reductions in greenhouse gas emissions. In Europe, as part of a strategy for achieving substantial reductions in greenhouse gas emissions, the EU Member States agreed in spring 2007 that by 2020 renewable sources of energy should provide 20% of the *total* EU energy consumption, which includes the energy used for heating and transportation. The European Commission subsequently proposed that the UK should commit to obtaining 15% of the its total energy consumption from renewables by 2020, the lower than average target for the UK reflecting the fact that in 2007 renewables supplied less than 2% of its energy, well below the EU average.[39] The UK will consequently need to greatly increase its use of renewable energy sources, and the Government has indicated that much of this increase will have to be provided by electricity generating technologies. Initial estimates suggest that to meet the 2020 target renewables would need to provide somewhere between 30% and 35% of UK electricity needs, with approximately two-thirds provided by wind energy. This would require a total installed wind capacity by 2020 of just under 30 000 MW, with offshore wind farms contributing about half.[40] A subsequent assessment by the Carbon Trust concluded that by 2020 wind power alone could provide over 30% of UK electricity needs from a total installed wind power capacity of 40 000 MW, most of which would be offshore.[41] And in addition to the benefits already stated, of reduced dependence on gas imports and reduced greenhouse gas emissions, the Carbon Trust noted that the programme they proposed would create many tens of thousands of jobs.

In the United States a detailed assessment published by the Department of Energy in mid 2008 noted that the on-land economic wind resource was 'sufficient to supply the electrical needs of the country several times over', and indicated that 300 000 MW of wind power capacity could be installed by 2030.[42] This would be sufficient to provide 20% of the electricity consumed in the US and would reduce emissions of the greenhouse gas carbon dioxide by about 820 million tonnes per year, which is equal to 14% of the present US emissions. It would also reduce by more than half the number of new coal-fired power stations that would need to be built by 2030 and it would reduce – by an estimated 60% – the volume of liquefied natural gas (LNG) that would need to be imported. Water supply is a growing problem in many parts of the US, and the power industry is a major water consumer; the installation of 300 000 MW of wind power capacity would reduce consumption by about 450 billion gallons per year, which is equal to about 17% of the power industry's forecast 2030 consumption. The 20% wind programme would also be a major source of employment; it is calculated that by 2030 it would create more than 170 000 direct jobs, and if indirect and induced jobs are included the total would be over 500 000. Though there has been no commitment so far to proceed with the 20% wind energy by 2030 programme that has been outlined, the new Administration of President Barack Obama has signalled very clearly that it intends to progress the development and deployment of renewable sources of energy, such as wind energy, much more vigorously than in the past.

In 2008 wind power supplied just over 1% of global electricity consumption, up from barely 0.1% a decade earlier. Its rapid growth is continuing and it seems probable that by 2020 wind power will supply at least 10% of global electricity needs. The ultimate practicable limit beyond 2020 is several times larger. Wind power has come a long way from the corn-grinding traditional windmills that first flourished in southern and eastern England and northern France in the late twelfth century. The pages that follow tell the fascinating story of the development of wind power technology from its medieval beginnings through many centuries of success to its nineteenth-century decline and its late-twentieth-century revival, and look forward to the important role that wind power seems likely to play – as a clean, competitive and abundant energy source – in helping to meet twenty-first-century energy needs.

2

The first windmills

Hero's toy windmill

Though watermills have been used since Roman times, both for grinding grain and pumping water,[1] there is no evidence that they used windmills. There is just one intriguing reference to the use of wind power in the writings of Hero of Alexandria, dating from the first century AD. He describes a wind-driven device for operating a toy organ which has an axle with two discs on it; one of these is fitted with vanes and the other has radial pegs, and the axle can be turned to face into the wind. As Landels points out the word translated as 'vanes' is used elsewhere to mean oar-blades, which suggests they were wooden and rigid.[2] The action of the wind on the vanes turns the axle, and the radial pegs on the second disc then act on a lever to raise a piston in a cylinder; then when each peg moves past the end of the lever the piston falls under its own weight, pumping air into the organ. Figure 2.1, from Schmidt's 1899 translation of Hero's Pneumatica,[3] illustrates how this device *may* have looked. (The wind direction is *from* the far side of the disc with vanes, which turns anticlockwise.) There is no reason to doubt that Hero's windmill worked, but it is clearly a toy and no-one seemed to recognise that a scaled-up version could provide a useful power output.

Early Persian drag-driven windmills

Though the windmills that we now think of as traditional windmills, such as the one shown in Figure 1.1, first appeared in England, Flanders[4] and northern France in the late twelfth century, windmills of a very different design were used in eastern Iran at least two centuries earlier. Needham has researched the origin of these early Persian windmills, and has found clear contemporary written

15

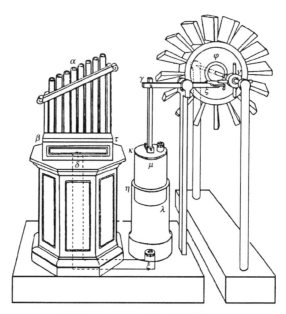

Figure 2.1 Hero's windmill toy. *Source*: Schmidt (1899).

evidence for their use from at least the tenth century in Seistan, a region in eastern Iran which is noted for its regular and very strong winds.[5] Needham also notes the story, first recorded by Ali al-Tabari in the late ninth century, that the Caliph Omar was murdered in Medina in 644 by a captured Persian technician who claimed to be able to build mills driven by the wind, and who was bitter about the high taxes he had to pay; this suggests that windmills were being used in Persia as early as the seventh century.

The earliest detailed description of a Persian windmill dates back to about 1300, and is provided by al-Dimashqi, a Syrian cosmographer.[6] His description and drawing make it clear that the Persian windmills he observed had sails attached radially to a vertical axle; in other words the sails turned around a vertical axis, (unlike a traditional English windmill, such as that shown in Figure 1.1, where the sails turn around an axis that is horizontal). Vertical-axis windmills have continued in use in eastern Iran through to modern times, and Figure 2.2 shows a typical example; Wulff noted a line of 50 such windmills still operating in the town of Neh when he was there in 1963.

Note a potential source of confusion; books and other publications that are primarily concerned with traditional windmills refer to windmills whose sails

Figure 2.2 Early Persian drag-driven windmill. *Source*: Wulff, Hans E., *Traditional Crafts of Persia*, figure 403, © 1967 Massachusetts Institute of Technology, by permission of The MIT Press.

rotate about a *horizontal* axis, such as the one shown in Figure 1.1, as *vertical* windmills, (because the sails turn in a plane that is vertical). Such publications then refer to windmills whose sails rotate about a *vertical* axis, like the one shown in Figure 2.2, as *horizontal* windmills. By contrast publications describing modern wind turbines categorise them according to their axis of rotation, and this convention has the advantage that the description is then unambiguous. For clarity this book describes all windmills and wind turbines, traditional or modern, by their axis of rotation.

The method of operation of the Persian vertical-axis windmills is shown in Figure 2.3. The massive structure which supports the rotating sails is shaped so that the wind is funnelled to one side of the rotor. The drag on the sails exposed to the wind then forces them downwind, so that the rotor shown in the figure turns anti-clockwise. However the support structure is shaped so as to shield the sails on the other side of the rotor from the force of the wind. The rotor therefore turns continuously in the wind, and the vertical shaft turns a millstone located one floor below.[7] The drag on the exposed sails results in considerable

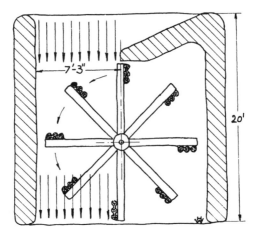

Figure 2.3 Plan view of early Persian drag-driven windmill. *Source*: Wulff, Hans E., *Traditional Crafts of Persia*, figure 402, © 1967 Massachusetts Institute of Technology, by permission of The MIT Press.

turbulence and loss of energy and drag-driven windmills have only a low efficiency. Bennett and Elton, based on their knowledge of several drag-driven vertical-axis windmills that had been built and operated in England in the eighteenth and nineteenth centuries, state that such windmills give only about one-fifth the power output of 'ordinary' windmills,[8] that is to say traditional windmills with sails turning around a horizontal axis; Persian vertical-axis windmills, like the one shown in Figure 2.3, were even less efficient.[9] Moreover the massive fixed structure of the Persian design of vertical-axis windmill meant that it was only suitable for use in locations where the wind comes from one predominant direction. In Seistan the winds are strong enough to compensate for the low efficiency, and they do blow consistently from one direction, the north. In the words of Colonel Kennion, who was British Consul there in the 1920s: 'The wind of Seistan! No one that has visited the country can ever think of it without the consciousness that there the word "wind" acquired for him a new significance … There is … the wind of a hundred and twenty days, that howls through the land during the summer; the freezing but shorter lived winds which rage in the winter; all from the same point of the compass … Everywhere in Seistan are evidences of the wind – the wind and man's struggle against it. Hollows scooped out of the earth, clay buffs carved into fantastic shapes, dunes with horns facing southwards, trees stunted and with a permanent list in the same direction, every building, the ruins even of buildings of the most ancient times, oriented southward.'[10]

Chinese drag-driven windmills

Winds that are consistently strong and unidirectional are very uncommon, so the Persian design of vertical-axis windmill was of little use elsewhere. However the Chinese have for centuries used a design that is in essence very similar, though cleverly adapted so that it works equally well whatever the wind direction.[11] A typical Chinese windmill, with sails similar to those on a sailing junk, is shown operating in Figure 2.4. On the left-hand side each sail is held by a rope so that it is kept broadside on to the wind (as in a sailing dinghy that is running downwind) and so is blown downwind, i.e. *towards* the camera. Then on the right-hand side, as each sail moves into the wind (and *away* from the camera) the rope goes slack and the sail turns to become roughly parallel to the wind. The drag on the sails going downwind, when they are held roughly square to the wind, is much larger than the drag on the sails going upwind, so the windmill is able to deliver a useful power output. As with all drag-driven windmills the efficiency is low, but is probably superior to the efficiency of the Persian windmill. Moreover the Chinese design has the great advantage that it is omni-directional, and the rope-supported structure is also relatively light weight and low cost. Windmills in China – as elsewhere – were widely used for grinding grain, but pumping water was another important application, and the windmill shown in the figure was being used to pump brine into shallow ponds, where it would be evaporated to make salt.

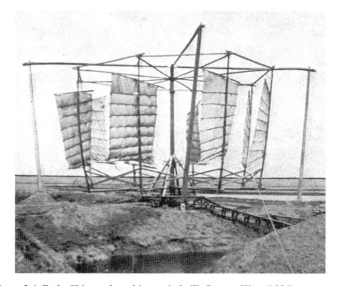

Figure 2.4 Early Chinese drag-driven windmill. *Source*: King (1926).

Though the Chinese have on occasion been credited with inventing the windmill, and with using it for over two thousand years, there is no evidence to support this. Needham, in his classic work on *Science and Civilisation in China*, indicates that the earliest significant Chinese reference dates from 1219, when a Chinese nobleman who had visited Samarkand (in what is today Uzbekistan) reported that he had seen windmills being used there to mill grain.[12] According to Needham later references all suggest that the concept travelled to China from lands to the west, rather than vice versa. The credit for inventing the earliest useful windmills would therefore seem to belong to the people living in the Seistan region of eastern Persia, which is consistent with the exceptional windiness of this region.[13]

The European windmill

As noted previously windmills made their first appearance in Europe in the late twelfth century and the earliest references point to their first use in the south and east of England, in Flanders and in northern France. Their use spread rapidly and illustrations in early manuscripts show that these earliest European windmills were post mills similar to that shown in Figure 1.1 and so-named because the whole body of the windmill is supported by – and can turn around – a massive vertical wooden post, about 0.7 m diameter, as shown in Figure 2.5.[14] At the top of this post a protruding peg, or pintle, fits into a hole in the underside of another massive timber – the crown tree – which extends across the full width of the windmill and carries the whole weight of the rotatable structure. This pintle bearing arrangement allows the whole body of the mill to be turned so that the sails can face into the wind, whatever its direction. Post mills – like that shown in Figure 1.1 – usually have four sails, and canvas is spread over the lattice wooden framework of each sail when the mill is working. The four sails are attached to – and turn – the windshaft, which carries the large diameter brakewheel, so-called because it usually has a brake on its circumference. Projecting pegs near the brakewheel's outer edge mesh with similar but vertical pegs on a smaller diameter wheel (as is shown more clearly in Figure 3.2) which is attached to the shaft that drives the upper of the two millstones. Post mills are clearly quite different from Persian or Chinese windmills, the very visible main difference being the fact that the sails of post mills rotate about a nearly horizontal axis, and so move in a direction that is roughly at right angles to the wind. Because the sails have to face into the wind, the structure that supports them must be turned when the wind changes direction; this the miller does by pushing on the tail pole (having first lifted the ladder so that it is clear of the ground).

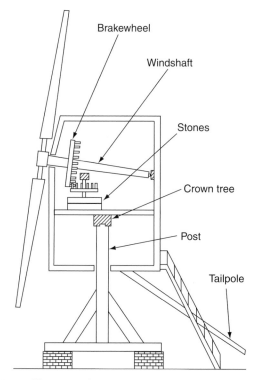

Figure 2.5 Post mill, cross section.

How post mills work

How do post mills work? The wooden lattice support structure of each sail (see Figure 3.2) is fixed to the windshaft so that each sail is set at an angle to the circle in which it is free to turn (called the plane of rotation) as is illustrated in Figure 2.6 (a), *which gives an end view of one sail looking from the tip in towards the centre;* for early windmills the sails were set at a constant angle of about 20° along their whole length. Initially, when the sail is stationary, the force due to the wind acts – as indicated – at an angle to the wind direction. Though the major part of this force acts in the downwind direction the sail is fixed to the windshaft and cannot move downwind. A smaller part of the wind force acts at right angles to the wind, and since the sail is free to move in this direction it starts to turn.

Figure 2.6(b) shows the situation a little later, when the sails are turning at their normal speed. At the tip the sail speed is faster than the wind speed, and for a traditional windmill is typically about 2½ times the wind speed. The wind direction *relative to the sail tip* is therefore angled forward, as shown in the figure, and if a small flag were attached to the sail tip it would be parallel to this

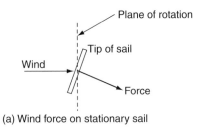

(a) Wind force on stationary sail

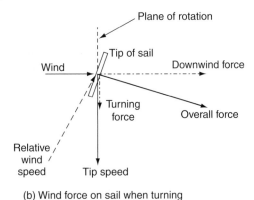

(b) Wind force on sail when turning

Figure 2.6 Wind forces on a windmill sail.

relative wind speed.[15] The flat sail of the windmill is a crude aerofoil, and the overall force on it, which is mostly due to aerodynamic lift,[16] acts nearly at right angles to the relative wind. As is illustrated in Figure 2.6(b) the overall force is equivalent to the combination of a large downwind force, which the structure supporting the sails and indeed the whole windmill must be strong enough to withstand, and a smaller turning force which drives the sails around and provides the useful power output. The overall and very important result is that the sails on a working post mill, like the blades on a modern wind turbine, are *pulled* around by aerodynamic lift,[17] unlike the Persian and Chinese windmills whose sails are *pushed* around by aerodynamic drag. Aerodynamic lift is much more efficient at converting the energy in the wind into a useful shaft-power output and, as noted earlier, the result is that horizontal-axis windmills (such as the post mill) are typically at least five times more powerful than drag-driven Persian or Chinese vertical-axis windmills of similar size.

Origins and economics

Exactly when and where the first European windmill was built is uncertain. Bennett and Elton review and dismiss a variety of claims made for windmills being used in England in the seventh, ninth and eleventh centuries,[18] and the clear evidence provided by the Domesday Book[19] is that by 1086 there were over 6000 watermills in use in England but no windmills. Though several authors and websites declare that the first windmill can be dated to an 1105 charter granted to the Abbott of Savigny, in France, Bennett and Elton point out that this document is known to be a monkish forgery written more than a century after its nominal date; the abbey did not even exist in 1105.[20] For many years the earliest authentic reference to a windmill was thought to be the one built in 1191 by Dean Herbert at Bury St. Edmunds, about 40 km east of Cambridge; as recorded in *The Chronicle of Jocelin of Brakelond* this infringed the manorial rights of the abbey and the enraged Abbott Samson ordered that it be demolished.[21] The novelty of the windmill at this time is confirmed by the decree issued by Pope Celestine III (1191–8) and addressed to the Abbot of Ramsey and the Archdeacon of Ely (both locations in the vicinity of Cambridge),[22] who had complained to him that a local knight had built a windmill and had refused to pay tithes[23] on it; the decree affirmed that tithes had to be paid on windmills.

More recently Kealey, in his book *Harvesting the Air: Windmill Pioneers in Twelfth-century England*, stated that in searching through twelfth- and thirteenth-century charters he found evidence for the existence of at least 56 windmills in England dating before 1200, with the oldest dated to 1137. However Holt, in his subsequent book *The Mills of Medieval England*, argued that the evidence for at least half of these is dubious. As noted by Hills the earliest reasonably reliable date comes from a document dated 1155 which reports on the gift of a windmill in Iford, East Sussex by Hugo de Plaiz, in memory of his father, to monks in the nearby town of Lewes.[24] On the other side of the English Channel there is evidence, see Holt, for five windmills dated before 1200, three in Flanders and two in France, with the earliest reliably dated to 1183.[25] And, intriguingly, White notes a reference to a windmill in Syria dated to 1190, at the time of the Third Crusade; he quotes an eyewitness who observed that 'the German soldiers used their skill to build the very first windmill that Syria had ever known'.[26] It would therefore seem that the windmill was already in use in Germany by this date. Given the passage of time we will probably never know with certainty where the first European windmill was built, but the weight of the evidence suggests an English origin. It is perhaps worth noting that when the windmill first appears England and France were politically very closely linked; Henry II, who was King of England from 1154 until 1189, also ruled much of France including Normandy.

What prompted the birth of the European windmill? As Holt points out windmills proliferated in England following their first appearance, and by 1200 they were in use along the whole length of eastern England; many hundreds more were built in the following decades.[27] The English feudal system was such that the right to erect windmills, like the right to erect watermills, belonged to the lord of the manor, and his customary tenants (which included most of the people who lived within the manor boundaries) were obliged to have their grain milled in the manorial mill.[28] For this service the miller – on behalf of the owner – took a proportion of the corn milled, the so-called *multure*,[29] which was typically about one-twentieth. The value of this was sufficient to make a well-sited windmill (or watermill) very profitable; for example a windmill which in the thirteenth century cost about £10 to build, see Holt, could be rented out for about £2 per year.[30] To put these figures in perspective a labourer's wage would then have been about one penny per day, with £1 = 240 pennies. Maintenance costs, mostly related to the periodic replacement of millstones, averaged about one-third of the income but the net return on the investment was still very attractive.

Watermills also represented a good profitable investment; however most of the rivers and streams that could be used already had mills on them by the time the Domesday Book was written in 1086. Once the first windmills had been shown to be practicable they offered an attractive new investment, especially in those relatively flat southern and eastern parts of England where the waterpower resource is poor but where the wind resource is good. The lords of many hundreds of manors were then willing to make the required quite substantial investment, through the course of the thirteenth century, to build the many hundreds of windmills that have since been documented. Investment risk was minimised by the fact that most tenants had to use the manorial mill; even the option of using a hand-mill in the home was forbidden, and anyone caught could expect to be fined sixpence, then a very substantial sum. And though their profitability explains the rapid expansion in the use of windmills in the thirteenth century, their sudden appearance in the latter part of the twelfth century is intriguing. Where did the idea come from? It has been suggested that twelfth-century Crusaders saw windmills in use in the eastern Mediterranean and brought the design back with them, but there is no evidence that windmills were used at that time in the eastern Mediterranean; as noted previously the first windmill in Syria was built in 1190 by German soldiers on the Third Crusade. Moreover the only windmills known to pre-date the horizontal-axis post mills that first appear in England, Flanders and France in the late twelfth century were the fundamentally different and much less efficient drag-driven windmills from eastern Persia, which were only suitable for use where the wind blew consistently from just one direction.

It seems unlikely that we will ever discover who invented and built the first post mill, and what prompted its invention. My guess is that there could well be a link to the Crusades,[31] but only insofar as it is seems likely that Crusaders in the eastern Mediterranean would have heard traveller's tales of wind-driven mills

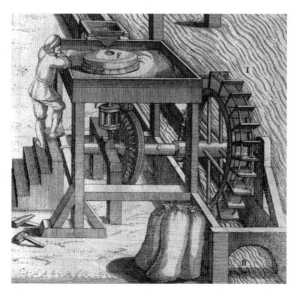

Figure 2.7 Traditional watermill. *Source*: Böckler (1661).

being used to grind corn in countries much further to the east.[32] It is unlikely that traveller's tales would have provided any technical detail (such as the fact that the axis of rotation of these eastern windmills was vertical) but they might well have given some indication of their size, such as 'several times taller than a man'. On his return home to England my putative Crusader, encouraged perhaps by the absence of sites suitable for watermills on his own lands, would instruct his local master carpenter to make him a mill powered by the wind and would make available to him the necessary resources – mostly timber. With no other information to go on apart from the approximate size, perhaps the fact that the eastern windmills used long flat blades driven round by the wind, but most importantly the knowledge that it could be done, the carpenter might look for inspiration to the design of watermills, which had been widely used since Roman times.

The traditional watermill, as first described by Vitruvius in about 30 BC, is shown in Figure 2.7.[33] A large diameter wheel supported by a horizontal axle, and with vanes or paddles around its circumference, has its lower part immersed in a stream so that the force of the water on the paddles turns the wheel. Two meshing toothed wheels are then used, as shown in the figure, to turn a shaft which rotates about a vertical axis; this passes through a hole in the centre of the lower – fixed – millstone and drives (via a short, horizontal, metal rod at its upper end) the upper – rotating – millstone. This millstone turns much faster than the waterwheel as the toothed wheel on the horizontal axle has many times the number of teeth on the lantern pinion with which it meshes. Grain is fed in via a central hole in the upper millstone, is ground between the two stones and the mixture of flour and bran then exits at their outer edge.

As has been pointed out by Wailes and by others the machinery within an early post mill, see Figure 2.5, is very similar to that of an early watermill turned upside down.[34] The unknown inventor of the first post mill, having inverted the machinery of a watermill, then had to find a means of using the wind to turn the horizontal shaft, and must have discovered that planks of wood attached radially would do the job,[35] provided that they were angled to the shaft (as indicated in Figure 2.6). It was serendipitous that as rotation commenced the arrangement became more efficient and more powerful. Trial and error would have determined that four planks were sufficient, as well as the best length and width for these planks. And at some stage early in the evolution of the design it would have been discovered that cloth covering a relatively lightweight wooden lattice structure, see Figure 1.1, was just as effective as solid wood.[36] The final major inventive step, which may well have occurred some years later, was to mount the whole structure containing the windshaft, the meshing toothed wheels and the millstones on a central post (see Figure 2.5) so that it could be turned to face into the wind. The post mill which was the end result of this creative process was a truly outstanding invention, much more efficient and versatile than the earlier Persian windmills, and it is unfortunate that we cannot identify and give credit to the individuals who were responsible.

3

Seven centuries of service

Post mills

Diffusion through Europe

In the decades following the first recorded appearance of the windmill in England in 1155 its use grew steadily. It seems reasonable to assume that the initial and probably rather crude design was gradually improved until, by the end of the twelfth century, a very practical design had evolved. By this date there is documentary evidence for at least 20 windmills in England, plus 3 in Flanders and 2 in northern France; given the fragmentary survival of records from this era actual numbers would have been substantially larger. Through the thirteenth century their use expanded rapidly, and by 1300 there were about 4000 in England alone.[1] Their use also expanded across Europe. They are recorded in Germany by 1222,[2] in Denmark by 1259 and in Holland by 1274; they were erected in large numbers across the plains of northern Europe, where unlike watermills they could continue to operate in freezing weather, and they reached Russia by 1330 and Finland by 1400. To the south they are recorded as reaching Italy by 1237, Portugal by 1261, Spain by 1330, Rhodes by 1389 and Gallipoli, in western Turkey, by 1420.[3] These early windmills were all post mills, similar to those in the foreground of Figure 3.1, which shows windmills on the outskirts of a city in the Low Countries in the late sixteenth century.[4] They were used initially just for milling cereal grains, especially wheat, though other applications – including land drainage – followed. Food choice was very limited in the Middle Ages, and since bread was relatively low cost it provided almost three-quarters of the calories in the diet for most people;[5] a typical family of two adults and three children would therefore consume about 1.2 bushels (about 32 kg) of wheat and barley per week.[6] To mill this quantity of grain in a hand mill, or quern, would take about 9 hours, requiring a tedious $1\frac{1}{2}$ hour long

Figure 3.1 Windmills in the Low Countries. *Source*: Stradanus (1584).

'daily grind', though as noted in Chapter 2 the English feudal system was such that if the Lord of the Manor provided a mill his tenants had to use it.[7]

Medieval milling capacity

Though later post mills were larger, Langdon indicates that in medieval times the length of the sails from tip to tip was about 13 m.[8] The power produced would clearly depend on the strength of the wind but for most of southern and eastern England the average wind speed over open fields is typically in the range 5 m/s to 6 m/s,[9] and the several thousand windmills built in the thirteenth century would have had an average power output of about 1 kW (as is discussed later). Though this is small by today's standards it is equivalent to the muscle power output of about 15 to 20 men, or $2\frac{1}{2}$ horses.[10] Hiring labourers to do the work that the mill could do would have cost about £15 to £20 per year, plus the cost of their equipment; alternatively each horse mill would have cost about £5 to build and there was the ongoing expense of horse fodder;[11] by comparison the post mill costing – as noted in Chapter 2 – about £10 was an attractive investment.

It takes about 14 W to mill 1 kg/hr of grain, see Appendix B; the post mill's 1 kW average power output would therefore be sufficient to mill about 70 kg of grain per hour. Assuming an average of 8 operating hours per day, and using Langdon's estimate that mills would be worked for about 240 days per year,[12] a post mill could therefore grind about 140 000 kg of grain per year, enough to

Table 3.1 *The Beaufort scale*
The wind speeds listed below correspond to a height of 10 m above the ground

Force	Description	Specification (over land)	Mean speed m/s; (knots)	Speed range (m/s)
0	Calm	Smoke rises vertically	0; (0)	0.0–0.2
1	Light air	Wind direction shown by smoke	0.8; (2)	0.3–1.5
2	Light breeze	Wind felt on face; leaves rustle	2.4; (5)	1.6–3.3
3	Gentle breeze	Leaves in motion; small flags lift	4.3; (9)	3.4–5.4
4	Moderate breeze	Raises dust and loose paper; small branches move	6.7; (13)	5.5–7.9
5	Fresh Breeze	Small trees begin to sway	9.3; (19)	8.0–10.7
6	Strong breeze	Large branches in motion; umbrellas difficult to use	12.3; (24)	10.8–13.8
7	Near gale	Whole trees in motion	15.5; (30)	13.9–17.1
8	Gale	Twigs break; walking impeded	18.9; (37)	17.2–20.7
9	Severe gale	Slight structural damage; (chimneys & slates removed)	22.6; (44)	20.8–24.4
10	Storm	Trees uprooted; considerable structural damage	26.4; (52)	24.5–28.4
11	Violent storm	Widespread damage	30.5; (60)	28.5–32.6
12	Hurricane	Devastation & loss of life	– ; (–)	> 32.7

meet the needs of just over 80 families, which corresponds to about 400 people. Langdon estimates that there were about 10 000 mills in 1300, including water-mills as well as windmills, and his data indicates that they had broadly similar outputs; the total milling capacity was therefore sufficient to meet the needs of about 4 million people, which is consistent with population estimates.[13]

Controlling the mill

Though the average power output of a medieval post mill was just 1 kW, corresponding to an average wind speed within the range 5 to 6 m/s, the power increased substantially in higher wind speeds. Traditional windmills would start to turn and run slowly when the wind speed reached about 3 to 4 m/s, which is a Gentle Breeze (Force 3) on the Beaufort scale,[14] see Table 3.1, and as noted by Hills would 'run steadily' in a wind speed of about 15 m.p.h. (6.7 m/s),[15] which equates to a Moderate Breeze (Force 4). The maximum output was delivered in a wind speed of about 11 m/s, which on the Beaufort scale is a Strong Breeze (Force 6); for a 13 m diameter post mill this power output would be about 5 kW. In higher wind speeds the sails would need to be reefed, in other words the amount of canvas spread over the lattice framework of the sails would need to be progressively reduced. To do this the windmill had first to be stopped with one sail in the '6 o'clock' position; the miller could then reduce the amount of canvas spread over this sail, sometimes having to use the lattice framework as a ladder if he needed access to where the canvas was secured close to the hub. He then had to allow the sails to slowly turn through one quarter of a revolution, stop the next sail at the '6 o'clock' position and reduce its spread of canvas, and repeat the process for the third and fourth sails. This was tedious and time-consuming work which might have to be repeated many times in the course of a very windy day, and would sometimes have to be done with rain streaming off the canvas. In the earliest post mills the only way to stop the mill was to turn it out of the wind. To do this one had to lift the outer end of the ladder, see Figure 3.1, off the ground using a rope running from the top of the mill to the foot of the ladder, which is hinged where it meets the body of the mill. The miller then had to push on the end of the ladder to turn the mill; as the mill weighed about 6 or 7 tons this required considerable force.[16] Later post mills were larger and heavier and a tailpole, shown in Figure 2.5 and also in Figure 3.4, was used to give increased leverage.

The need for a mechanical brake to stop the sails from turning would soon have become evident and the obvious place to put a brake was on the large diameter toothed wheel attached to the windshaft.[17] Figure 3.2, from Abraham Rees' 1819 *Cyclopaedia*, shows a typical brake arrangement using linked timber blocks that can be pulled onto the circumference of the gear wheel – which is consequently usually referred to as the brakewheel. When the sails are turning the brake lever *mk*, which is pivoted at *k*, is latched in the raised position so that there is a small gap between the brake blocks and the brakewheel; the brake is engaged by unlatching the brake lever and pulling down on its left-hand end. Once the amount of sailcloth had been adjusted to suit the wind conditions the windmill would be turned to face the wind and the ladder would be lowered to the ground, as it helped to stabilise the mill as well as providing access. The brake would then be released and the sails allowed to turn.

Sail developments

The canvas used for the sails was woven from hemp fibres (the word *canvas* in fact derives from the Latin word for hemp, *cannabis*) and most windmills had four sails, attached to two main spars – called stocks in the terminology of traditional windmills – which were mortised at right angles through the end of the windshaft, as shown in Figure 3.2, and wedged securely in place. Early windmills, like those shown in Figure 3.1, had the main spar located near the

Figure 3.2 Post mill internal details. *Source*: Rees (1819).

middle of the sails in the reasonable expectation that the force due to the wind would be about equal on the two sides. However as the design of windmills evolved, and sails became progressively larger, windmill builders learnt by experience what aerodynamicists only rediscovered in the early twentieth century, which is that the wind force on the sail acts not through its centre line, but through a line that is about one quarter the width of the sail behind the leading edge.[18] Putting the spar near this quarter-width position eliminates the twisting force that it would otherwise experience, and this made it possible to use the longer spars that were needed by the larger-diameter windmills. Once the spar had been moved to the quarter-width position, a change which took place somewhere around 1600, wooden boards were commonly used to provide the sail area in front of the spar, as is illustrated in Figure 3.2; canvas supported on a wooden lattice framework continued to be used to cover the rear part of the sail.

Early windmills had sails that were effectively flat, like those shown in Figure 3.1, and Smeaton indicates that some windmills still used flat sails in the mid eighteenth century.[19] The angle between the sail and the plane of rotation is known as the *weather*, and for flat sails Smeaton states that this angle was usually between 15° and 18°. However around 1600 it was discovered that windmills with twisted sails gave better performance. This effect was presumably first noticed by chance on a windmill on which one of the sail spars had twisted after installation, and was then optimised by trial and error. Smeaton refers to such sails as 'Dutch' so it seems probable that this innovation was pioneered in Holland. Smeaton also tells us that English windmills that used 'Dutch' sails usually had their sail tips parallel to the plane of rotation, i.e. zero weather angle, and the weather then steadily increased towards the hub, where it would be about 15°; his windmill model tests, which are discussed in Appendix B, indicated that for the best performance the weather angle should increase from about $7\frac{1}{2}°$ at the sail tip to about $22\frac{1}{2}°$ near the hub. The Flemish windmills on which the eminent French scientist Charles Coulomb[20] made performance measurements in the late eighteenth century had weather angles at the sail tips that were in the range 6° to 12°, increasing to about 30° near the hub.

The main reason why twist along the length of the sail improves a windmill's performance is that it reduces the drag on the sail.[21] For the drag to be low the angle between the relative wind and the sail, see Figure 2.6(b), should be no more than about 15°. For traditional windmills the best performance is given when the tip speed is about $2\frac{1}{2}$ times the wind speed; the requirement that the angle between the relative wind and the sail should be no more than about 15° can therefore be met with just a few degrees of weather at the tip. However towards the hub the sail's circumferential speed reduces and the angle between the relative wind direction and the plane of rotation increases very substantially. Increasing the weather angle towards the hub makes it possible to keep the angle between the relative wind and the sail below about 15°, and so reduces the drag.

The windshaft which carries the sails – see Figure 3.2 – is typically about 0.5 m to 0.6 m diameter and, like most of the main structure of the mill, made from oak. The weight of the windshaft and the sails is supported, close to the sails, by a simple bearing, which is usually faced with iron; a second bearing at the windshaft's narrower downwind end takes a small part of the weight but a large part of the downwind thrust produced by the sails. As the figure indicates the windshaft was usually inclined to the horizontal by a small angle, usually between about 5° and 15°; in part this was done to increase the clearance between the sails and the body of the mill but it also helped to prevent damage which might otherwise be caused by a sudden lull in the wind. The inertia of the sails is such that they would take many seconds to slow down so for this short period the sails would be turning too fast for the wind; in extreme cases this could result in the sails acting like a propeller and trying to pull the sails and the windshaft forward into the wind, and out of the bearings which are designed to take thrust only in the downwind direction.[22] The simplest way to avoid this potential problem is to incline the windshaft, so that part of the weight of the sails and the windshaft always acts along the length of the windshaft and prevents its forward movement.

Milling the grain

The toothed brakewheel on the windshaft meshes with the lantern pinion, shown at E in Figure 3.2, and so drives the two-piece iron spindle GN which passes through the two millstones and their cylindrical wooden casing. Only the upper millstone, the runner stone, rotates and it is turned by the upper half of the spindle whose forked lower end engages with a metal cross-piece, or *rynd*, which is recessed into the lower surface of the runner stone, see Figure 3.3.[23] (The top end of the upper spindle, just above the lantern pinion, is held in position by a bearing attached to a sturdy cross-beam.) The rynd sits in the slotted top of the lower half of the spindle, which supports the 1 to $1\frac{1}{2}$ tonne weight of the runner stone. This lower part of the spindle passes through a close-fitting sleeve (often made from a hard wood such as lignum vitae) fitted centrally in the stationary lower millstone, and is supported at its lower end by a bearing – usually made of brass – attached to a wooden beam called the bridge-tree. This bridge-tree is pivoted at one end and the other end can be raised or lowered by the system of ropes and levers shown in Figure 3.2; the miller can in this way adjust the very small gap between the two stones.

Sharp-edged grooves are cut to a depth of about 20 mm into the surface of both millstones, and the pattern of these grooves for the runner stone can be seen in Figure 3.3; the apparently flat surfaces between the grooves are in fact covered with closely spaced, narrow grooves that run parallel to the deeper and more visible grooves. The pattern of grooves for the lower millstone, the bed stone, is identical so that when the runner stone is placed face down over the

Figure 3.3 Upper millstone, with rynd.

bed stone – and rotated – the opposing grooves give a scissor-like action. Grain is fed into the hole at the centre of the runner stone by a chute, see Figure 3.2, which is pivoted where it joins the grain bin so that the miller can – by pulling a string – adjust the slope of the chute and so control the rate at which the grain flows to the stones; when the wind speed increases, and more power becomes available, the flow of grain must also be increased. The spindle has a square cross-section where it goes through the lower end of the chute, and by design the side of the chute touches the spindle; then as the spindle turns it makes the chute vibrate, which helps to ensure that the grain flows freely. Though the bed stone is flat the runner is slightly 'dished' so that the gap between the stones gradually narrows as the grain moves out from the eye to the periphery, and the grain is broken down by a combination of shearing and crushing. The 'meal' that emerges at the stones' periphery is quite warm to the touch, and collects in the annular gap between the stones and the surrounding wooden enclosure. A 'sweeper' attached to the runner stone sweeps the meal round to a hole in the floor through which the meal flows, via the meal spout, to the bin on the floor below.

The British stone most commonly used for millstones was – as its name suggests – millstone grit, a hard sandstone found in the Peak District and elsewhere. Stones from this material were relatively inexpensive, and were mostly used in the north of England. In the south and east millers preferred to use the higher quality but more expensive millstones imported from Germany or

France.[24] German millstones had been imported for watermills since Roman times, and came from a lava stone source to the west of Koblenz; since they were shipped along the Rhine via Cologne they became known as Cullins. French burr stones were particularly highly favoured as they did not contaminate the flour with sandy particles. The porous quartz stone came from a source to the east of Paris, but came in lumps that were too small for a one-piece millstone; a number of pieces had therefore to be matched, shaped and cemented together, and the completed stone was then bound with iron hoops, see Figure 3.3. Millstones, from whatever source, were typically about 1.3 m diameter and the best speed for milling was about 100 to 125 rpm. Measurements made on traditional windmills in Flanders by Coulomb in the 1780s, and by Dutch engineers in the 1930s, have shown that they give their best performance when the tip speed of the sails is about $2\frac{1}{2}$ times the wind speed.[25] For a medieval post mill with a diameter from sail tip to sail tip of about 13 m this means that the sails need to turn at about 33 rpm in a good milling wind of about 9 m/s. The number of gear teeth on the brakewheel must therefore be about three times the number of staves in the lantern pinion, so as to give the required speed of about 100 rpm at the runner stone. Later, larger windmills would need to turn more slowly to give their best performance: for example a windmill that was double the size, 26 m diameter, would only need to turn at half the speed – $16\frac{1}{2}$ rpm – to give the same optimum tip speed ratio of about $2\frac{1}{2}$ in a wind speed of 9 m/s; to keep the runner stone at the correct speed the number of teeth on the brakewheel would then need to be about *six* times the number of staves in the lantern pinion.

Design evolution

Over the centuries following their first use in the late twelfth century the basic design of the post mill changed little, though the sails, the bearings, the gearing and other details were all gradually improved. The most noticeable change was that post mills grew steadily larger, so as to provide more power. In the diary of his *Journey to the Low Countries 1755* Smeaton records seeing a number of post mills near Dunkirk with diameters of about 27 m, but these seem to have been unusually large.[26] Post mills continued to be built through to the nineteenth century and Bennett and Elton, in their classic 1899 *History of Corn Milling*, observe that post mill diameters were typically in the range 15 m to 18 m.[27] The power output of a windmill is proportional to the area swept by the sails (see Appendix A) and so is proportional to the square of the windmill's diameter. Later post mills with diameters of about 18 m, such as the seventeenth-century Outwood mill shown in Figure 3.4, would therefore give

Figure 3.4 Outwood windmill, 1665, Surrey, England.

about double the power output of their 13 m diameter medieval predecessors and the internal arrangement of the gearing would then be modified so that two pairs of millstones could be used, either individually or – in stronger winds – simultaneously.[28] Outwood mill, built in 1665, is the oldest working windmill in Britain and its main body (or *buck*) weighs 25 tons. Though early post mills, such as those shown in Figure 3.1, had the trestle structure supporting the central post open to the weather it was usual by the eighteenth century to enclose the trestle with a roundhouse, as can be seen in Figure 3.4; this gives protection from the weather and also provides additional space for storage.

Following Abraham Darby's breakthrough in using coke to smelt iron at Coalbrookdale, in the early eighteenth century, the use of iron grew rapidly and Smeaton pioneered the use of cast iron in watermills and windmills in the 1750s.[29] In windmills iron was increasingly used for the windshaft and for the gearing,[30] as well as for the *poll end* or *canister*, both traditional windmill terms for a cast iron fitting at the end of the windshaft to which the sails were attached. Cast iron end fittings could be used in conjunction with either a wooden wind-shaft or a cast iron one, and they made it possible for windmills to have more than four sails. From their origins in the twelfth century traditional windmills had used four sails because it was relatively straightforward to make two perpendicular, rectangular holes in the end of the windshaft – see Figure 3.2 – through which the two sail spars could be passed and wedged in position, and on which four sails could then be built. Making additional holes through a wooden windshaft to allow the use of 6 or even 8 sails would have required that the windshaft be extended and would have led to structural weakness. However a cast iron end fitting could easily be made with any number of equally spaced short radial arms, to which the spars for individual sails could be bolted. Figure 3.5 shows one of these cast iron end fittings being used on a windmill with four sails, but Smeaton favoured the use of five sails, and windmills with six sails – as well as a few with eight – were also built. Presumably there was the expectation that increasing the number of sails would give an increase in the windmill's power output. In fact, as is discussed in Appendix A, a windmill's power output is almost unaffected by the number of sails. The power in the wind flowing through the area swept by the sails can be efficiently intercepted using any number of sails, from one upwards, but as the number of sails increases the optimum speed of rotation decreases. The maximum power output of an eight-sailed windmill would be about the same as a four-sailed windmill, but it would turn more slowly. The main advantage of having more sails is that they give a higher starting torque, so that the windmill will start to turn in lighter winds; there is however little power available in these light winds. The main disadvantage of having more sails is their extra cost, and the extra work required when sails have to be reefed in stronger winds. Most windmills built after the use of cast iron became widespread continued the tradition of using four sails; there was evidently no clear advantage to be gained by using more.

Smock mills and tower mills

Building a post mill such that the whole structure could be easily turned about the centre post required a high level of woodworking skill, and quite early in the evolution of the windmill a significant variant – the smock mill – was developed. In this most of the structure was fixed and only the upper part, the cap, had to be turned to face into the wind. Figure 3.5 shows the internal arrangement of a late-eighteenth-century smock mill; the figure is reproduced from Abraham Rees' 1819 *Cyclopaedia* but shows, with minor modifications, Smeaton's 1782 design for a 21 m diameter corn grinding windmill that was built on the outskirts of Newcastle.[31] Four conventional canvas-covered sails (only two, at A and A, can be seen in the section) drive the slender cast iron windshaft B. The brakewheel C on this windshaft meshes with the smaller gear wheel D and so

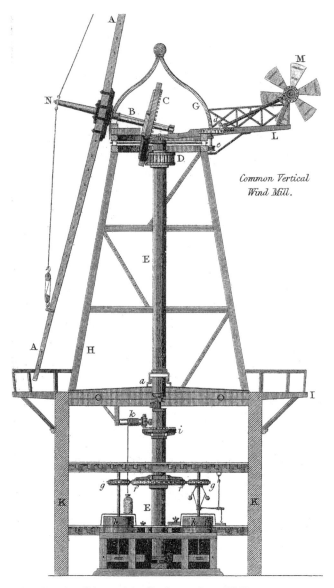

Figure 3.5 Smock mill internal details. *Source*: Rees (1819).

drives the central vertical shaft EE; at the lower end of this the large-diameter
gear wheel *ff* meshes with the two smaller gearwheels *g* each of which drives a
spindle passing through a pair of millstones. The arrangements for feeding the
grain to the centre of each pair of millstones, and for collecting the meal that

exits from their periphery, are the same as has been described for the post mill – and as is shown in Figure 3.2 – but are omitted here for clarity.

The lower part of the windmill's structure has brick walls KK which support an octagonal wooden tower H. Fixed to the top of this is a circular curb, and the underside of the wooden cap has a matching circular curb. Rollers between these two curbs support the weight of the cap, but also allow it to be easily turned so that the sails face into the wind. The cap is kept in position, as it turns, by wheels which turn on vertical axles attached to the cap; these wheels run on – and thrust against – the inside of the lower curb, which is fixed to the top of the tower.[32] The top bearing of the vertical shaft is attached to the centre of a horizontal beam which forms part of the main structure of the cap; the axis of this bearing must be accurately in line with the centre of the two circular curbs, so that there is no lateral movement when the cap turns.

Smock mills are so-named because of the superficial resemblance that their tapered wooden towers have to the smocks that used to be worn by most of the people living in rural areas. Tower mills are in all essentials the same as smock mills, except that their whole structure up to the cap is made from stone or brick. Tower mills and smock mills have been used since at least the fourteenth century,[33] and some early tower mills can be seen in the left background of the illustration by Stradanus, Figure 3.1. Tower mills are more durable than post mills, but more expensive, and until the end of the eighteenth century most English windmills were post mills. However tower mills – and smock mills – can be made larger than post mills and the growing demand for power through the eighteenth and nineteenth centuries led to the construction of an increasing number of tower and smock mills. In early smock and tower mills the cap was turned using a long tail pole, extending from the cap to the ground, and this arrangement continued in use – particularly in Holland – well into the nineteenth century. However Smeaton's 1782 smock mill, shown in Figure 3.5, turns the cap into the wind using a fantail, which was invented by Edmund Lee in 1745. When the windmill faces into the wind the air flow is parallel to the fantail's plane of rotation, and it therefore does not turn. However when the wind direction changes the air flow approaches the fantail at a small angle and the fantail starts to turn. Through reduction gearing it then turns a small gear wheel *e* which meshes with a large gear ring fixed to the outside of the lower curb; the cap is consequently slowly turned until the sails again face into the wind and the fantail stops turning. Fantails became widely used, but far from universal.[34]

One unusual feature of Smeaton's 1782 windmill was the forward extension of the windshaft BN, reminiscent of a ship's bowsprit, from which bracing ropes were attached to the sail spars. Taking part of the downwind load on the sails through these ropes would have allowed the use of thinner spars, but this innovation never became popular in England, where the wood required for un-braced spars continued to be available. And in the Low Countries iron spars became widely used in the nineteenth century. However in countries around the Mediterranean, where wood suitable for large spars was not so readily available, the use of 'bowsprits' on windmills became widespread.[35] Figure 3.5 also shows, at *k*, a sack hoist driven via gearing from the main vertical shaft. By tradition a sack contained 4 bushels, which is 32 gallons, and weighed about 110 kg.[6] Millers had to manhandle these heavy sacks, but understandably chose to use the shaft power available within the windmill to lift the sacks from ground level to the floor above the millstones where grain was stored prior to milling. It is not clear when sack hoists were first used, but one can be seen lifting a sack in the illustration of sixteenth-century windmills in the Low Countries, see Figure 3.1.

Control refinements

One final feature of interest in the windmill shown in Figure 3.5 is the centrifugal governor, shown on the spindle above the right-hand pair of mill-stones. Iron balls are attached via hinged iron rods to the underside of the gearwheel *g*. When the wind speed increases, and the sails start to turn faster, the higher speed of the gearwheel and spindle increases the centrifugal force on the iron balls and makes them swing radially outwards. Hinged metal links attach the rods holding the iron balls to a circular sleeve which is free to move up and down around the spindle, so that when increased spindle speed makes the balls move outwards the sleeve is raised. As the sleeve moves up (or down) it moves the forked end of a lever which forms part of the system of levers (as previously described, and as shown more clearly in Figure 3.2) used to raise or lower the runner stone. The centrifugal governor consequently makes it possible for the gap between the stones to be adjusted automatically in response to changes in the wind speed.[36] Several variants of centrifugal governor were developed for windmills in the 1780s and their use inspired Boulton and Watt in 1788 to adapt the principle to control the steam supply, and hence the speed, of their steam engines.

Reefing the canvas sails in strong winds was – as noted earlier – tedious and time consuming, and several attempts were made in the eighteenth century to devise less tedious means for adjusting the sails. One of the most successful was the system of spring sails invented by Andrew Meikle in 1772 which can be seen in Figure 3.4 on Outwood mill. The conventional arrangement with sailcloth covering a lattice wooden framework (in windmill terminology known as a *common* sail) is replaced by a series of rectangular shutters, rather like a Venetian blind. The shutters are pivoted at their ends and can be opened or closed by an iron rod running the length of the sail. At the hub end of each sail is an adjustable spring (their black and bowed outline is clearly visible in Figure 3.4 on the two upper sails) which holds the shutters closed when the windmill is working over its normal range of wind speeds, that is up to about 11 m/s. However in strong winds the force of the airflow on the shutters is sufficient to overcome the resistance of the spring; the shutters then open, which limits both the power produced and the potentially damaging downwind force on the sails.

Though they made life for the miller easier and safer, spring sails were somewhat less efficient than common sails, and as a compromise it was not uncommon for windmills to be fitted with one pair of spring sails and one pair of common sails. A later refinement, patented by William Cubitt in 1807, used a *striking rod* which passed through the centre of a hollow windshaft to actuate, via a central *spider* and system of levers, the rods along the length of the sails which opened or closed the shutters.[37] The striking rod extended out of the back end of the cap, and could be moved backwards or forwards using a rack and pinion arrangement, with the rack fitted on the exposed length of the striking rod and the pinion on a horizontal axle attached (usually) to the underside of the frame supporting the fantail; a chain wheel on the same axle supported an endless chain which reached almost to the ground. The miller would pull on this chain to move the striking rod and close the shutters when he wanted to start the mill. A weight hung on the endless chain would hold the shutters closed until in stronger winds the air flow forced them open, and by adjusting this weight the miller could pre-determine the wind speed (and power level) at which this happened. When he wanted to stop the mill all he had to do was remove the weight and pull on the chain to open the shutters. It was very much easier to control a windmill equipped with a fantail and Cubitt's *patent* sails, and this combination was used on many nineteenth-century windmills. Since common sails were more efficient many mills used one pair of patent sails together with one pair of common sails, and this arrangement can be seen on Wilton windmill, see Figure 3.6, a 20 m diameter tower mill built in 1821 on an exposed location in Wiltshire.

The power output from traditional windmills

Tower and smock mills could be made substantially taller than post mills, and could carry longer sails. Their greater height allowed them to rise above obstacles, such as houses, which might otherwise interfere with the air flow past the windmill; and even if the windmill could be sited clear of nearby obstacles there was benefit from their greater height, since the wind speed increases with distance above the ground (see Appendix C). As noted

Figure 3.6 Wilton windmill, 1821, Wiltshire, England.

previously a windmill's power output is proportional to the area swept by the sails, so larger sail diameters give substantially larger power outputs. More formally (see Appendix A) the power output P from a windmill is given by the expression $P = C_P(\frac{1}{2}\rho A V^3)$, where V is the wind speed, $A = \pi D^2/4$ is the rotor swept area corresponding to the sail diameter D, ρ (rho) is the air density[38] which at sea level and a temperature of 15 °C is equal to about 1.23 kg/m³, and C_P is a number – known as the power coefficient – whose value depends on the detailed design of the windmill. The maximum theoretical value of the power coefficient is 0.59, which is known as the Betz limit after the German aerodynamicist Albert Betz who in 1920 first formulated the theoretical model from which it derives.

Modern wind turbines can achieve power coefficients of $C_P \approx 0.45$; as is discussed in Appendix A this means that they are able to extract about 67% of the energy in the air flowing through their rotors. However the sails of traditional windmills give less lift and more drag than the blades of modern wind turbines and measurements made on traditional windmills (which are reviewed and discussed in Appendix B) indicate that by the eighteenth and nineteenth centuries their performance was equivalent to a peak power coefficient $C_P \approx 0.07$; this lower power coefficient indicates that they could only extract about 10% of the energy in the air flowing through the area swept by their sails. Medieval windmills with their untwisted sails and their more primitive bearings and gearing would have been less efficient, with a peak power coefficient $C_P \approx 0.05$. Importantly, the formula given above indicates that the power in the wind is proportional to the cube of the wind speed, so at a location where the wind speed averages 6 m/s the power in the wind is 8 times higher than at a location where the wind speed averages 3 m/s. Hence the tendency to site windmills where possible on exposed local hill tops; even if the wind speed is only 20 to 30% higher than at less exposed locations nearby, the average power output will be approximately double.

The wind speed over open countryside in southern and eastern England typically averages between 5 and 6 m/s for heights in the range 10 to 20 m (above ground level),[9] which are typical of the hub heights for traditional windmills. Because the power in the wind is proportional to the cube of the wind speed the average power output P_{Av} is substantially higher than the power output at the average wind speed[39] V_{Av}. This can be seen from Figure 3.7. The upper half of this figure shows, for a location where the average wind speed is 5 m/s, how many hours per year the wind blows at different wind speeds.[40] It shows, for example, that the wind blows for about 1300 hours per year at wind speeds in the range 4 m/s to 5 m/s; wind speeds in the range 8 m/s to 9 m/s are however only experienced for about 490 hours per year. The figure also shows that wind speeds higher than three times the average wind speed, in this case 15 m/s, are rarely encountered.

Traditional windmills would start to turn in a wind speed of about 3 to 4 m/s, and developed their maximum power output, with their sails fully spread, in a wind speed of about 11 m/s;[41] above this speed the power output would be held approximately constant by reefing the sails, but if the wind speed exceeded about 15 m/s the mill would be stopped. The lower half of Figure 3.7 shows how the energy output for a traditional windmill with a diameter of 26 m varies with wind speed, again for a location where the average wind speed is 5 m/s. Though wind speeds are in the range 4 m/s to 5 m/s for 1300 hours per year the power level is low (≈ 2 kW) and the contribution to the overall annual energy output is

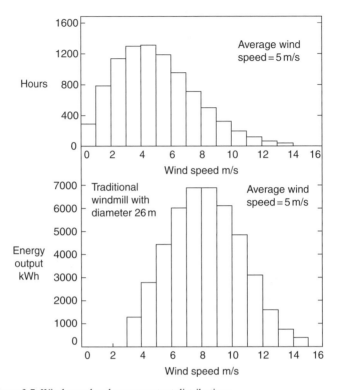

Figure 3.7 Wind speed and energy output distributions.

only about 2700 kWh. By contrast though the wind blows for just 490 hours per year at wind speeds between 8 m/s and 9 m/s the power level is high ($\approx$ 14 kW) and the contribution to the overall annual energy is over 6800 kWh. The figure also shows that though the windmill operates in wind speeds above the 11 m/s that give its maximum power level ($\approx$ 30 kW) for only about 2% of the time ($\approx$ 190 hours) these conditions make a significant ($\approx$ 13%) contribution to the annual energy output. The windmill's overall energy output is obtained by adding up the contributions in each of the wind speed bands and amounts to 44 800 kWh per year, which gives an average power output over the whole year (8760 hours) of 5.1 kW. Though the site's *average wind speed* is 5 m/s this *average power level* corresponds to the power produced by the windmill when the wind speed is just over 6 m/s. Note that most of the energy is produced around a wind speed that is about 60% higher than the average wind speed.

Similar calculations for a location where the wind speed averages 6 m/s again show that most of the energy is produced around a wind speed ($\approx$ 10 m/s) that is

Table 3.2 *The power output from traditional windmills*

Sail diameter tip to tip	13 m (medieval)	18 m	26 m	30 m
Average power, P_{Av} (where $V_{Av} \approx 5$ m/s)	0.8 kW	2.4 kW	5 kW	7 kW
Average power, P_{Av} (where $V_{Av} \approx 6$ m/s)	1.3 kW	3.8 kW	8 kW	10 kW
Maximum power (when $V = 11$ m/s)	5 kW	15 kW	30 kW	40 kW

60% higher than the average wind speed, and that the *average power level* of about 8 kW corresponds to the power produced by the windmill when the wind speed is just over 7 m/s. For windmills larger or smaller than the 26 m diameter assumed in the above discussion the power output is proportional to the area swept by the sails, in other words to the square of the diameter. Table 3.2 summarises both the average power output and the maximum power output for windmills varying in size from the 13 m diameter of medieval post mills up to the 30 m diameter of the largest tower and smock mills that were built in the eighteenth and nineteenth centuries. For the smaller medieval windmills a lower value of the power coefficient has been used ($C_P \approx 0.05$, compared with $C_P \approx 0.07$ for the later and larger windmills) reflecting, as indicated previously, their lower efficiency.

As noted earlier the overall average power output of the several thousand medieval windmills would have been intermediate between the values of 0.8 kW and 1.3 kW shown in the table, and would therefore have been about 1 kW, equivalent to the muscle power output of about 15 to 20 people and sufficient to grind enough grain to meet the needs of about 400 people. However by the late eighteenth century the average size of windmills had approximately doubled (to about 26 m diameter) and their overall average power output would have been about 6 kW, which is equivalent to the muscle power output of about 100 people; it is also sufficient to grind enough grain to meet the needs of about 2400 people, an estimation which is consistent with the statement made by Frederick Stokhuyzen, the chairman of De Hollandsche Molen (the Dutch Association for the Preservation of Windmills) that 'for every 2000 inhabitants there had to be one windmill to ensure an uninterrupted supply of meal for the population'.[42] The very largest traditional windmills had a diameter of about 30 m and at coastal locations or on very exposed hill tops, where wind speeds averaging 6 m/s or more can be found, these large windmills could give average power outputs up to about 10 kW, corresponding to the muscle power output of about 150 to 200 people. Maximum power outputs ranged from about 5 kW for medieval windmills up to about 40 kW for the largest eighteenth- and nineteenth-century windmills.

Land drainage and other applications

Windmills were used for a variety of other purposes, besides grinding grain, and one of the most important was land drainage, an application for which the Dutch became renowned. The first evidence for a windmill in what is now the Netherlands dates back to Haarlem in 1274, more than a century later than in England. Windmills were used initially, as elsewhere, to grind grain but the need to drain low lying land led to the use of windmills for this purpose by the early fifteenth century. Over the following centuries thousands of windmills were used to drain and reclaim land, as is discussed in some detail both by Stokhuyzen and by Hills.[43] Both post mills and smock or tower mills were used for drainage. In smock (or tower) mills the central, vertical drive shaft – see Figure 3.5 – would extend almost to ground level, where a lantern pinion on the vertical shaft would mesh with a larger-diameter geared wheel fixed to a horizontal shaft, at the other end of which there was a large-diameter scoop wheel. The arrangement at ground level in a drainage mill is essentially the same as in the watermill shown in Figure 2.7, except that there are of course no millstones and the drive is *from* the vertical shaft via the gears and *to* the scoop wheel, which for drainage mills is housed in a close-fitting channel so that as it is turned by the power from the windmill it lifts the water from the level of the inlet channel to the higher level of the outlet channel. The maximum lift is given by the radius of the wheel, but it is usually about half of this, typically in the range 1 to 1.5 m. If the height through which the water had to be raised was larger than this the Dutch would use a *molengang*, a number of drainage mills close to one another with the water output from the first channelled so that it became the water input to the second, and so on.

   With post mills used for drainage there was the problem that the scoop wheel was fixed in position by the drainage channel, and so could not turn with the body of the mill when the wind direction changed. The ingenious solution took advantage of the fact that the central post was usually completely supported by the quarter bars, see Figure 2.5, and the bottom end of the post was a short distance above ground level. Post mills for water pumping were therefore made with a hollow central post so that the windshaft could (through the usual gearing) drive a vertical shaft which was located inside the post and extended down to ground level.[44] There, as for the drainage smock mill, a lantern pinion meshing with a geared wheel similar to that shown in Figure 2.7 drove a horizontal shaft which turned the scoop wheel. These mills, known as *wipmolen*, date back to at least 1526 and though smaller than the drainage smock mills they were very widely used.[45]

By the time that Smeaton made his Journey to the Low Countries in 1755 the Dutch were acknowledged leaders in the construction of windmills for corn milling, land drainage and a variety of other applications, including sawing wood and crushing oil seeds.[46] The French scientist Charles Coulomb[47] made his pioneering eighteenth-century study of how the power output from windmills varied with wind speed using one of the many windmills built near Lille to crush rapeseed and extract the rapeseed oil, and the operation of these mills, as well as Coulomb's measurements, is described and discussed in Appendix B. Though in England as elsewhere most windmills were built to grind grain they were also used from the sixteenth century onwards for land drainage, especially in the area around the Wash and the Norfolk Broads, and this application is described in some detail by Hills;[48] the industrial uses of windmills are comprehensively covered by Gregory's *The Industrial Windmill in Britain*.

The windmill's nineteenth-century peak

As noted earlier windmill numbers in England grew rapidly through the thirteenth century, and by the year 1300 about 4000 post mills had been built, each about 13 m diameter and with an average power output of about 1 kW. These were sufficient, together with about 6000 watermills, to meet the corn milling needs of the population, which was then about $4\frac{1}{2}$ million. However the Black Death of 1348–9 killed nearly half the population,[49] which by 1350 was down to about $2\frac{1}{2}$ million, and one of its many consequences was that windmill numbers subsequently declined. The population did not recover to its level before the plague until about 1600. Thereafter it grew, fairly slowly at first to about 6 million by 1700 and to about 8 million by 1800. Then, through the nineteenth century it grew rapidly, reaching 17 million by 1850 and 30 million by 1900. And as the population grew so did the requirement for milling capacity.

Late-eighteenth-century windmills were typically about 26 m diameter and would average between about 5 kW and 8 kW, see Table 3.2, depending on the windiness of their location. Overall their average power output was about 6 kW, sufficient – as discussed previously – to meet the milling needs of about 2000 people. So if in 1800 all the milling needs of the population of 8 million were met by windmills about 4000 would be required. By that date the dominant manufacturer of steam engines, Boulton and Watt, had built fewer than 500 and these, with a power output averaging about 12 kW (barely double that of a windmill),[50] were mostly used in mines, metal plants, textile factories and breweries. Milling was therefore done predominantly by windmills and watermills, though many of the latter were used in the textile industry and for other industrial applications. This therefore suggests that the number of windmills used for milling grain in England in 1800 was no more than about 4000, to which can be added about 100 industrial windmills and up to one thousand drainage mills.[51] There are, perhaps surprisingly, no accurate figures for the number of windmills in England in the nineteenth century. Hills quotes two estimates for about 1820, (when the population was 11 million) one suggesting there were 5000 windmills and the other suggesting 10 000; it would seem, from the figures given above, that the lower of these estimates is more probable.[52]

Causes of decline

Steam power

Though James Watt is best known for his invention of the separate condenser (patented 1769) which greatly improved the efficiency of steam engines, their applications were at first limited to pumping water; and it was only after his further innovations in the early 1780s that steam engines able to drive rotating machinery became both possible and commercially attractive.[53] These innovations were all embodied in the Albion Mill which was built by Boulton and Watt in London, next to the Thames and close to the southern end of Blackfriars Bridge, with one steam engine powering (initially) 6 pairs of millstones. It was the first building designed to use rotary steam power and was completed in 1786; a second engine and additional millstones were added in 1789. The two double-acting 50 hp steam engines then powered up to 20 pairs of millstones, each able to grind 8 bushels of wheat per hour, plus the machinery for lifting flour and grain into and out of barges and for dressing (that is sieving) the meal.[54] The volume of flour produced was sufficient to reduce the price of flour in London and put a number of windmills out of business. Albion Mill was destroyed by fire in March 1791, with arson suspected. It was not rebuilt, but it pointed clearly towards the changes that would affect milling (and many industrial processes) in the next century.

James Watt's patents expired in 1800, allowing other manufacturers to compete with Boulton and Watt, and the pace of steam engine development quickened. Boulton and Watt had been reluctant to use 'strong' steam, that is steam at a pressure much above atmospheric, because of the risk of explosions. Given the materials available at the time the risk was real, but materials improved, as did boiler designs, and engines that could operate safely at higher pressures were developed. This led to higher powers, more compact engines, and higher efficiencies. And as steam engines improved and became more economical they were used increasingly to meet the milling needs of the growing population. One of the short stories in Daudet's *Letters from my Windmill*, which was first published in Le Figaro in 1866, poignantly describes how the construction of a steam-powered flour mill near Tarascon, in Provence, in the 1840s put out of business most of the surrounding windmills, an episode that would have been repeated many times in England, as in France, through the nineteenth century.[55]

The Corn Laws, which had restricted imports and kept prices high, were finally repealed in 1846 and the increased corn imports which followed helped to feed the growing population. The practicality of using steamships for long-distance transportation was first demonstrated by the transatlantic crossing of Brunel's *Great Western* in 1838, and by the 1870s iron-hulled, screw-propelled ships powered by high-pressure steam engines had greatly improved the economics of marine transportation. This – helped by a succession of poor harvests in England through the 1870s – led to a further surge in the volume of grain imports, which came mostly from the United States, Russia and India.[56] By the 1880s wheat imports averaged 3.5 million tons per year, up by a factor of 30 from their level in the 1820s and 1830s, and sufficient to meet about half the total consumption. Much of this imported grain was milled at the ports, using purpose-built steam-powered flour mills, and distribution throughout the country was facilitated by the network of railways built since the 1830s. Windmills struggled to compete with the high-volume steam-powered flour mills, and they had the added disadvantage that they could not be relied upon to mill grain every day; there would be periods, especially in the summer, when for days on end wind speeds would be too low for them to operate. In earlier centuries customers had to put up with this inconvenience; now they no longer had to. The windmill's difficulties were made worse by the fact that the flood of imports had halved the price of wheat and encouraged many English farmers to switch from growing cereals to market gardening and dairying.[57] But an even greater problem for the hard-pressed windmill owner was the public's growing demand for white bread.

White bread

In traditional milling, also known as *low milling*, wheat grains are reduced to meal in a single pass between closely spaced millstones. Each grain is a few millimetres long and has three components, a brownish protective outer coating of bran, a pin-head sized germ which is the plant embryo and the white starchy endosperm which accounts for about 85% of the grain and which – when ground – provides the flour. The *wholemeal flour* that results from low milling contains all three components, and until the nineteenth century most people ate bread baked from this;[58] as the flour contained the bran, the bread was brown. To get white bread the bran must be removed. In medieval times this was achieved by sieving the meal, a task that was at first performed by the baker, not the miller. With most of the bran removed the baker could produce a softer, whiter bread which was favoured by the few who could afford it, and white bread became a symbol of high social status. In the City of London in the fourteenth century there was already a White-Bakers Guild as well as a Brown-Bakers Guild, and by the seventeenth century the white bread bakers were dominant.[59]

In later centuries sieving was more usually done by millers, and the process of sieving was improved by the invention of the *bolter* in the sixteenth century. This was a cylindrical wooden frame covered (except for its ends) in fine woven cloth and inclined at an angle of about 30° to the horizontal. The meal chute from the stones channelled the meal into the upper end of the cylinder, which was rotated about its axis using some of the power available from the main shaft in the windmill, either by belt drive or by gearing. The weight of the meal made the cloth sag, and as the cylinder turned the cloth was pressed against longitudinal rods fixed outside but parallel to the cylinder. This forced the white flour through the cloth, where it was collected in a wooden bin. The bran and the *middlings* (i.e. particles of endosperm, many with flakes of bran still attached) fell out from the lower end of the cylinder, and were collected in a separate bin. Over the years more elaborate and more efficient arrangements for separating out the white flour were developed,[60] but it was hard to prevent small brownish particles of bran and small yellowish particles of germ from being mixed in with the flour.

White bread slowly became more affordable, and its consumption spread down the social scale. This process was helped by the introduction of *high milling*, first developed in France in the eighteenth century and greatly improved through the nineteenth century in Austria-Hungary. In high milling the grain was first passed through one pair of stones with a relatively large gap between them, so that only partial splitting and breaking of the grain occurred. Some flour and bran was produced, and was separated out, but most of the grain was reduced to *semolina*, relatively large particles which are mostly the white, starchy endosperm. This was then passed through several additional pairs of stones, set progressively closer together, with careful sieving to give some removal of flour and bran after each stage. This high milling process gave nearly three times the yield of fine white flour.[61]

Roller mills

The final major step in the evolution of milling technology was to use a succession of pairs of chilled iron rollers instead of millstones.[62] This allowed the whole milling process to be controlled more closely, was much less labour intensive, and could more readily be adapted to cope with the very different milling characteristics of the *hard* wheat which was being imported from the United States in increasing volumes during the second half of the nineteenth century. Hard wheat generally has a higher gluten content, and it is gluten which is responsible for the light and spongy texture that is characteristic of bread. This is because when water is added to the flour the gluten gives a pliable but elastic mixture; this traps as small bubbles the carbon dioxide produced by fermentation when a small quantity of yeast is also added. The higher gluten content of the hard wheat imported from the US gave a lighter loaf, which consumers preferred.

The first complete roller-milling system was installed in a McDougall Brothers mill in Manchester in 1879 and by 1895 the use of millstones had almost finished. With roller milling about 75% of the grain could be converted to fine, white flour, far in excess of what low milling between stones could provide. And the high volume of grain processed at the large, steam-powered, port-based, roller mills, combined with the lower prices that had resulted from massive grain imports, meant that white bread could at last be afforded by virtually the whole population. Windmills just could not compete. The capital cost of a small-throughput roller-milling system, complete with the building required to house it, was far too high for small millers; and the power required for roller-milling was more than double the power required for traditional low-milling between stones.[63] And despite a campaign in favour of wholemeal bread[64] by the Bread Reform League and some members of the medical profession it was clear that the public preferred white bread.

The few windmills that survived in operation beyond 1900 were mostly used to grind animal feed. At their peak, in the period 1825 to 1850, there had been somewhere between 5000 and 10 000 windmills in Britain, but their time had passed. They had served people well for a full seven centuries, but they were now succeeded by a more cost-effective technology. Exposed to the weather, and made of wood, a windmill in operation required continual maintenance and occasional repairs; when it ceased to be used it decayed, and within a generation or two it would usually be little more than a ruin. The Society for the Protection of Ancient Buildings became concerned about the rate at which windmills were being lost, and in 1931 formed a Windmill Section which has been instrumental in protecting and preserving many old windmills and, where possible, encouraging their restoration to working order.[65] Today there are about 150 surviving windmills in the UK with about 35 still occasionally grinding grain, cherished reminders of the contribution that windmills once made to meeting our need for our daily bread.

Windmill numbers in other countries also declined from the mid nineteenth century onwards, and for similar reasons, though the pace of decline was generally not so rapid as in the UK. Germany had about 16 000 corn-grinding windmills in 1875, with 4000 still in use in 1933.[66] In the Netherlands there were about 9000 windmills in 1850, used for land drainage, corn grinding and for a wide variety of industrial applications, and 2500 were still working in 1900.[67] Attempts were made in the 1920s, most notably by Adriaan Dekker, to apply to the design of windmill sails some of the lessons that had been learnt in the design of aeroplane wings, and significant performance improvements were achieved.[68] It was however too late to have any material effect on the steady decline in the use of windmills. By then windmills had additional competition from stationary diesel and petrol engines, and the spread of electricity distribution systems into rural areas gave an even more convenient option. A society for the preservation of windmills in the Netherlands, De Hollandsche Molen, was formed in 1923 and has helped to ensure that nearly 1000 remain in working order; of these just under 400 are drainage mills and the remainder are corn mills and industrial mills.[69]

The American multi-bladed windpump

Though the latter half of the nineteenth century was in general a period of decline for windmills, there were two notable exceptions. The closing years saw the beginnings of the use of windmills for electricity generation, a story that will be told in the next chapter. The second half of the nineteenth century also saw the remarkable success of the multi-bladed American farm windmill, or windpump. Walter Prescott Webb argued in his influential book *The Great Plains*, published in 1931, that success in settling the Great Plains of the United States – to the west of the Mississippi river – in the late nineteenth century was the result of three technological innovations, the Colt 45 revolver, barbed wire and the windpump. The revolver helped to defeat the Native Americans, who were the previous occupants of the land; barbed wire could be used to contain cattle and other livestock on land where there were too few trees for wooden fences; and though the land was semi-arid, there was abundant water underground, which the windpump could pump to the surface to meet the needs of the settlers and their animals.

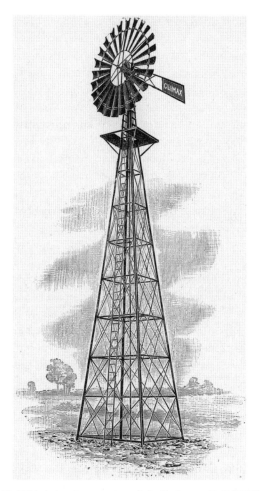

Figure 3.8 Multi-bladed water pumping windmill. *Source*: Ball (1908).

The development of windmills to pump up water from the underground aquifers commenced in the 1850s, as is detailed by Hills[70] and by Baker,[71] and by the 1870s multi-bladed windmills similar to the one shown in Figure 3.8 were being made by dozens of companies across the US. The most striking feature is the large number of blades, which in early windpumps were flat and made from wood; later machines used blades with a concave profile (as can be seen at the blade tips in the figure) made from sheet steel, which were more durable as well as being aerodynamically more efficient. As noted previously, the number of blades has little effect on the power output, for the power in the wind flowing through the rotor can be efficiently intercepted using any number of blades. The use of many blades, and a high *solidity*,[72] does however give a relatively high starting torque, even in light winds, and a relatively low rotational speed.

The high starting torque was particularly important as the windmill was used to operate a reciprocating pump in a borehole, which was typically 10 m to 100 m deep. The boreholes were drilled by itinerant teams of well drillers and were usually about 0.15 m diameter, with a sheet-iron casing.[73] Within the borehole was a drop pipe and at the bottom of this there was a cylinder pump of matching diameter. A crank-arm arrangement converted the rotary motion of the windshaft to a reciprocating motion, and so actuated a pump rod which extended down through the drop pipe to the cylinder pump at the bottom. During every down stroke of the pump rod water flowed through a non-return valve into the volume above the pump's plunger; during every subsequent lift stroke the plunger would then lift the column of water above it, and water would flow out from the top of the drop pipe.[74] The torque produced by the windmill had to be sufficient to lift the weight of the column of water in the drop pipe, so for deep boreholes one would have to use either a smaller diameter drop pipe or a larger diameter windmill. Typical windmill sizes for use on farms ranged from about 1.8 m to 4.9 m diameter (6 ft to 16 ft). Another important application was to pump water into storage tanks on towers by the side of railway lines, so that the steam locomotives could replenish their water tanks; railway windpumps were usually larger and could be up to about 9.1 m diameter (30 ft).

It was essential that the windpump could operate unattended. The use of a tail vane, as is shown in Figure 3.8, was sufficient to ensure that in light-to-moderate winds the rotor always faced into the wind. The bigger problem was how to furl the windmill in stronger winds. Different manufacturers found different solutions but the one that in due course was most widely adopted was to offset the windshaft slightly from the vertical axis about which the rotor turned to face the wind;[75] the tail vane was also hinged where it joined the body of the windmill so that the vane could fold round to be parallel to the plane of the rotor. A spring between the tail vane and the windmill body would in light-to-moderate winds hold the tail vane in the position shown in Figure 3.8, so that it could do its job and keep the windmill facing into the wind. However in stronger winds the downwind force on the offset rotor (which was trying to turn the rotor out of the wind) combined with the force exerted by the tail vane (which was trying to turn the rotor back into the wind) would be sufficient to overcome the spring and allow the tail vane to fold parallel to the plane of the rotor; the plane of the rotor would then lie approximately parallel to the wind direction, and could safely withstand storm force winds. When wind speeds moderated the spring would make the tail vane open out; the rotor would then turn to face into the wind, and pumping would recommence.

Having enough water to meet daily needs was the prime concern of the farmers and ranchers who bought the windmills.[76] The performance in light winds was therefore particularly important and the best windmills, such as the Aermotor, would start to pump water in wind speeds below 3 m/s. Though most of the companies that made windmills in America in the late nineteenth century were locally based, and had only a few employees, a few grew to dominate the market. Aermotor was one of these. Its first factory in Chicago was established in 1888 and in that year it sold 45 windmills, but by 1891 it was selling over 20 000 windmills per year. This volume of sales allowed Aermotor to make and sell their windmills at very competitive prices, and they went on to capture over 50% of the market. Aermotor windmills had curved steel blades, and their detailed rotor design was based on the work of Thomas Perry, who in the early 1880s had systematically tested a wide range of multi-bladed rotor configurations using a steam-powered whirling-arm test rig.[77] Aermotor windmills also used gearing, which reduced the windshaft speed by a factor of about three before converting the rotary motion to the reciprocating motion required to drive the pump. This effectively gave a three-fold increase in the starting torque, and so helped to give the windmills their good performance in light winds.

A multi-bladed farm windmill with a diameter of 3.0 m (10 ft) would, in a 5 m/s wind, give a power output of approximately 56 W;[78] about equal to the muscle power output of one man.[79] Modest though this may seem, it is sufficient to pump about 11 litres per minute from a depth of 30 m. Average wind speeds over the Great Plains are typically in the range 5 m/s to 6 m/s,[80] so a 3.0 m (10 ft) windmill could be expected to average a power output of this magnitude in most locations. An average of 11 litres per minute would then become 16.5 m^3/day, or 4350 US gallons/day (equal to 3620 Imperial gallons/day), which was sufficient for most applications. However to reduce the risk of running short of water during sustained periods of light winds it was prudent, and customary, to have a water tank close to the windpump which was large enough to store the equivalent of several days consumption.

Many hundreds of thousands of multi-bladed farm windmills were made in the United States in the late nineteenth century, and their volume production continued through to the 1930s. Some estimates put the total built at over 6 million, but numbers subsequently declined as electricity networks spread into rural areas.[81] Then when a windmill needed repair it would usually be cheaper to buy an electric pump. However Gipe suggests that about 60 000 are still in use in the US,[82] mostly in the southern Plains area, and sales of new windmills continue at just a few thousand per year. Not surprisingly, given its success in the US, the multi-bladed farm windmill has also been used very successfully in other parts of the world where semi-arid land conditions exist in combination with abundant underground water supplies. Argentina, Australia and South Africa have all used windpumps in large numbers, with imports often succeeded by local manufacture and continuing volume production well into the second half of the twenteeth century. Walton, for example, notes that though production of multi-bladed farm windmills did not commence in South Africa until 1942 (when the Second World War disrupted imports) in the next 30 years over 150 000 were built, with windmill number 150 000 being presented to Kruger National Park.[83]

The continuing spread of electricity networks into rural areas has reduced the need for windpumps in other areas, as well as the US, but Gipe indicates that about 600 000 remain in use in Argentina, with about 250 000 still in use in Australia and about 100 000 still in use in South Africa.[82] Windmills cannot compete with electric pumps in areas served by national or regional electricity networks, so these numbers can be expected to decline as electricity networks continue to spread into rural areas. However rural electrification schemes are costly and it seems likely that the multi-bladed farm windmill will continue to meet local needs for water in many areas, for many decades.

4

Generating electricity: the experimental years, 1887 to 1973

Electricity is so fundamental to the way we live today that we tend to take for granted the many services it provides. As the very occasional power cut serves to remind us, one of the most important of these is having lighting available to us at the touch of a switch. The practical use of electric lighting commenced in the late 1870s, when carbon arc lamps were used to illuminate streets in Paris, London and other major cities.[1] The fact that a brilliant light was produced when a current of electricity flowed across a small gap between two carbon rods had been discovered several decades earlier, and a number of early arc lamps had been shown at the Great Exhibition of 1851. However it was only after the development of practical, steam-powered, electric generators in the 1860s that electric lighting could start to challenge the dominance of gas lighting. Though carbon arc lamps gave a very bright light the white-hot tips of the carbon rods would slowly burn away, and a complex mechanism was required to move the rods so as to keep the gap constant and sustain the arc. This made carbon arc lamps very expensive; their light was also far too bright for use in homes.

The problem of producing an electric light that could be made more cheaply, and whose light was suitable for use in homes, was solved independently by Thomas Edison in the United States and Joseph Swan in England in the late 1870s. Both developed a practical incandescent lamp in which a slender carbon filament was surrounded by a glass envelope from which the air had been evacuated; when an electrical current flowed through the filament, via wires that passed through the insulating neck of the glass bulb, it would glow white hot.[2] These light bulbs, as they came to be called, were in essence the same as the incandescent light bulbs still widely used today, the main difference being the use today of filaments made from tungsten instead of carbon. Edison and Swan avoided a potentially very costly legal dispute over UK patent rights by setting up a jointly owned company to manufacture their electric light bulbs, which were made in a range of sizes from 1 candlepower up to 1000 candlepower; 16

candlepower was the most popular, roughly equivalent to a modern 25 W tungsten filament bulb or to a 6 W energy-saving compact fluorescent lamp.

Early demonstrations of electric lighting systems included an evening football match played under arc lights at the Bramall Lane ground in Sheffield in October 1878; and the Victoria Embankment, beside the Thames in London, was illuminated by arc lights from December 1878. These and other demonstrations aroused considerable public interest, and the promise of electric lighting – without the smell, the heat or the risk of leaks associated with gas lighting – was soon recognised.[3] To have electric lighting in the home was at first only possible if, like the industrialist Sir William Armstrong, you were wealthy enough to afford your own complete installation.[4] However public supply systems, offering electricity to shops and to homes (usually as additions to an electric street lighting installation) soon followed. The very first such system was installed in Godalming, a small town about 50 km south-west of London, in November 1881, and was followed in January 1882 by the larger and steam-powered Edison installation at Holborn Viaduct in London. A similar Edison installation built at Pearl Street, New York, opened in September 1882 and gave the first public electricity supply in the United States.[5]

The first wind-powered electricity generation

It was against this background of growing interest in the use of electricity that Sir William Thomson (who later became Lord Kelvin) suggested in his address to the British Association meeting in York in 1881 that as fossil fuel resources were consumed, and became more expensive, wind power might be used to generate electricity.[6] This comment was widely noted and certainly encouraged Professor James Blyth, of Anderson's college, Glasgow (a predecessor of today's Strathclyde University) to build and test in 1887 the first windmill to be used for electricity generation.[7] As Blyth describes his first electricity generating windmill was of traditional smock mill design (broadly similar to that shown in Figure 3.5) with four canvas-covered sails and a diameter of about 8 m, and adjoined a holiday cottage that he had in Marykirk, a small village about 45 km north-east of Dundee.[8] Bevel gearing was used at the foot of the vertical drive shaft to give a horizontal output shaft, on which was a 3 m diameter flywheel; a rope drive from the rim of this then turned the shaft of a dynamo, which charged a battery of 12 storage cells. When the wind speed dropped and the dynamo output voltage became too low a cut-out arrangement would disconnect it from the battery's charging circuit. The power stored in the battery was used to light up to ten 8 candlepower, 25 volt, incandescent lamps in Blyth's cottage.

Figure 4.1 Blyth windmill, 1891, Marykirk, Scotland. *Source*: Ball (1908).

Blyth clearly had considerable expertise in electrical matters, and had been publishing papers on electrical topics through the 1880s, but his windmill would seem to have been fairly basic. It had to be turned into the wind by hand, and the canvas sails also required manual reefing in strong winds. Blyth recognised the need for a windmill that could operate unattended and, after a period of experimentation with an American-style multi-bladed rotor, opted in favour of a vertical axis configuration and built the windmill shown in Figure 4.1 at Marykirk in 1891. The railway embankment and bridge over the river North Esk can be seen in the background, and the lady standing next to the open door of the hut (which contains the dynamo and storage battery) gives an indication of scale. The diameter was about 10 m and the 8 semi-cylindrical boxes, as Blyth called them, are each about 1.8 m wide and 1.8 m high.[9] The drag on a box is greater when the wind is blowing in to its open side – as on the right in the figure – than when the wind is blowing – as on the left – on to the curved outer surface. This drag difference is what makes the rotor turn but as noted previously drag-driven windmills are less efficient by a factor of about 5 than traditional windmills,[10] and Blyth's machine would have been no exception.[11] The power output from the rotating central vertical shaft was taken via gearing at its lower end, followed by a belt drive which further increased the speed, to a dynamo which – as with Blyth's 1887 windmill – was used to charge a battery of storage cells. Blyth states that 'in a fair wind' the power output was about 3 kW, but given the low efficiency the 'fair wind' must have been about 20 m/s, which corresponds to a Force 8 Gale.

In 1895 Blyth licensed Mavor & Coulson, a Glasgow firm of 'Engineers and Electricians', to make a modified version of his design. They built only one, which went to provide an electricity supply to Sunnyside Asylum, Montrose (just a few miles from Marykirk). A Mavor & Coulson handbill in the Blyth archive at Strathclyde University indicates that it was slightly larger than the Marykirk machine, with a diameter of 12 m, and was more robustly supported; it was rated, somewhat optimistically, at 10 hp (7.5 kW).[12] There is no information on its actual performance; Robert Ball, in his 1908 book *Natural Sources of Power*, simply notes that it suffered a fracture in the main 4 inch (102 mm) diameter vertical driving-shaft and was not re-erected. Blyth's windmill at Marykirk, shown in Figure 4.1, continued to generate electricity until it was dismantled in 1914.[13]

In the paper Blyth presented in January 1892 to the Royal Scottish Society of Arts he made reference to a windmill being used in 1891 to light two lighthouses near Le Havre in France, and there is evidence that Charles de Goyon (a French aristocrat with the title Duc de Feltre) first tried to use a 12 m diameter American multi-bladed windmill to provide electricity for the lighthouses there in the summer of 1887. However it took him at least two years to overcome the difficulties he encountered, and the only evidence we have that he eventually succeeded is Blyth's brief reference.[14] After Blyth the next person to succeed in using a windmill to generate electricity would seem to have been the American, Charles Brush who in 1888 used a large multi-bladed windmill to illuminate his Cleveland mansion. Brush had pioneered the development of practical arc lamps and the dynamos required to power them, and had become very wealthy through making and selling electric street lighting systems to cities ranging from New York to San Francisco.[15] Brush's windmill is described in considerable detail in a December 1890 issue of *Scientific American* which includes a picture, see Figure 4.2; the Brush mansion can be seen in the background.[16] The rotor had 144 blades and a diameter of 17 m and with the tower weighed about 36 tonnes. The tower was supported on – and free to turn around – a central wrought iron post (an arrangement similar in principle to that used in post mills) and the post extended 2.1 m below ground level into solid masonry. The small stabilising wheels at the end of outriggers near the base of the tower would in normal operation *not* touch the circular track on the ground; however in strong winds they would touch down as the structure deflected and would reduce the load on the central post. The large tail vane would point the rotor into the wind; the smaller auxiliary vane (which extends to the side of the rotor) was part of the furling system and helped to turn the rotor out of the wind during periods of high wind speed.[17]

Within the tower a two-stage belt drive arrangement transferred the power output from the windshaft to the dynamo's drive shaft, and increased the dynamo's shaft speed to 50 times the speed of the rotor. At full load the dynamo would then turn at 500 rpm and give an output of 12 kW.[18] Underground

Figure 4.2 Brush windmill, 1888, Cleveland, US. *Source*: Scientific American, vol. 43, p. 383, 20 December 1890.

conductors took this output to the basement of the Brush mansion where 12 storage batteries, each having 34 separate storage cells, gave an overall storage capacity of about 80 kWh. The house had in total about 350 incandescent lamps, with about 100 in everyday use; most of these were of 16 to 25 candlepower, equivalent in their light output to modern 25 W to 40 W tungsten filament bulbs.[19] When the *Scientific American* report appeared in December 1890 Brush's wind-powered electric lighting system had been operational for two years, and had 'proved in every respect a complete success'. It continued to operate for another decade, until the more convenient option of a public electricity supply became available.[20]

Though Blyth and Brush were successful in using wind-generated electricity to light their homes neither had developed a system good enough, and inexpensive enough, to encourage replication. Through the 1890s many more experimented with the use of windmills to generate electricity,[21] but the practical problems that they had to overcome were substantial. As Robert Ball noted in his 1908 book, 'it might be supposed that cheap electricity would be possible with the windmill at our disposal to drive our dynamos, and that the acquisition of "power for nothing" would solve the problem of electric lighting, and banish the small isolated steam plant and internal combustion engine. But such is by no means the case, for, with few exceptions, and these expensive ones, the windmill electric installation has proved to be a failure so far.'[22]

Ball gave as a typical example the windmill installed in the 1890s by George Cadbury (founder of the chocolate company that bears his name) to provide electricity to his home. As shown in Figure 4.3 it was a typical multi-bladed

Figure 4.3 Cadbury's windmill, ca. 1895, England. *Source*: Ball (1908).

design, with a diameter of 10.7 m, and was installed at an exposed location within the grounds surrounding the house. The output was used to drive a dynamo and charge storage batteries, however the batteries at that time had problems with the continual variations in the windmill's power output.[23] As Ball states 'the accumulators were found to wear out rapidly owing to the treatment to which they were exposed, and the plates buckled. The engineer in charge of the plant informed the writer that every possible expedient had been tried to make the plant a success, but without avail.' So after several years' experimentation Cadbury's windmill was relegated to the task of pumping water for the house, and a gas engine was installed to drive the dynamo and provide the house with electricity.[24]

The problems experienced by Cadbury were clearly quite common and were caused by continual variations in the wind speed. *Average* wind speeds change only slowly, over periods of several hours, as the large scale weather systems that can be clearly seen in satellite photos track across the country; the air in motion over a region such as southern England weighs many *billions* of tonnes and – like a supertanker – can only slowly start or stop.[25] However friction with the ground causes turbulence and eddies that are superimposed on the average wind speed and lead to significant short duration fluctuations in the wind speed (over time scales ranging from a few seconds to a few minutes),[26] with consequent variations in the speed at which the windmills turned.[27] These rotor speed variations were of relatively minor consequence for corn grinding or water pumping,[28] but with early stand-alone electricity-generating windmills they led to continual variations in the dynamo's output and were the cause of significant problems with battery storage systems.

Danish beginnings: Poul la Cour

Ball noted that one experimenter had been successful in overcoming these battery storage problems.[29] He was the Dane, Poul la Cour, whose work is credited with laying the foundation for what became, in the late twentieth century, Danish dominance in the design and manufacture of electricity generating windmills. La Cour's early career had been in the Danish Meteorological Office, where he became deputy director, but he was also a prolific inventor who made some notable contributions to the still-novel electric telegraph. In 1878 he became a teacher at Askov Folk High School, and was very concerned to improve living conditions in rural areas. In particular he wanted to bring to people living in the countryside the benefits of electricity, so that they could have a good light for reading and power for working barn machinery in the long and dark Danish winter evenings; he saw the electricity-generating windmill as a suitable means for achieving his objective. He therefore sought and secured Danish government funding and built his first windmill at Askov (in Jutland, about 40 km east of Esbjerg) in 1891. This, shown on the right in Figure 4.4, had a diameter of 11.6 m and four sails each 2 m wide. To provide more power he built in 1897 the larger windmill shown on the left in the figure; this had a diameter of 22.8 m with sails that were 2.5 m wide.

Both windmills were fully automatic in operation, with fantails to point them into the wind and with self-reefing shuttered sails.[30] And both had the benefit of a device la Cour invented in 1891, to smooth out the power fluctuations that result from the turbulence in the wind, which he called a *kratostat*. This was a mechanical device in the belt drive between the windshaft and the dynamo which – in a controlled way – allowed some belt-slipping during gusts of wind.[31] Offices and laboratories were housed on the ground floor in the larger 1897 windmill and the output from the power-smoothing kratostat was used to drive two dynamos, each giving an output of up to 9 kW. One of these was used to charge a 60 cell storage battery which had a capacity of 43 kWh; power from this battery was then used for lighting and to drive electric motors. The other

Figure 4.4 La Cour 1897 and 1891 windmills, Askov, Denmark. *Source*: Image courtesy of the Poul la Cour Foundation.

dynamo was used to produce hydrogen and oxygen by the electrolysis of water; these two gases were stored in separate gas-holders and the hydrogen was used in gas lamps to illuminate the school rooms.[32]

La Cour also used a very early wind tunnel to investigate the performance of model windmills and concluded that the optimum configuration for electricity generation was a four bladed rotor with a solidity of about 20%, operating at a tip speed ratio of 2.4. His tests indicated that the sail profile, especially near the tips, should be curved (rather like the curvature used on the late-nineteenth-century multi-bladed water-pumping windmills, such as that shown in Figure 3.8) and the weather angle should vary from about 10° at the sails tips to about 25° near the hub.[33]

To spread knowledge of how to generate and use electricity from the wind la Cour founded a Society of Wind Electricians in 1903 which ran courses for electricians in rural areas and helped to establish rural power stations. By the time la Cour died in 1908 it had been involved in 95 such installations, and 32 used wind power to provide part of the electricity supplied. However the spread of public electricity systems using oil burning engines to power electrical generators made electricity more affordable, and interest in electricity-generating windmills then diminished until the outbreak of World War I in 1914. This led to severe restrictions on Danish fuel imports and prompted a resurgence of interest in electricity-generating windmills. When the war ended in 1918 there were about 120 in operation, each with a diameter of approximately 16 m and a maximum power output of about 30 kW, helping to provide the electricity supplied by rural power stations. Fuel imports then returned to normal and, as electricity produced by power stations burning fossil fuels was cheaper than wind-generated electricity, the use of windmills again declined.[34]

Public electricity supply systems that need to supply large areas use alternating current (AC), primarily because the high voltages that are required for low-cost transmission over long distances can be easily reduced using transformers to the much lower voltages at which – for safety's sake – power is supplied to consumers.[35] The growing public electricity supply system in Denmark was therefore, as in other countries, mostly an AC system, though localised direct current (DC) systems continued to be used in some of the more remote locations, including islands. Electricity-generating windmills at that time were designed only to provide a DC power output, which limited the areas where they might be used to those that still had DC supply systems. Local circumstances would occasionally favour the use of an electricity-generating windmill, and Lykkegaard made and installed about 20 during the inter-war years, with outputs up to 45 kW. Windmills made by this company leant heavily on la Cour's pioneering work, and when his 1897 windmill was destroyed by fire in 1929 it was replaced by one made by Lykkegaard, see Figure 4.5. Another small Danish company that made and sold electricity-generating windmills in the inter-war years was Agricco; taking advantage of developments that had been made in aircraft wing design this company was one of the first to make windmills with metal-covered blades that had a streamlined, aerofoil, cross-section.[36]

The inter-war years

Though the level of wind power activity in Denmark was low during the inter-war years, there was only limited progress elsewhere. One very novel project was the windmill built by Flettner in Germany in 1926.[37] A spinning object moving through air (or any other fluid) experiences a side force known as the Magnus effect, named after the nineteenth-century German physicist who is credited with first describing it.[38] This effect is well known on spinning balls, whether golf balls, tennis balls, footballs or any other, because of the way it makes them swerve through the air. A spinning cylinder experiences a similar side force, and Flettner built a ship with two such cylinders mounted vertically on the deck; motors below the deck made it possible to control both the spin rate and the spin direction and Flettner's rotor ship made a successful crossing of the Atlantic in 1925. The

Figure 4.5 Lykkegaard windmill, 1929, Askov, Denmark. *Source*: Image courtesy of the Poul la Cour Foundation.

windmill he built the following year had four spinning cylinders (instead of blades or sails) which were turned by electric motors in the hub. This prototype machine had a diameter of 20 m and was rated at 30 kW in a wind speed of 10.3 m/s, which corresponds to a somewhat disappointing power coefficient $C_P \approx 0.14$. Flettner had plans to build a second machine in 1927 with a rotor diameter of about 100 m but – perhaps unsurprisingly – finance was not forthcoming.

Savonius, a Finnish engineer, invented a windmill in 1922 which in operation resembles a rotating cylinder, though its cross-section approximates to an S-shape.[39] It turns readily in the wind but tests in the 1920s were only moderately encouraging. There was a resurgence of interest in Savonius rotors in the 1970s, largely due to the ease with which enthusiasts could make their own by cutting an oil drum lengthwise into two semi-cylinders and then joining the two halves to give the required S-shape. By the standards of today's wind turbines their efficiency is low, with a maximum power coefficient $C_P \approx 0.2$, but a bigger problem results from the fact that their solidity[40] is 100% and their design is such that they cannot be turned out of the wind. Their structure and foundation

Figure 4.6 Balaclava, 1931, 30 m dia., 100 kW, Crimea. *Source*: Putnam (1948).

must therefore be able to withstand very high once-in-a-lifetime wind speeds; they are consequently more costly – relative to their output – than conventional horizontal-axis windmills.

Probably the most significant wind project in the inter-war years was the 100 kW rated, 30 m diameter, three bladed, Russian wind turbine shown in Figure 4.6, which was constructed in 1931 near Balaclava, on the north coast of the Black Sea.[41] The downwind force on the rotor was reacted by the inclined strut, whose lower end sat in a trolley that could ride round a circular track to point the rotor into the wind. The significance of this project lies in the fact that it was the first wind turbine[42] designed to supply AC power to the local electricity distribution system and, like most of the wind turbines deployed commercially through the 1980s and subsequently, it was grid-connected via speed-increasing gearing and an induction generator that were housed within a metal-clad nacelle at the tower top. Induction generators[43] are simple and reliable machines that turn at a speed determined by the grid frequency[44] and they consequently keep the wind turbine's rotational speed *almost* constant. The amount of speed variation depends on the detailed design of the induction generator, but as the power output of the wind turbine goes from zero to its rated maximum the rotational speed will typically increase by about 1%. Small though this is it is sufficient to allow the wind turbine's rotor to act as a flywheel and partially smooth the power fluctuations that result from turbulence in the wind.

Though its blade construction was relatively crude the Balaclava machine achieved its rated output of 100 kW in a wind speed of 11 m/s, corresponding to a power coefficient $C_P \approx 0.17$. The rotor turned at 30 rpm, giving a blade tip speed of 47 m/s, and the multi-stage gearing (which – somewhat surprisingly – used wooden teeth) increased the output shaft speed to the 600 rpm required by the induction generator. Prior to grid connection the rotor speed was controlled by varying the blade pitch,[45] and the force to change the blade pitch was provided by the offset ailerons that are visible halfway along the length of the blades. The wind turbine operated successfully for ten years, until it was damaged in World War II, and though successor machines were planned there is no evidence that any were built.

Small wind-electric systems

On a smaller scale, but of considerable importance in many rural areas during the inter-war years, was the development and volume production of small electricity-generating windmills. Public electricity supply systems did not then extend into most rural areas and though lighting needs could be met reasonably well without electricity, by using oil lamps with incandescent mantles,[46] the rapid growth of radio broadcasting through the 1920s and 1930s – and the concurrent development of more advanced (and more power demanding) radio receivers – created a demand for electricity for which there was no substitute. In the United States and elsewhere there were renewed attempts to adapt multi-bladed water pumping windmills for electricity generation, but with only limited success. The fundamental problem was that multi-bladed windmills like those shown in Figures 3.8 and 4.3 had been developed to provide a high starting torque in light winds, and with a tip speed ratio near unity their rotational speed was low. For electricity generation the starting torque is low, and what is needed is a fast-turning rotor that can efficiently capture the energy in a wider range of wind speeds. Several manufacturers developed such windmills and the most successful, such as the Wincharger and the Jacobs, were sold in their tens of thousands in the United States in the 1930s.[47] Though Jacobs windmills were expensive they acquired a reputation for reliability and durability, and to this day are highly prized by enthusiasts.

Marcellus Jacobs has given an interesting account of his windmill development through the late 1920s.[48] He started by using a two-bladed rotor but found

Figure 4.7 Jacobs battery charging windmill, 1930s, 2.5 kW, US. *Source*: Hau E.,
Wind Turbines (2006) figure 2.9, © Springer-Verlag Berlin Heidelberg 2006.

that excessive vibrations occurred whenever the wind direction changed and the
tail vane turned the rotor to face the new wind direction. He eventually realised
that the problem resulted from the fact that when the two blades were vertical
there was very little inertial resistance to turning, but one quarter of a revolution
later – when the two blades were horizontal – the inertial resistance to turning
was much greater.[49] Consequently the tail vane was 'forced to follow wind
changes by a series of jerks, causing considerable serious vibration to the plant'.
The solution was to use three blades, as can be seen in Figure 4.7, which shows a
4.6 m diameter, 2.5 kW Jacobs windmill. The blades were made from spruce, a
light but strong wood which was then widely used for making aeroplanes, and
had an aerofoil cross-section.

Power output from the Jacobs windmill was controlled by varying the blade
pitch, using an adaptation of the centrifugal governor described in Chapter 3.
Higher wind speeds made the rotor turn faster and the consequent increased
centrifugal force made three weights within the hub enclosure move radially
outwards; these weights were connected through a mechanical linkage to the
blades so that as they moved out they changed the blade pitch and limited the
rotor's speed increase.[50] Rotational speeds were typically in the range 125 to

225 rpm, and the DC generator was designed for low-speed operation so that it could be coupled directly to the rotor, without the need for a gearbox; the usual output voltage was 32 V. Jacobs windmills in the 1930s cost $490, plus $175 for a 15.2 m (50 foot) steel tower, and were usually sold with a 21 kWh storage battery which cost an additional $365. And Marcellus Jacobs indicated that at a typical location in the western half of the United States their average power output was about 450 kWh per month. A wide variety of Jacobs-branded domestic electrical appliances was made, ranging from vacuum cleaners and laundry irons to refrigerators and freezers, and for families who could afford a Jacobs (and whose land provided a suitably windy site close to the home) they performed a valuable service. However in the Depression years of the 1930s their $1030 total cost was beyond the reach of many potential customers, who had to be content with the less costly – and generally less reliable – windmills made by other manufacturers.

The small electricity-generating windmills made by Jacobs and others brought the benefits of electricity to tens of thousands of families in rural areas of the US through the 1930s and into the 1940s. Similar windmills, such as the Lucas Freelite in the UK and the Dunlite in Australia, performed a similar service elsewhere. However with the spread of electricity power lines into rural areas – a process that was accelerated in the United States by the creation in 1936 of the Rural Electrification Administration (REA) – these grid-independent small wind systems fell into disuse; few, when given the option, could resist the convenience of connection to a regional electricity grid system.

Grandpa's Knob, Vermont: the first megawatt-plus machine

As the inter-war years came to a close one very ambitious – and very large – wind project was initiated in the United States. The prime mover was Palmer C. Putnam, an engineer whose interest in wind energy commenced in 1934 when he 'had built a house on Cape Cod and had found both the winds and the electric rates surprisingly high'.[51] His initially quite modest plans gradually evolved until in 1939 he persuaded the S. Morgan Smith Company to finance the design and construction of a megawatt-sized wind turbine that could generate electricity and supply it at a competitive price to the Central Vermont Public Service Corporation. Almost all the electricity generated by this utility came from hydroelectric power stations, and they recognised that the output from wind turbines would allow them – during windy periods – to reduce the volume of water used; this retained water could then be used to help meet demand at other

Figure 4.8   Smith–Putnam, 1941, 53 m dia., 1250 kW, Vermont, US. *Source*: Putnam (1948).

times. If the prototype large wind turbine proved a success the utility expected to purchase several more.

Putnam describes the development of what became known as the Smith–Putnam wind turbine in his 1948 book *Power From The Wind*. The turbine, shown in Figure 4.8, was installed on an exposed hill-top site about 20 km west of Rutland, Vermont, in 1941 and commenced generation in the October of that year. With a rotor diameter of 53 m it was far larger than any previous wind turbine, and had a maximum output of 1250 kW (= 1.25 MW). The designers chose to use two blades, rather than three, in the expectation that a third blade would add substantially to the overall cost but would provide only 2% more energy. The blades were made with steel spars and ribs, covered with thin sheets of stainless steel, and each weighed about 8 tonnes. Unusually they were positioned downwind from the tower and were hinged near the hub so that they were individually free to swing – or *cone* – in the downwind direction, an arrangement designed to reduce stresses in the blades.[52] To help control the power output in strong winds the blade pitch could also be varied, that is to say each blade could – within limits – be rotated about its lengthwise axis so as to increase (or decrease) the level of interaction with the air flow.[53]

The wind turbine's design rotational speed was 28.7 rpm, giving a blade tip speed of 80 m/s. A speed increasing gearbox was then used to increase the output shaft speed to 600 rpm and this was connected not to an induction generator (as used by the Balaclava wind turbine shown in Figure 4.6) but to a synchronous generator.[54] As noted previously an induction generator allows a small variation in the speed of the wind turbine, which helps to smooth power fluctuations (caused by the wind's turbulence) and – importantly – this power smoothing substantially reduces the fluctuating loads in the transmission system. With a synchronous generator there is no speed variation whatsoever, and therefore no smoothing of the power output and no smoothing of the loads in the transmission system. To overcome this disadvantage an hydraulic coupling was positioned between the gearbox and the synchronous generator; though expensive this allowed a small variation in the rotor speed and so smoothed both the power output and the loads in the transmission system.

The gearbox, hydraulic coupling and electrical generator, together with the low-speed shaft bearings, oil pumps and other necessary machinery were enclosed in the box-like nacelle adjoining the rotor, and the complete nacelle and rotor assembly could be turned about a vertical shaft so as to face the rotor into the wind. The arrangement for turning the nacelle and rotor to face the wind had much in common with that used on smock mills (or tower mills) such as the one designed by Smeaton and shown in Figure 3.5. In both cases a large gear ring fixed to the top of the tower meshed with a small gearwheel attached to the nacelle or, in the case of the smock mill, to the cap. In the smock mill any misalignment between the wind direction and the rotor would make the fantail turn, and through reduction gearing this would drive the small gear and turn the rotor back into the wind. With the Smith–Putnam wind turbine any misalignment between the wind direction and the rotor was sensed by a wind vane placed on top of the nacelle; the signal from this went to an electrical control system which powered an hydraulic motor to drive the small gearwheel and turn – or *yaw* – the rotor back into the wind. Almost all the large wind turbines built in recent years have used a similar yaw orientation system.

The 610 m high ridge-top location chosen for the Smith–Putnam wind turbine had no specific name but the family who owned it always referred to it simply as 'Grandpa's', and as it had a well-rounded profile it became known as Grandpa's Knob. The site was selected with the expectation that the average wind speed would be a very high 10.7 m/s (24 m.p.h.) though subsequent measurements indicated that it was much less windy, with an annual average wind speed of just 7.6 m/s (17 m.p.h.). Wind speeds above the annual average contain more than 90% of the energy available in the wind (see Appendix C), so with the expectation that the average wind speed would be nearly 11 m/s the turbine was designed to start generating when the wind speed reached 8.9 m/s (20 m.p.h.). At higher wind speeds the power output increased rapidly, reaching the usual operational limit of 1000 kW in a wind speed of 13.4 m/s (30 m.p.h.). In wind speeds above 13.4 m/s the blade pitch was adjusted so as to hold the power output constant at 1000 kW; and on the infrequent occasions when the wind speed exceeded 26.8 m/s the wind turbine would be shut down.

A variety of mechanical problems interrupted the operational testing of the Smith–Putnam wind turbine following its first generation of electricity on 19 October 1941, and the failure of a main bearing in February 1943 led to a prolonged shut-down.[55] The United States had entered World War II on 8 December 1941, following the Japanese attack on Pearl Harbour, and wartime priorities were such that it took more than two years before a replacement bearing could be obtained and installed. Operation re-commenced on 3 March 1945 but 23 days later, when the wind turbine was operating in a quite moderate 11 m/s wind, one of the blades failed near its hub end and was thrown about 230 m. No-one was hurt but the machine did not operate again. The cause of the blade failure was identified and could have been avoided in subsequent wind turbines.[56] However a comprehensive design review concluded that though a batch of six successor machines of similar size could be built and installed on a nearby ridge (where wind speeds were substantially higher) for $190/kW the value of their output to the local utility was only $125/ kW. Smith–Putnam's attempt to build a cost-effective, megawatt-scale, grid-connected wind turbine therefore came to an end.

Wartime shortages rekindle Danish interest

One might anticipate that in Europe during the war years there would be no significant wind turbine developments, as priorities would lie elsewhere. However in Denmark the war years again saw restrictions on diesel fuel imports, with a consequent resurgence of interest in electricity-generating windmills. Between 1940 and 1943 Lykkegaard made and installed about 60 of their four-bladed windmills,[57] similar – see Figure 4.5 – to those they had made during the inter-war years and showing very clearly la Cour's influence. They were mostly about 18 m diameter, with 30 kW generators, and though their DC output restricted their use to the islands and small rural communities that still used DC these were the communities most affected by the diesel fuel shortage. Wartime restrictions on their usual export business also encouraged another Danish

Figure 4.9 FLS Aeromotor, 1941, 17.5 m dia., 60 kW, Denmark. *Source*: Claudi-Westh (1976).

engineering company, F. L. Smidth & Co., to develop – in 1941 – a 17.5 m diameter two-bladed electricity generating windmill with a 60 kW DC generator, see Figure 4.9; they followed this in 1942 with a 24 m diameter three-bladed windmill with a 70 kW DC generator.[58] F. L. Smidth had a subsidiary company that made small aeroplanes; their windmills benefited from this experience and had aerofoil section blades that were made with laminated wooden spars, and covered with a plywood or metal skin. The power in high wind speeds was limited by using spoilers that were normally recessed within the blades but could be deployed when required. As with most windmills (though not Smith–Putnam's) the blades were upwind of the tower, and twin fantails were used to point the rotor into the wind. F. L. Smidth had another subsidiary company that specialised in building reinforced concrete chimneys and grain silos; most of their windmills consequently had reinforced concrete towers, as can be seen in Figure 4.9 which shows one that was installed on the island of Bornholm. F. L. Smidth built 12 of their two-bladed windmills and 7 of their three-bladed windmills in the period 1941–3 and they remained in service typically for about 10 years. When the war ended in 1945 diesel fuel supplies returned to normal, and the use of electricity-generating windmills once again declined.

Post-war energy concerns, and Denmark's Gedser mill

There was however concern, even in the early post-war years, about the longer-term availability of energy supplies. For example Golding, author of the classic 1955 book *The Generation of Electricity by Wind Power*, noted that 'coal, oil and other fuels are being used up at an alarming rate'.[59] In Denmark concern was heightened by the fact that it had no domestic sources of coal or oil, and was dependent on fuel imports to meet the growing demand for electricity. With this in mind Johannes Juul – who had been a student of Poul la Cour – persuaded his employer, the Danish utility SEAS, to investigate the use of wind power in helping to meet Danish electricity needs. Until then windmills had only been used in Denmark to provide electricity in the relatively few and isolated areas that still had a DC electricity supply. Most consumers in Denmark, as else-where, had an AC electricity supply and Juul recognised that if windmills were ever to make a significant contribution they would need to provide AC power. From 1947, with SEAS support, Juul led a research and development pro-gramme with this objective and by May 1950 the first Danish windmill designed to produce AC power was operational at Vester Egesborg, on the south-west coast of Zealand and about 75 km south-west of Copenhagen.

This first Danish AC power-producing windmill was relatively small, with a rotor diameter of only 8 m and a maximum power output of 15 kW. However this small size allowed Juul to test different configurations speedily and inex-pensively. For example it was first operated, like the Smith–Putnam machine, with its 2 blades downwind of the tower but within a few months one of the blades failed. It was therefore rebuilt with new blades that were positioned upwind of the tower and was back in operation in this new configuration before the end of 1950.[60] It was grid-connected using an induction generator and was used by Juul for a variety of tests through 1951, after which it was left in service until 1960 when – having served its purpose – it was dismantled. The success of this small windmill encouraged Juul to build a larger machine in 1952 on the small island of Bogø, which is just to the south of Zealand and about 90 km south of Copenhagen. One of the F. L. Smidth DC generating windmills had been installed there in 1942 and Juul was able to re-use its concrete tower and gearbox. It was given a new 13 m diameter three-bladed rotor to Juul's design (with the blades in their usual upwind location) and used to drive a 45 kW induction generator connected to the local AC grid. It was designed, like the smaller Vester Egesborg windmill, to operate unattended and gave many years of trouble-free operation.[61] And although Juul's AC-generating Bogø windmill had a substantially smaller rotor diameter than its DC-generating predecessor its annual electricity output was three times higher.

Figure 4.10 Gedser, 1957, 24 m dia., 200 kW, Denmark. *Source*: Image courtesy of
the Danish Museum of Electricity, Bjerringbro.

The successful operation of the Vester Egesborg and Bogø windmills enabled
Juul to secure Danish Government funding for the next major step, which was to
design and construct a 24 m diameter, 200 kW rated, windmill – shown in
Figure 4.10 – at Gedser, about 130 km south of Copenhagen. This was completed in the summer of 1957 and after a period of measurement and adjustment
it was released for continual unattended operation in the summer of 1958.[62] Its
rather ungainly appearance results from the use of a steel box spar as the main
source of strength within each blade. Experience with the Vester Egesborg
windmill had shown the need to reduce the fatigue-inducing loads in the steel
spars by using a forward extension of the windshaft and stays to help carry the
downwind loads on the blades (Smeaton had used a similar arrangement on his
windmills in the eighteenth century, as is shown in Figure 3.5).[63] The Gedser
mill also had bracing between the blades to help reduce the bending load in each
blade that results from its own weight, and which reverses every revolution as it
turns from the '3 o'clock' position to the '9 o'clock' position and then on again
to the '3 o'clock' position. The blades of modern wind turbines are made from

fibreglass or other fibre-reinforced materials and with these materials the need for stays and bracing wires can be more easily avoided.

The Gedser mill had a concrete tower and the blades, positioned upwind of the tower, turned at 30 rpm; the output shaft speed was then increased using a 2-stage chain drive to the 750 rpm required by the 8-pole induction generator. Freely hanging cables then carried the 380 V power output from the generator,[64] which was in the nacelle, to a transformer at the base of the tower where the voltage was increased to 10 000 V; the power then flowed into the local electricity distribution network. The mill was turned to face into the wind using the same type of yaw orientation system as Smith–Putnam's, with a wind vane mounted on top of the nacelle giving a signal to the electrical controller when the wind direction changed; a small yaw motor would then slowly turn the rotor back into the wind.

The blades were given an aerofoil section by attaching shaped wooden ribs to the steel spar; the ribs were then covered with thin (1 mm) aluminium alloy sheeting. And like the sails on traditional windmills the blades were twisted along their length, with a weather angle of 3° at the tips which increased to 16° at the hub ends. As with Juul's earlier machines the blades on the Gedser mill were fixed to the hub. Compared with the variable pitch blades used by the Balaclava and Smith–Putnam wind turbines this substantially reduces the cost and complexity, but it also removes the ability to limit the power output in high wind speeds by changing the pitch.[65] However Juul knew from his experience with the smaller Vester Egesborg and Bogø windmills that with the blade pitch fixed, and with the rotor speed held constant by the induction generator, the power output would be naturally self-limiting in higher wind speeds. This is because as the wind speed increases the angle between the airflow and each blade steadily increases until the flow over the blade stalls,[66] just as the flow over an aeroplane wing stalls if the angle between the airflow and the wings becomes too large. For an aeroplane the loss of lift and the increase in drag that result from stalling can be dangerous; for wind turbines the principal consequence is the very beneficial limitation of the power output. This method of power limitation in high winds is often referred to as *stall control* so as to distinguish it – and the turbines that use it – from the *pitch control* used by turbines, such as Smith–Putnam's, that have blades with variable pitch.

One potential problem with all grid-connected wind turbines is what happens if the grid connection is lost, as can happen when storms bring down power lines. The resistance provided by the generator is then lost and if no action were taken the wind turbine would start to accelerate, and within a few tens of seconds it would seriously overspeed and fail. Pitch-controlled machines can respond by rapidly pitching the blades through a large angle so that they act as air brakes and the rotor speed is soon safely reduced to just a few revolutions per minute. With fixed blades, as on the Gedser mill, one cannot do this and the obvious solution, which is to provide a brake on the windshaft that is able to stop turbine runaway, is difficult and expensive (particularly so for larger turbines) because of the large amount of kinetic energy that has to be dissipated. Juul's simple but very effective solution to this problem was to make the blades with just the blade tips able to rotate, when required, through a large angle so that they would act as air brakes; and though usually all three blade tips would be actuated simultaneously Figure 4.10 shows the Gedser mill under test with just one tip turned to act as an air brake.

Each blade tip was attached to the adjoining main blade by a circular rod which passed through a tubular sleeve fitted lengthwise within the adjoining blade main structure.[67] In normal operation a cable running from this rod to the hub of the rotor held each blade tip tightly against the main blade, so that there was no discontinuity between the tip and the blade. However if the grid connection was lost the cables would be automatically released and centrifugal force would then drive the blade tips radially outwards; a helical groove within the tubular sleeve ensured that as each tip moved outwards it was turned through 60°. Even though their area was relatively small the braking effect of the blade tips was quite sufficient to reduce the rotor speed to a safe low level. This blade tip actuation arrangement also allowed the windmill to be shut down safely whenever required; then when re-starting the windmill the cables were used to pull the blade tips back to their normal running position. For operational convenience there was also a parking brake on the windshaft, but this could only be used after the tip brakes had reduced the rotor speed to a low level.

The Gedser mill had a solidity of 12% and its blade tip speed was 38 m/s. With its aerofoil profile blades Juul predicted a maximum power coefficient $C_P \approx 0.45$ in a wind speed of 8 m/s and it was designed to give its maximum output of 200 kW in a wind speed of 15 m/s, when the power coefficient would have been $C_P \approx 0.20$. Later measurements[68] indicate that the actual power coefficients were somewhat lower but it operated reliably for ten years following its 1957 completion. By then it needed a major overhaul. However by the mid 1960s the early post-war concerns about the longer-term availability of energy supplies had subsided; the full magnitude of the Middle East oil reserves had become apparent and increasing supplies, both from this region and elsewhere, had resulted in a substantial reduction in the price of oil.[69] This had led to a significant reduction in the price of electricity, and the decision was therefore made not to refurbish the Gedser machine.[70] Juul's Gedser mill was later to prove hugely influential in the development of the Danish wind industry, and many thousands of the Danish wind turbines installed in the decades after 1973 used the same basic stall-controlled rotor configuration, with three fixed-pitch blades located upwind of the tower, grid-connected using an induction generator and with overspeed protection provided by rotatable blade tip brakes.

French and German post-war programmes

Denmark's early post-war concern about energy supply security was shared by other European countries and several supported modest wind turbine research and development programmes. In France three experimental grid-connected wind turbines were built and tested between 1958 and 1966 by the electricity utility EDF (Electricité de France) in collaboration with two companies, BEST and Neyrpic.[71] All were three-bladed grid-connected machines with steel lattice towers and with self-aligning blades mounted downwind of the tower.[72] First to be completed was the BEST-Romani machine installed in 1958 at Nogent-le-Roi, about 60 km west of Paris, and shown in Figure 4.11. It had fixed-pitch aluminium alloy blades with a diameter of 30 m and was rated at 800 kW; it ran

Figure 4.11 BEST-Romani, 1958, 30 m dia., 800 kW, France. *Source*: Hau E., *Wind Turbines* (2006) Figure 2.13, © Springer-Verlag Berlin Heidelberg 2006.

experimentally for five years until a blade failed in 1963. The other two machines were built by Neyrpic and installed at Saint-Rémy-des-Landes, near the English Channel. The smaller of these had a rotor diameter of 21 m, variable-pitch fibreglass blades, and a rated output of 132 kW; it was completed in 1962 and ran reliably for more than three years. The larger Neyrpic wind turbine had a diameter of 35 m and a rated output of 1000 kW, and it too had variable-pitch fibreglass blades. It was completed in 1963 but ran for only seven months before the main shaft broke. By then declining oil prices had led to reduced interest in wind energy; it was consequently not repaired and the French wind energy programme ended when the smaller Neyrpic wind turbine was shut down in 1966.

In Germany in the early post-war years Ulrich Hütter developed a 10 m diameter 6 kW AC wind turbine, with three variable-pitch blades, which was used successfully for water pumping and other applications in several developing countries. He then went on to develop the 34 m diameter, 100 kW rated Hütter-Allgaier wind turbine shown in Figure 4.12 which was completed in 1958 at Stötten, about 80 km east of Stuttgart.[73] Hütter favoured low-solidity rotors designed to operate at high tip speed ratios, and his Stötten machine had a solidity of just 3%. It was grid-connected using a synchronous generator and the rotor turned at 42 rpm; this gave a blade tip speed of 75 m/s, nearly double the tip speed of the Gedser mill. The contrast between the 1000 kW rating of the 35 m diameter Neyrpic turbine and the 100 kW rating of the only slightly smaller Hütter-Allgaier machine is very marked, but it should not be assumed that the much higher rating of the Neyrpic machine would necessarily give a higher annual energy output. A high rated power output, relative to the swept area, can even be counter-productive as the extra energy output during relatively infrequent periods of high wind speed has to be offset by reduced energy output in the much more frequent periods of moderate winds, due to the reduced gearbox and generator efficiencies when they are operating well below their maximum output.[65] The Neyrpic turbine's high rated wind speed of 17 m/s would only be appropriate for very high wind speed locations. Modern wind turbines – such as those shown in Figure 1.3 – are designed for windy locations where hub height average wind speeds are usually in the range 7 to 9 m/s and they typically have rated wind speeds in the range 12 to 14 m/s.

Hütter's 100 kW wind turbine was one of the first to have blades made from fibreglass, and the two variable-pitch blades were located downwind of the tower. To help reduce stresses the blades were attached to a hub that could *teeter*,[74] that is to say that the hub with its two blades was hinged where it attached to the windshaft so that the blade experiencing the greater downwind

Figure 4.12 Hütter-Allgaier, 1958, 34 m dia., 100 kW, Germany. *Source*: Hau E., *Wind Turbines* (2006) Figure 2.16, © Springer-Verlag Berlin Heidelberg 2006.

force could tilt backwards by a few degrees; and see-saw fashion the other blade would then tilt forwards. Though there were some flutter[75] problems with the long slender blades the turbine operated experimentally until the late 1960s. However in Germany, as elsewhere, low oil prices led to lack of interest in alternative energy sources and the wind turbine development programme was ended.

The UK post-war programme

UK concern about energy supply security in the early post-war years led to a wind power programme that continued through to the early 1960s.[76] It was

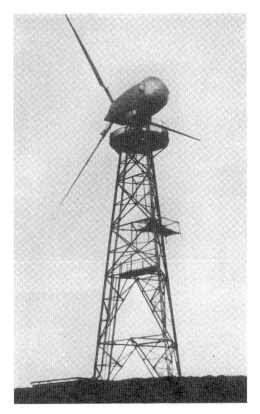

Figure 4.13 John Brown, 1952, 15 m dia., 100 kW, Orkney, Scotland. *Source*: Golding (1955).

started in 1948 when a national wind-power committee was set up, co-ordinated by the Electrical Research Association (ERA) under the supervision of E. W. Golding, to carry out a UK wind survey, to select sites and to design and test prototype wind turbines with ratings up to 100 kW. Its survey work led to measurements being made of the wind resource at 101 locations, most of which were hill tops; three of these were then chosen to be the test sites for three very different wind turbines, each with a rated power output of 100 kW.

The first of these machines, shown in Figure 4.13, was installed in 1952 by John Brown and Co. on Costa Hill, Orkney, an island group about 10 km north of the Scottish mainland. The Costa Hill site was extremely windy – measurements indicated that the average wind speed was 11 m/s – and the wind turbine's rated wind speed was therefore also a relatively high 16 m/s. Its three fabricated wooden blades were located downwind of the tower, and it was grid-connected using an induction generator; the gear ratio between the rotor's low-speed drive shaft and the generator was then chosen so as to give a rotor speed of 130 rpm. The blades were variable pitch and, as on the Smith–Putnam machine, free to cone[52]; following helicopter practice the blades also had so-called drag hinges which gave them some freedom of movement in the plane of rotation, but all this rotor complexity led to stability problems. As first installed the wind turbine had a diameter of 18.3 m, and the blade tip speed was – at 124 m/s – very high. It was however insufficient to give the blades the necessary rigidity[77] and when operating in a high wind one of the blades hit the tower. New and shorter blades were fitted, reducing the rotor diameter to 15.2 m, and tests continued. Problems also continued and attempts to correct them and improve the turbine's reliability were then hindered by the site's remoteness. The Costa Hill wind turbine was shut down in 1956.

The second of the UK 100 kW wind turbines to be built in the 1950s was the novel two-bladed, 24 m diameter, Enfield–Andreau machine shown in Figure 4.14 with some of the equipment used for erection still in place.

Figure 4.14 Enfield–Andreau, 1955, 24 m dia., St. Albans, England.

Designed by the French engineer J. Andreau and built by the de Havilland aircraft company (under contract to Enfield Cables) it had hollow blades which – as they turned – centrifuged air out from their open tips.[78] Air was sucked up inside the tower to replace the air expelled from the blade tips, and the rectangular vents where the air entered can be seen just over half way up the cylindrical lower end of the tower. As the air flowed up inside the tower it passed through a multi-bladed turbine wheel within the lower end of the tower, just above the air inlet vents, and the turbine's shaft power output was directly coupled to a synchronous generator. Eliminating the gearbox, and locating the generator close to ground level, would – it was claimed – reduce maintenance costs; and the absence of any direct connection between the rotor and the generator meant that the rotor speed could be allowed to vary. The two variable-pitch blades were made from aluminium alloy, and located downwind of the tower. The rotor was self-aligning and could turn at speeds up to 95 rpm, corresponding to blade-tip speeds up to 120 m/s.

The wind turbine was initially erected in 1955 near St. Albans, about 30 km north-west of London. This low wind speed location was chosen for its convenience close to where the turbine was built, and it was intended that after initial testing at St. Albans it would be moved to a much windier site at Mynydd Anelog, in North Wales, where it was to be operated by the British Electricity Authority.[79] However planning permission was refused and in 1957 it was bought by Électricité et Gaz d'Algérie and installed on a hill-top site called – appropriately – Grand Vent, near Algiers. Though it was designed to give its 100 kW rated output in a wind speed of 13.5 m/s tests in Algeria indicated that a wind speed of 15 m/s was required, corresponding to a power coefficient $C_P \approx 0.11$ at a tip speed ratio of 8.[80] This low power coefficient – little more than half that achieved by the much less complex Gedser mill at the same wind speed – was largely the result of high internal flow losses. The high speed of the air expelled from the blade tips added to these losses and also made the Enfield–Andreau machine very noisy. It was operated experimentally in Algeria for in total about 180 hours,[81] and when the bearings at the hub ends of the blades failed the wind turbine was not repaired. It was dismantled in the early 1960s.

The last of the three UK 100 kW wind turbines to be built in the 1950s, shown in Figure 4.15, had a diameter of 15.2 m and three fixed-pitch blades and was installed on a hill-top site in the Isle of Man in 1959. It was designed to be a simple and low-cost machine and the blades, which were positioned upwind of the tower, were made from extruded aluminium, with steel bracing rods providing additional strength.[82] The blades were made without twist or taper and their width, or chord, was 0.56 m, giving a rotor solidity of 7%. A fantail was used to keep the rotor facing into the wind. The wind turbine was grid-connected using an induction generator, and the gearbox ratio was chosen so as to give a rotor

Figure 4.15 Smith (Horley), 1959, 15 m dia., 100 kW, Isle of Man. *Source*: Reprinted from *Applied Energy*, **1**, Elliott D. E., Economic wind power, p. 192 figure 19(a). © 1975, with permission from Elsevier.

speed of 75 rpm; this gave a blade-tip speed of 60 m/s. With its fixed-pitch blades and constant-speed rotor the Isle of Man machine in some respects resembled the Gedser mill that had been built two years earlier; and like the Gedser mill its power output was limited in high wind speeds by stalling of the flow over the blades.

Stall-controlled wind turbines must have means of preventing overspeed if the grid connection fails and the Isle of Man machine had air brakes at the blade tips, which are clearly visible in Figure 4.15. These were hinged so that when the grid connection was lost they would open outwards and be perpendicular to the airflow; their drag would then be more than sufficient to prevent overspeed. However, and in contrast with the tip brakes used on the Gedser mill (see Figure 4.10), the tip brakes on the Isle of Man machine caused considerable drag even in the normal running position shown in Figure 4.15, and measurements made in zero wind conditions indicated that they caused a power loss of about 30 kW. The wind turbine was operated experimentally until 1963, when a blade broke after striking the tower. By then funding for the UK wind energy programme had ceased, so it was not repaired.

The Isle of Man machine had been the most successful of the three UK 100 kW wind turbine prototypes, and though it was less advanced than Juul's Gedser mill it pointed in the same development direction,[83] that is towards a robust, three-bladed, stall-controlled rotor with fixed-pitch blades located upwind of the tower, with grid connection using an induction generator and with air brakes at the blade tips for overspeed protection. However the early post-war worries about future energy supplies that had prompted the exploration of wind energy's potential in Denmark, France, Germany and the UK had, by the mid 1960s, been superseded by complacency. As already noted the price of oil had steadily declined from its early post-war peak,[69] and as people became increasingly aware of the magnitude of the Middle East oil deposits there was the expectation that cheap oil would continue to be available for the foreseeable future. There was, in addition, confidence that nuclear power would provide an additional and abundant source of cheap energy. After all Lewis Strauss, the Chairman of the US Atomic Energy Commission, had in September 1954 told an audience of science writers in New York that 'our children will enjoy in their homes electrical energy too cheap to meter'; and only two years later the first commercial nuclear power station, at Calder Hall in Cumbria, was opened by the Queen. Through the 1960s orders for nuclear power stations proliferated, both in Europe and the United States, and for a few brief years it seemed reasonable to anticipate a future in which energy supplies would be both abundant and inexpensive. There was – so it seemed – no need to put resources into the development of wind energy or any other renewable energy source.

5

The evolution of the modern wind turbine, 1973 to 1990

The 1973 oil crisis

Though people were confident, through the 1960s, that low-cost energy supplies would continue to be readily available for decades[1] there was growing concern in Europe, the United States and elsewhere about environmental issues. Emissions from coal-fired power stations, and the acid rain that these produced, were of particular concern. And the American Space programme that had its climax in 1969 with the lunar landing had produced dramatic images of the Earth – as seen from the surface of the Moon – which reminded people just how isolated our planet is, and encouraged them to question the wisdom of continuing to use the atmosphere and the oceans as dumping grounds for all our wastes. The growing environmental movement led to discussion of the possible use of alternative energy sources, but the prevailing low energy prices were a barrier to action. This all changed in late 1973 when a crisis in the Middle East caused oil prices to rise sharply. Though the background to the 1973 oil crisis is complex[2] what happened then to the oil price, and what has happened since, has had a profound effect on the subsequent development of renewable energy technologies in general, and wind energy in particular; forecasts of future oil prices continue to exert a strong influence on the programmes that many countries now have to increase their use of renewable energy sources.

Prior to 1973 the oil-producing countries had for many years complained that they were not getting a fair price for their oil. They had formed OPEC, the Organisation of Petroleum Exporting Countries, in 1960 to help press for higher oil prices but through the 1960s oil supplies exceeded demand and prices remained low. However the 1960s was a decade of rapid worldwide economic growth. Car ownership almost doubled from a global total of 98 million cars in 1960 to 193 million in 1970,[3] and free world oil demand more than doubled from 19 million barrels per day in 1960 to 40 million barrels per day in 1970;[4] two-thirds of this increase in demand was met by producers in the Middle East. By 1970 American oil production had reached its all-time peak, at 11 million barrels per day, and would subsequently slowly decline; meanwhile its demand for oil continued to grow strongly and between 1970 and 1973 its net oil imports doubled to 6 million barrels per day. By 1970 worldwide demand for oil was close to exceeding supply and the oil-producing countries could at last start to raise the official posted price of oil, slowly at first from $1.80 per barrel in 1970 to $2.90 by mid 1973.[5] Political pressures in the Middle East then led to the Yom Kippur[6] war when, on 6 October 1973, Egyptian and Syrian forces attacked Israel. By mid October OPEC had raised the posted price of oil to $5.1 per barrel, so as to match prices on the extremely nervous spot market, but the announcement soon afterwards by the Arab oil producers that they would cut back their oil production (with a complete ban on exports to the United States and the Netherlands) led to considerable alarm and further price volatility.

Though the Yom Kippur war ended in late October the oil production cuts continued for some months and had a very considerable political impact, which was enhanced in many countries by fuel shortages and long queues of motorists seeking to refuel their cars. In late December OPEC established a new price for oil, $11.7 per barrel; oil supplies thereafter slowly returned to normal, but prices remained high. High oil prices continued through the 1970s, and the overthrow of the Shah of Iran by Ayatollah Khomeini's Islamic Revolution in 1979, followed by the Iran–Iraq war in 1980, pushed prices higher still; by October 1981 OPEC's posted price for oil reached $34 per barrel.[7]

The fourfold increase in oil prices in late 1973 had a profound effect on attitudes to energy supply over the next decade, and countries sought to reduce their dependence on oil in a variety of ways. Measures to increase the efficiency of energy use were introduced and electricity generators turned away from oil to other sources, most notably coal and nuclear power; however construction delays, soaring costs and growing environmental concerns – reinforced in 1979 by the near-catastrophic accident at the Three Mile Island nuclear power station – soon reduced enthusiasm for the latter.[8] The need to reduce dependency on oil also led several countries, including the United States, to examine the prospects for using renewable energy sources, including wind energy. America moved more speedily than most and though wind was not initially thought to be the most promising renewable energy option a Federal wind power program was soon initiated.[9]

The American Federal wind power program

First generation machines, Mod-0 and Mod-1

The 1.25 MW Smith–Putnam wind turbine installed at Grandpa's Knob in 1941, shown in Figure 4.8, had operated for only about 1100 hours spread over 3½ years before it failed in 1945; and unlike Juul's Gedser mill, shown in Figure 4.10, it had never been allowed to operate unattended. However Putnam's book, *Power from the Wind*, gave the impression that the Grandpa's Knob machine was almost a success. And if near-success could be achieved in the early 1940s with the first attempt to build a megawatt-scale wind turbine then three decades later, with a much improved understanding of aerodynamics, materials and structural dynamics, the next attempt would surely succeed. This expectation, and the desire to make a contribution to meeting American electricity needs as speedily as possible, encouraged the Federal wind program managers to make the development of large grid-connected wind turbines their main priority. The decision was therefore made to move speedily forward with the design and construction of a series of large wind turbines which, like Smith–Putnam's, would have two blades located downwind from a lattice tower.

The first of these,[10] known as Mod-0 and with a rotor diameter of 38.1 m (125 ft), was designed and built by the Lewis Research Center of NASA (the National Aeronautics and Space Administration) and was installed near Sandusky, Ohio, in 1975; see Figure 5.1. Because the site wind speed was low the rated power output was just 100 kW. Four very similar machines, each with the same two-bladed, 38 m diameter, downwind rotor configuration but with a 200 kW rated output, were then built by Westinghouse and – designated Mod-0A – they were installed between 1977 and 1980 in locations ranging from Hawaii to Rhode Island.[11] The last of the Federal wind program's first generation machines, shown in Figure 5.2, was designed and built by GE (the General Electric Company) and installed at a location called Howard's Knob near Boone, North Carolina, in 1979.[12] Designated Mod-1 it also had its two-bladed rotor downwind from the tower; with a diameter of 61 m (200 ft) and a rated power output of 2 MW it was larger than the Grandpa's Knob wind turbine, and its blades were longer than the wings of a Boeing 747 'Jumbo Jet'.[13]

All six of the Mod-0 and Mod-1 wind turbines were, like the Grandpa's Knob machine, grid-connected using synchronous generators, all had lattice towers and all had full-span variable-pitch blades for power control. For yaw orientation they all used the arrangement pioneered on the Grandpa's Knob machine (and also used on the Gedser mill) with a wind direction indicator on

Figure 5.1 Mod-0, 1975, 38 m dia., 100 kW, US. *Source*: NASA.

top of the nacelle sending a signal to an electrical controller when the wind
direction changed; if the change was sustained the controller then activated a
yaw motor to slowly turn the rotor so that it again faced into the wind. The
Mod-0 and Mod-0A machines at first had blades built like aircraft wings, with
spars, ribs and skins made from aluminium alloy, but the cyclic loads they
experienced as they passed through the wake of the tower were greater than
anticipated and soon caused fatigue cracks. The aluminium blades were

Figure 5.2 Mod-1, 1979, 61 m dia., 2 MW, US.

therefore replaced with blades made from fibreglass or laminated wood-epoxy, and both these materials proved satisfactory.

The Ohio Mod-0 was used to test a variety of rotor and tower designs before being dismantled in 1987. The four Mod-0A machines were operated as and when they were able but averaged an annual energy output of only about 220 kWh per unit of rotor swept area, barely half what was then being achieved by small Danish wind turbines;[14] their cost of maintenance was greater than the value of the electricity they produced and they were shut down and dismantled in 1982.[15] The Mod-1 machine had blades made from welded steel by the Boeing Aerospace Company but the cyclic loads they experienced as they passed through the tower's wake were again greater than expected, and they would have required replacement if testing had not been curtailed for other reasons. A more pressing problem for this machine was the low-frequency aerodynamic noise caused by the blades passing through the localised wakes produced by the tower's tubular legs. The problem was largely resolved by reducing the rotor speed, which was achieved by changing the generator. However a mechanical failure where the hub attached to the rotor main shaft (the windshaft) led to the Mod-1 being dismantled in 1983, after just four years intermittent operation.

Figure 5.3 Mod-2, 1982, 91 m dia., 2.5 MW, US.

Second- and third-generation machines, Mod-2 and Mod-5

It was by now apparent that the design of reliable megawatt-scale wind turbines was much more challenging than had been anticipated, but the American programme to develop large wind turbines had acquired its own momentum. Before Mod-1 had even commenced testing Boeing had been contracted to design a 'second-generation' machine, designated Mod-2, which was to have a diameter of 91.4 m (300 ft) and a power output of 2.5 MW. The problems that had been experienced on Mod-0 with the blades downwind of the tower encouraged Boeing to design Mod-2 with its two blades upwind of the tower; and to further reduce the cyclic loads produced by interaction between the blades and the tower the Mod-2 was designed with a slender tubular steel tower.[16] Three Mod-2 wind turbines were completed and installed at Goodnoe Hills, near Goldendale, in Washington State between September 1980 and May 1981 and two of these are shown in Figure 5.3.[17] Two further Mod-2's were installed in early 1982, one – bought by the Bureau of Reclamation – in Medicine Bow, Wyoming, and the other – bought by the Pacific Gas and Electricity utility – in Solano County, to the east of San Francisco.

The Mod-2's two-bladed rotor was fabricated from steel plate and to save cost only the outermost 30% of each blade could have its pitch varied, for power control. Though the Mod-0, Mod-0A and Mod-1 wind turbines all had their blades attached rigidly to the rotor main shaft Hütter had successfully demonstrated, on his 1958 Stötten machine (see Figure 4.12), that blade stresses could be substantially reduced by using a *teetering* hub;[18] a similar arrangement was therefore used on the Mod-2, that is to say the central section of the blades was hinged where it attached to the rotor main shaft so that whichever blade was temporarily experiencing the greater downwind force could rock backwards by a few degrees. The main shaft speed was 17.5 rpm, giving a blade tip speed of 84 m/s, and a compact gearbox increased the output shaft speed to the 1800 rpm required by the synchronous generator. The rotor solidity was just 4% and the wind turbines were designed to give their maximum power output when the wind speed at the hub height of 61 m (200 ft) reached 12.2 m/s.[19] A succession of operational problems included the fatigue failure of the rotor main shaft, blade bearing failures and failures of the bolts which joined sections of the blades together. These and other problems were gradually overcome but through to the end of 1986 the three Goldendale Mod-2's had operated in total for barely 10% of the time available, and they were sold for scrap in 1987. The other two Mod-2's had similarly troubled operational histories and were also sold for scrap in the late 1980s.

The final step in the American Federal program to design and build cost-effective large wind turbines was the construction in 1987 on Oahu, Hawaii, of the 'third-generation' Boeing Mod-5B. The design of this was already well under way by the time the Mod-2's commenced operation, and Mod-5B was in essence a stretched version of Mod-2 with pitchable blade tips that were each 3 m longer; the rotor diameter was therefore 97.5 m (320 ft) and the rated power output was 3.2 MW.[20] And instead of the synchronous generator used by the Mod-0, Mod-1 and Mod-2 turbines the Mod-5B used a wound rotor induction generator,[21] which allowed the rotor speed to be varied in operation between 13 and 17 rpm. Mod-5B was more successful than its predecessors in the American large wind turbine programme and in its first 55 months of service it generated 24.5 million kWh of electricity, which corresponds to a *capacity factor* of 19%.[22] However it could not compete with the Danish wind turbines (discussed later in this chapter) which were then available; though these were smaller they were much more reliable and much more cost-effective. Mod-5B continued in operation intermittently until late 1996; it was then shut down and its gearbox and generator removed, and after standing idle for a few more years it was finally demolished in 2003.

Though the American large wind turbine programme had cost in total about $350 million it had, in the judgement of the American wind power expert and author Paul Gipe, delivered very little.[23] Lessons were of course learnt from the problems experienced with the dozen machines that were built, but they were lessons that could have been learnt much less expensively if there had been less haste in progressing to multi-megawatt wind turbines. However one useful and lasting result was the development of the laminated wood-epoxy method for making wind turbine blades. After the aluminium blades initially used on the Mod-0 and Mod-0A turbines proved unsatisfactory Gougeon Brothers of Bay City, Michigan, who had developed the skills necessary to make laminated wooden hulls for fast racing boats, were contracted to make replacement blades. The epoxy resin used to glue together the thin veneers of wood (each just a few millimetres thick) made the wood-epoxy composite structure waterproof and dimensionally stable, and enabled full advantage to be taken of wood's high strength-to-weight ratio and fatigue resistance. The wood-epoxy replacement blades for the Mod-0A's were very successful,[24] and when Westinghouse sought to commercialise their wind experience by building a modified and enlarged version of the Mod-0A the blades for the 43 m diameter rotor were made using wood-epoxy; unlike the Mod-0A this new turbine had its two-bladed rotor upwind of the tower. A batch of fifteen of these 600 kW rated Westinghouse wind turbines was installed on Oahu, Hawaii, in 1985 and the blades performed well for more than a decade.[25] However problems with the pitch-change mechanism impaired the overall turbine performance and Westinghouse, unable to compete commercially with the Danish wind turbines that were then available, withdrew from the market.

The American/Swedish WTS-4

One other noteworthy large wind turbine was erected in the United States in the early 1980s, outside the main NASA/Department of Energy Mod programme, and this was the 4 MW rated WTS-4 built by Hamilton Standard in co-operation with the Swedish company Swedyards. These two companies collaborated in the design of a 78 m diameter wind turbine with a two-bladed, teetered, down-wind rotor for the National Swedish Board for Energy Source Development (NE), and Hamilton Standard made the fibreglass variable-pitch blades using a novel filament winding process. Two almost identical prototypes were built and the first, designated WTS-3 and with a rated power output of 3 MW, was installed in mid-1982 for NE at Maglarp, near Malmo, on the southernmost tip of Sweden; see Figure 5.4.[26] The second prototype was the 4 MW rated WTS-4 which was bought by the US Bureau of Reclamation and installed in late 1982 at Medicine Bow, Wyoming, close to the site of the Bureau's Mod-2. Both turbines were grid-connected using synchronous generators, and the different power ratings were a consequence of the fact that the grid frequency in Sweden – as in the rest of Europe – is lower than in the United States (50 Hz as compared with 60 Hz); the WTS-3 rotor consequently turned more slowly

Figure 5.4 WTS-3, 1982, 78 m dia., 3 MW, Sweden.

than the rotor of the WTS-4 (25 rpm as compared with 30 rpm) and its rated power output was therefore also reduced. Like other multi-megawatt wind turbines of this period the WTS-4 performed poorly. In its first four years it generated only 8 million kWh, corresponding to a capacity factor of barely 6%. When its generator failed in 1986 the Bureau of Reclamation decided it was not worth repairing and it was sold at scrap value to a local wind power enthusiast. He succeeded in repairing it, and it then ran intermittently until 1994 when a failure of the pitch control mechanism caused a blade to strike the tower and break. This led to other component failures and the turbine, damaged beyond repair, was scrapped.

The Darrieus vertical-axis wind turbine

Though the American Federal wind power program spent most of its budget on the development of large wind turbines, many tens of millions of dollars were also spent on small wind turbines and on a variety of innovative wind turbine designs, most notably the Darrieus wind turbine. Grid-connected small wind turbines are less cost-effective than larger ones; hence the trend in Denmark through the 1980s from small to progressively larger wind turbines (as is discussed later in this chapter). The American small wind turbine programme will therefore not be covered here, but has been reviewed by Divone[27] and – somewhat more critically – by Gipe and by Righter.[28] Most of the innovative designs considered were soon dropped as their low efficiency and/ or high cost became evident,[29] but the Darrieus concept was more durable and for more than a decade held promise that it might challenge the dominance of the conventional two- or three-bladed horizontal-axis wind turbine.

Georges Darrieus was a French wind turbine pioneer who was responsible for the design of a number of conventional horizontal-axis wind turbines (HAWTs) in the 1920s and 1930s, but he is best known today for his invention in 1925 of the modern vertical axis wind turbine (VAWT).[30] It made little impression at the time but Raj Rangi and Peter South, working in Canada at the laboratories of the National Research Council in Ottawa, re-discovered this turbine design in the late 1960s and – especially after the 1973 oil crisis – their work stimulated a great deal of interest in this novel configuration.[31] In the United States work on Darrieus wind turbines was centred on the Sandia National Laboratories in Albuquerque, New Mexico, and Figure 5.5 shows a Darrieus wind turbine on test at this location. The two curved blades are attached to the central vertical column, which is held in place by bearings at the top and bottom, and as the blades turn about the vertical axis of rotation they take power from the wind, irrespective of its direction.

How they do this is shown in Figure 5.6 which gives a view from above of the central, vertical, portion of the blades. In operation the circumferential blade speed is several times the wind speed, and with the wind blowing from the left side of the page the relative wind – as seen from the blade – meets the upwind blade at the small angle shown.[32] Provided this is less than the angle (of about 12° to 15°) at which the flow over the blade stalls[33] the *overall force* on the blade, which is mostly due to aerodynamic *lift*, will as shown be almost at right angles to the relative wind.[34] The major part of this overall force acts in the downwind direction but there is a small component that acts in the direction the blade is moving and this contributes to the shaft power output. On Darrieus wind turbines the blades' aerofoil cross-section is symmetrical and consequently, as the figure shows, the overall force on the downwind blade is comparable with the force on the upwind blade; this overall force too has a small component in the direction the blade is moving, and also contributes to the shaft power output. One quarter of a revolution later the two blades are

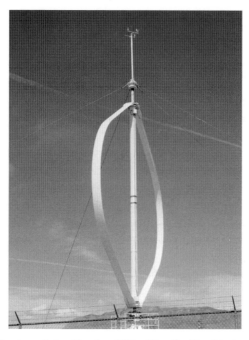

Figure 5.5 Darrieus wind turbine, late 1970s, 17 m dia., New Mexico, US. *Source*: Image courtesy of Sandia National Laboratories.

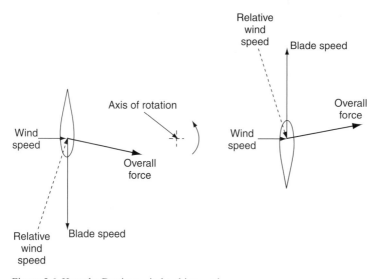

Figure 5.6 How the Darrieus wind turbine works.

moving parallel to the wind direction and there is no useful power output; in fact drag on the two blades results in some loss of power, though with a good aerofoil section the drag is low and the power loss is small. After another quarter revolution the blade that was upwind is now downwind, and vice versa, but the situation is identical with that shown in Figure 5.6 with both blades again delivering a useful shaft power output. It can be seen therefore that as the blades turn around the vertical axis the output power varies cyclically from a maximum when the blades are crossing the wind to a minimum – which corresponds to a small power loss – when the blades are running parallel to the wind. This cyclic power variation can be smoothed by having three blades, instead of two, but it is usually more cost-effective to smooth the power output by having some compliance in the drive-train so that the rotor can act as a flywheel.[35]

As is indicated above the blades on a Darrieus wind turbine interact with the wind in essentially the same way as the sails on a traditional windmill, or the blades on a conventional horizontal-axis wind turbine, in that the blades are *pulled* around by aerodynamic lift. The Darrieus wind turbine is consequently much more efficient than the Persian and Chinese vertical-axis windmills described in Chapter 2 whose sails are *pushed* around by aerodynamic drag, and as Divone notes the peak power coefficients achieved by Darrieus turbines are not far short of those achieved by modern horizontal-axis wind turbines.[36] As the Darrieus wind turbine is omni-directional the cost and complexity of a yaw orientation mechanism is avoided, and another advantage of the Darrieus configuration is that the vertical drive shaft allows the gearbox, generator and other equipment to be located at ground level, making maintenance easier. Disadvantages include the fact that the optimum tip speed ratio is about half that of a comparable horizontal-axis wind turbine so that for a given power level the torque input to the gearbox is about double, which increases the gearbox cost.[37] The guyed support of the central column is also a source of difficulty on all but very flat terrain, and though some Darrieus wind turbines have been built with un-guyed, cantilevered, central columns they tend to be more expensive.

Another frequently cited disadvantage of the Darrieus design is that it is not self-starting. For grid-connected applications (and most applications of all but the smallest wind turbines are grid-connected) this is in fact a minor issue. Taking power from the grid for a few tens of seconds to start a Darrieus wind turbine which will then run – typically – for many hours has little impact on either the overall energy output or the economics. A greater problem is the complex blade shape. As Darrieus noted in his 1925 patent there is benefit in making the blades to the same shape as a skipping rope, as blade stresses due to centrifugal forces are then zero. However making blades to this shape, which has been named a *troposkien* (from the Greek for 'turning rope'), is difficult and in order to keep blade costs down to an acceptable level most American Darrieus wind turbines have used extruded aluminium blades,[38] notwithstanding aluminium's poor fatigue resistance.

Darrieus wind turbines with power ratings up to 500 kW were developed in the United States and a Department of Energy review panel concluded in 1979 that the cost of energy from vertical-axis wind turbines, when in production, would be competitive with the cost of energy from comparable horizontal-axis wind turbines.[39] Two companies, Flowind and VAWTpower, subsequently went into production with Darrieus wind turbines derived from a Sandia Laboratories 17 m diameter 100 kW rated design and several hundred were installed in the Californian windfarms (discussed later) that flourished in the early 1980s. Though more successful than most American designs they proved much less reliable than the conventional Danish three-bladed horizontal-axis wind turbines that were then available. Subsequent commercial developments have naturally focussed on the more successful horizontal-axis wind turbines[40].

The American wind power programme has been described in some detail as it was initiated very speedily after the 1973 oil crisis, was well funded and dominated the international wind power scene through the 1970s and into the 1980s. As has been seen most resources were directed towards the speedy development of megawatt-scale wind turbines but the results were disappointing, with little of commercial value to show for the expenditure of several hundred million dollars. Unfortunately when European countries followed with their own wind power programmes most made the same mistake of underestimating the task of designing and building reliable megawatt-scale wind turbines.

European national wind power programmes

Sweden

The Swedish programme initiated in 1975 was dominated by the early commitment to build and test two different designs of multi-megawatt wind turbines, which were both completed in 1982. The smaller of these was the 75 m diameter, 2 MW rated, WTS-75 which was installed at Nasudden on the island of Gotland, about 250 km south of Stockholm, see Figure 5.7. Built by the Swedish company Kvaerner in collaboration with the German company ERNO (a subsidiary of VFW-Fokker) its two-bladed, upwind, rigid rotor turned at 25 rpm and the full-span variable-pitch blades each had a load-carrying central steel spar, with fibreglass leading edge and trailing edge fairings. The concrete

Figure 5.7 WTS-75, 1982, 75 m dia., 2 MW, Sweden.

tower was dynamically 'stiff' and, unusually, the fully assembled nacelle and rotor was raised into position using a temporary trackway – which is clearly visible in the figure – up the side of the tower. The annual average wind speed at the hub height of 77 m was 7.5 m/s and the forecast annual energy output was 6 million kWh, however in the five years that it was in operation it averaged less than half this; it was dismantled in 1991.[41] The other Swedish multi-megawatt wind turbine was the 3 MW rated WTS-3, shown in Figure 5.4, which as discussed was built by the Swedish company Swedyards in co-operation with the American company Hamilton Standard. This had a two-bladed downwind rotor with a diameter of 78 m and a slender, tubular steel tower that was dynamically 'very soft'.[26] It was sited near Malmo, on the southernmost tip of Sweden where the wind exposure was very good and the average wind speed at the hub height of 80 m was 8.1 m/s. Though the WTS-3 was more successful than most multi-megawatt machines of this era, and was in operation for 10 years, a succession of problems reduced the energy output to less than half the 8 million kWh per year that was forecast and it was dismantled in 1993.

Germany

The German wind power programme was initiated in 1976 and was strongly influenced by the work done by Hütter in the 1950s and 1960s. The decision was soon made to proceed directly with the design and construction of a 3 MW machine which would have a diameter of 100 m, much larger than any other wind turbine that had then been built, and it was designated Growian – short for *Gro*sse *Wi*ndkraft-*An*lage, or large wind turbine. Design and construction was undertaken by MAN and Growian – shown in Figure 5.8 – was installed late in 1982 at Kaiser-Wilhelm-Koog, by Germany's North Sea coast and about 70 km north-west of Hamburg. Like Hütter's Stötten machine (see Figure 4.12) Growian had a two-bladed teetered rotor, with the blades downwind of the slender guy-supported tower, but the full-span variable-pitch blades were made with a central load-carrying steel spar surrounded by an aerofoil-shaped

Figure 5.8 Growian, 1982, 100 m dia., 3 MW, Germany. *Source*: Hau E., *Wind Turbines* (2006) figure 2.25, © Springer-Verlag Berlin Heidelberg 2006.

fibreglass shell. It was grid-connected using a doubly fed induction generator[21] which allowed the rotor speed to be varied by up to 15% above or below the nominal 18.5 rpm; it was in fact the first wind turbine to use this arrangement, which has become increasingly popular in recent years. It commenced testing in the summer of 1983 but after only 100 hours operation serious cracks appeared in the teetering hub structure that joined the blades to the rotor's main shaft. Though repairs were attempted they were ultimately unsuccessful and the turbine was finally shut down in 1987 after just 500 hours of intermittent operation; it was dismantled the following year.[42] Growian had cost nearly 90 million DM (about £21 million) and its very public failure was a source of considerable embarrassment in Germany; some even suggested that its failure was the result of a deliberate attempt by the utilities and 'big industry' – who favoured nuclear power – to discredit wind power. The reality is that those involved, like their counterparts in the United States and elsewhere, had greatly underestimated the challenge of building reliable megawatt-scale wind turbines.

Though the German wind programme was dominated by Growian a second, smaller, wind turbine that was built is also of interest. This was the single-bladed Monopteros built by MBB (Messerschmitt-Bolkow-Blohm) and completed in 1981, see Figure 5.9.[43] As has been noted previously the power in the wind flowing through the rotor of a wind turbine can be efficiently intercepted with any number of blades, but single-bladed machines are about 10% less efficient than two-bladed machines, which in turn are about 2 to 4% less efficient than three-bladed machines; and for optimum performance single-bladed wind turbines need to turn more rapidly than two- or three-bladed machines. MBB's Monopteros had a 24 m long fibre-composite blade balanced by a heavy counterweight, and delivered its output of up to 370 kW to the grid through an AC/DC/AC power converter which allowed variable-speed operation in the range 39 to 50 rpm. This meant that the blade tip speed was in the range 98 to 126 m/s and since noise levels increase rapidly with tip speed Monopteros, like most single-bladed wind turbines, was relatively noisy.[44] It performed well enough for MBB to build a succession of single-bladed machines, with diameters ranging from 12 m to 56 m, but they were not competitive with conventional three-bladed wind turbines and MBB ceased their development in the late 1980s.[45]

Denmark

As elsewhere Denmark was very concerned about its energy supply security following the 1973 oil crisis, and in 1976 it was one of the first European countries to initiate a wind power programme. Its first step was to refurbish and

Figure 5.9 Monopteros, 1981, 48 m dia., 370 kW, Germany.

test Juul's 24 m diameter, 200 kW rated, Gedser mill – see Figure 4.10 – which in 1967, after ten years' operation, had been dismantled and stored at a time when oil prices were low and expected to stay low.[46] The next step in the Danish Ministry of Energy's wind programme was to build two 40 m diameter, 630 kW rated, wind turbines at Nibe, close to the coast and about 15 km west of Aalborg. Both had three-bladed upwind rotors with a rotor speed of 34 rpm, both were grid-connected using induction generators, and both had blades made primarily from fibreglass but with their inboard ends reinforced internally by an 8 m long steel spar. Where they differed is that the Nibe A machine, which was completed in 1979, had – like the Gedser mill – externally stayed fixed-pitch blades and was stall controlled; by contrast the Nibe B machine, which was completed the following year and is shown in Figure 5.10, had cantilevered blades which for power control had variable pitch.[47] They were both operational for more than a decade, and the Nibe B machine in particular was probably the most successful government-funded large wind turbine anywhere. Though they were succeeded in the late 1980s by the 'second-generation' Tjaereborg machine,[48] which was

Figure 5.10 Nibe B, 1980, 40 m dia., 630 kW, Denmark.

Figure 5.11 Tvind, 1978, 54 m dia., 1 MW, Denmark. *Source*: Image courtesy of Benny Christensen.

61 m diameter and had a rated output of 2 MW, this was not the route by which Danish wind turbines became commercially pre-eminent. Much more significant was the construction in the late 1970s of the 54 m diameter 1 MW wind turbine, shown in Figure 5.11, which was built by the teacher group of the schools at Tvind, near Ulfborg on the west coast of Jutland.

A different Danish way forward

The Tvind machine

The 1973 oil crisis had encouraged a lively debate in Denmark as to how future energy needs should be supplied, and there were many in Government and the electricity utilities who favoured the construction of nuclear power stations. The teacher group at Tvind were amongst those who were strongly anti-nuclear and who favoured the use of renewable energy sources. They saw wind power as the most promising option for electricity generation but perceived the need to demonstrate that wind power could provide electricity on a scale appropriate to electricity utilities; led by Amdi Petersen they therefore made the bold decision, in 1975, to build their own megawatt-scale wind turbine.[49] To keep costs down they bought a second-hand 2200 kW eight-pole synchronous generator and a 1200 kW gearbox that was originally built for use with a mine hoist. The solid steel, 700 mm diameter, low-speed shaft was bought second-hand from a shipyard in Rotterdam. With the assistance of volunteers the Tvind teacher group made the 53 m high reinforced concrete tower; remarkably they also made the turbine's fibreglass blades, after going to Stuttgart to learn from Ulrich Hütter how he had made the fibreglass blades for his 34 m diameter Stötten wind turbine (shown in Figure 4.12). Influenced by Hütter the Tvind wind turbine has its three-bladed rotor downwind from the tower, and the power output is controlled in high winds by varying the blade pitch. Up to 400 kW can be supplied to the grid using a rectifier/converter that was purpose built for the turbine by students from the Technical University of Denmark, supervised by Ulrik Krabbe; any additional power is used locally to provide hot water for the Tvind schools. The rotor speed is allowed to vary in the range 14 to 24 rpm.

The Tvind machine was completed in early 1978 and remains in operation, having generated more than 15 million kWh of electricity. A blade failure in 1993 prompted the replacement of all three blades but there have been no other major component failures. Its success is a tribute to the thoroughness and commitment of the Tvind teacher group and their volunteer helpers, and is all the more remarkable when compared with the problems experienced by all the other first generation megawatt-scale wind turbines that were built – notwith-standing the eminence of the engineering companies that designed and built them.

Small grid-connected wind turbines

The Tvind wind turbine inspired many wind enthusiasts in Denmark, but even before its construction there were a few who were starting to build smaller and less complex machines. Christian Riisager was one of these early Danish pioneers who built his first 15 kW rated wind turbine next to his house in 1974; this was followed in 1976 by two 10 m diameter 22 kW machines, and then in 1977 by the 10 m diameter, 30 kW rated, wind turbine shown in Figure 5.12. The Riisager machines, like Juul's Gedser mill, had fixed-pitch blades and were connected to the grid using an induction generator; this meant that the rotor speed was held constant and, as with the Gedser mill, the power output in high wind speeds was then naturally self-limiting.[50] The

Figure 5.12 Riisager, 1977, 10 m dia., 30 kW, Denmark.

stay-supported blades were made from wood and as with earlier Danish wind turbines (except Tvind) they were located upwind of the tower; following the example set by la Cour, Lykkegaard and F. L. Smidth twin fantails were used to keep the rotor facing into the wind.

Another important design from the late 1970s was the 10 m diameter, 30 kW rated HVK machine developed by Karl Erik Jorgensen and Henrik Stiesdal and shown in Figure 5.13. This also had three fixed-pitch blades located upwind of the tower and was grid-connected through an induction generator. In contrast to Riisager's use of fantails for yaw orientation Jorgensen and Stiesdal followed the Gedser mill's precedent and used an active yaw control system, with a wind vane on the nacelle sensing any misalignment between the wind direction and the rotor axis. Vestas, a leading Danish manufacturer of mobile cranes and agricultural machinery that then employed about 100 people, bought the HVK

Figure 5.13 HVK/Vestas, 1979, 10 m dia., 30 kW, Denmark. *Source*: Image courtesy of Benny Christensen.

manufacturing rights and commenced production of these wind turbines in 1979. From this modest beginning Vestas has grown to become the world's largest manufacturer of wind turbines with, in 2007, over 13 000 employees and a multi-billion pound annual turnover.[51]

One of Juul's significant innovations was the use of blade-tip brakes, see Figure 4.10, which were automatically activated if the grid-connection was lost. Such tip brakes add to the complexity and cost of blades and the first dozen or so small Danish wind turbines were made without this feature. However storms in September 1978 led to a number of incidents where the mechanical brakes that were fitted failed to stop the rotor; as the brakes overheated the rotor speed then rapidly increased until failure occurred. Blade designs were consequently modified so as to include some means of aerodynamic braking, and all subsequent Danish wind turbines were required to have aerodynamic as well as mechanical brakes. The two main options were either the rotatable tips pioneered by Juul or spoilers,[52] which usually took the form of a low fence that could be raised along the length of the outer half of each blade but which would normally lie flush with the blade's surface.[53] Some wind turbine makers chose to use spoilers; others used rotatable blade tips. Experience showed that the latter were more consistently effective, and were less expensive to implement, and they therefore became the preferred choice of most manufacturers.

To encourage small-scale wind turbine developments the Danish Parliament passed legislation in the summer of 1979 that gave purchasers a 30% subsidy,[54] provided the turbine was of an approved type. The approval process was administered by the Risø National laboratory at Roskilde, about 30 km west of Copenhagen, and covered only the basic safety systems (including the need for both aerodynamic and mechanical brakes) and general construction, but it helped to prevent the sale of machines that were poorly engineered. By 1980 there were a dozen manufacturers producing grid-connected wind turbines with rotor diameters in the range 5 m to 15 m and with rated power outputs in the range 5 kW to 55 kW.[55] Almost all were three-bladed stall-controlled machines with upwind rotors and with the grid connection provided by an induction generator, a combination of features that came to typify Danish wind turbines through the 1980s. Many used blades made by another Danish wind pioneer, Erik Grove-Nielsen, who had in 1977 set up a company – Økær Vind Energi – that specialised in the manufacture of fibreglass blades.[56] Following the storms and turbine failures in September 1978 he modified his blade design so that it incorporated a spring-loaded rotatable tip that in the event of overspeed would be activated simply by the increased centrifugal force. With this feature his blades were used in the early formative years of the 1980s by companies such as

Vestas, Bonus and Nordtank that subsequently grew to be world leaders in the design and manufacture of wind turbines.

The connection of small wind turbines to the grid was at first tolerated and then – in the late 1970s – formalised by Danske Elværkers Forening, the Danish Association of Electricity Supply Undertakings, in a way that gave protection to the grid and to other users without being unduly onerous for those who purchased the wind turbines.[57] This, plus the stimulus provided by the 30% subsidy and the generally positive experiences of early customers, encouraged a substantial increase in the number of turbine sales from 18 in 1979 to over 100 in 1980, and they remained at or above this level throughout the 1980s.[58]

In 1980 almost all the turbines sold had a rotor diameter of 10 m and were rated at 22 kW. However as manufacturers grew in experience they progressed to making larger turbines that were more cost-effective. By 1982 about half the turbines sold had 15 m diameter rotors and were rated at 55 kW, like the Bonus machine shown in Figure 5.14; this larger diameter gave a 125% increase in the rotor swept area, and therefore in the energy capture, for a cost increase that was only 60%.[59] At that time a 15 m diameter 55 kW rated wind turbine cost about 310 000 kroner (equal to £21 000 at the then current exchange rate of £1 = 15 kroner); and in what the Danes refer to as a 'class 1 roughness' location – that is to say open country areas with few trees or buildings – the average wind speed would be about 5.5 m/s and the turbine would give an annual energy output of about 90 thousand kWh.[60] A farmer buying such a turbine would use part of the output to meet his own needs, with the surplus during windy periods being sent back through the grid system to help meet the electrical needs of others; then when there was insufficient wind to meet his own needs he would buy in electricity from the grid in the usual way. And regardless of whether he was buying or selling electricity the grid could without difficulty absorb the short-period power fluctuations – caused by the turbulence in the wind – that had caused early experimenters such as Poul la Cour so many problems. The farmer would pay the usual rate of 0.8 kroner/kWh for his electricity purchases, so this would be the value to him of the electricity that his wind turbine provided directly to the farm; and he would be paid 0.4 kroner/kWh for the surplus energy that he sent back through the grid. Typically about 40% of the wind turbine's output would be used on the farm, with the remainder sold back to the utility, and the 90 thousand kWh produced would therefore be worth about 50 000 kroner; the farmer would then recover his capital expenditure in about 6 years.

Figure 5.14 Bonus, 1982, 15 m dia., 55 kW, Denmark. *Source*: Image courtesy of Siemens Wind Power.

This simplistic calculation ignores the cost of the concrete foundations, as well as the cost of grid connection and the turbine's operating and maintenance costs; however it also ignores the benefit of the 30% subsidy, which largely compensated for these omissions. The end result – as the calculations indicate – was that in locations that had good wind exposure the economics were sufficiently attractive to encourage a steady stream of farmers to purchase wind turbines. A further stimulus to the Danish market was legislation that allowed individuals who had shares in a co-operatively owned wind turbine that was in or adjacent to the district where they lived to pay no tax on the income from electricity sales, provided that their proportionate share of the turbine's annual output did not exceed about nine thousand kWh.[61] This made a well-sited and well-maintained turbine an attractive investment and led to many more turbine purchases.

Sales through the early 1980s were fairly steady at about 140 turbines per year, with the annual installed capacity rising from 3.5 MW in 1980 to 6.5 MW in 1984 as the more expensive but more cost-effective 55 kW machines gained market share. The companies which were most successful were agricultural equipment manufacturers, such as Vestas, Bonus and Nordtank, that chose to diversify into making wind turbines; they were used to dealing with farmers and recognised the need to provide equipment that was soundly engineered, robust and easy to maintain. The turbines were mostly installed singly or in small clusters, and ownership or co-ownership by people who lived locally helped to ensure that there were usually few problems in securing planning consent.

Though the first generation of Danish wind turbines, like those shown in Figures 5.12 and 5.13, had lattice steel towers the use of tubular steel towers, see Figure 5.14, soon became the norm; though somewhat more expensive experience was to show that they were visually much more acceptable to most people. They have the added advantage that, with their internal ladders, they give protection from the weather to the maintenance engineers who have to climb up to the nacelle to access the machinery within. Near the top of the tower a door allows access to the maintenance platform that can be seen in Figures 5.14 and 5.15; and the latter figure shows – with the cover raised – the arrangement of the equipment within the nacelle. Though the particular machine illustrated is a 100 kW rated, 20 m diameter, Vestas V20 of 1986 the layout within the nacelle is representative of most Danish wind turbines built through the 1980s and – with some changes[62] – later. The figure shows clearly how the blades (4) are bolted to the hub (3); from the hub the low speed shaft (2) – supported by bearings at both ends – transmits the power from the blades to the gearbox, which is almost completely hidden behind the disc brake (5) and its supporting structure. The smaller-diameter, high-speed, output shaft from the gearbox then drives the generator (6) and pendant cables (not shown) take the 3-phase, 415 V electrical output down to ground level. The shaft bearings, disc brake, gearbox and generator are all mounted on the bedplate (1). When the wind vane (seen in the figure above the disc brake) signals that the wind direction has changed, and that the rotor is no longer facing into the wind, the hydraulic unit (7) powers the yaw motors (8) which drive the yaw gearing (10) and turn small diameter yaw gears that mesh with the large diameter yaw ring (9) fixed to the top of the tower; the nacelle is consequently slowly turned until the rotor once again faces into the wind.

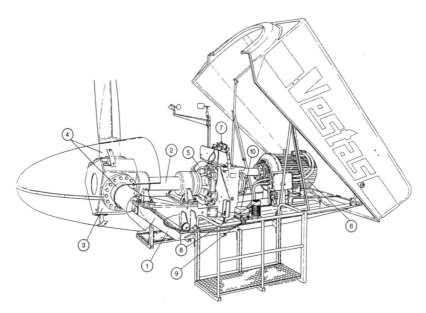

Figure 5.15 Vestas V20, 1986, 20 m dia., 100 kW, Denmark. *Source*: Image courtesy of Vestas.

The Californian wind boom

The importance of these Danish small wind turbine developments is – in retrospect – very clear, but it must be said that at the time they were little noticed outside Denmark. Through the late 1970s and early 1980s the main focus of attention, for wind power proponents in the UK and elsewhere, was the American Federal wind power program with its high hopes and successive disappointments. Then in 1981, and with little or no forewarning, the Californian wind boom began.[63] This had its beginnings in legislation passed by the Jimmy Carter administration in 1978 to encourage energy conservation and the use of indigenous energy sources. Part of this legislation was the Public Utilities Regulatory Policy Act, or PURPA, which amongst other provisions required utilities to buy electricity from 'qualifying facilities' (QFs) at a price that fully reflected the utilities' avoided costs. QFs were defined as those that produced their power from biomass or renewable energy sources, and it was stipulated that the avoided costs should include both the fixed costs and the running costs that the utilities avoided through their power purchases. The legislation did not attempt to put a price on these avoided costs; this was left to the supervisory Public Utilities Commissions for individual States and the

Californian PUC established that a rate of about 7 cents/kWh was appropriate (at the high end of the range of rates specified around the country by the various PUCs).

Earlier Federal legislation designed to help the economy recover from the recession of the mid 1970s had allowed a 10% tax credit for capital investment in any manufacturing sector. Concerns about energy supply security, heightened by the second round of sharp oil price rises that followed the overthrow of the Shah of Iran in 1979, then led in 1980 to a supplementary 15% tax credit for all energy-related capital investments made before the end of 1985. In addition California, with Jerry Brown as Governor, had passed legislation in 1978 that gave a 25% tax credit for solar energy investments made before the end of 1986; fortunately late in the legislative process this was amended so that wind energy investments were also included. These tax credits, totalling 50%, plus the certainty that the electricity output could be sold at a favourable price, combined to make investment in wind energy projects very attractive and prompted the installation of groups of wind turbines – or wind farms – in suitably windy locations.

The first wind farms

The wind farm concept was pioneered by U.S. Windpower, which was formed in the mid 1970s in Burlington, Massachusetts, and had put up its first wind farm of twenty machines of its own design on Crotched Mountain, New Hampshire, in 1980. U.S. Windpower saw the opportunity presented by the generous tax credits that were available in California and arranged to lease large tracts of land in the windy Altamont Pass, about 60 km east of San Francisco; they then secured zoning approval, contracted with the local utility for the grid connection and in 1981 installed one hundred of their wind turbines. By then their design had evolved into a 17 m diameter, 50 kW rated machine which had three fibreglass, variable-pitch, blades located downwind from the tower and was grid-connected through an induction generator. The design remained unchanged through most of the 1980s except that by 1983 the power rating had been increased to 100 kW, and Figure 5.16 shows some of the nearly four thousand U.S. Windpower turbines that were installed in the Altamont Pass through the 1980s.

Other entrepreneurs also recognised the opportunities created by the combination of Federal and Californian tax credits and secured land leases, zoning approval (equivalent to planning permission) and a grid connection, and made arrangements for the purchase, installation and operation of wind turbines. They, like U.S. Windpower, could then invite investment in their wind farm

Figure 5.16 U.S. Windpower, 1980s, 17 m dia., 100 kW, Altamont Pass, California, US.

projects. Typically a group of wind turbines would be bought through a project-specific limited partnership by many tens of investors each contributing several tens of thousands of dollars. Provided the investors were tax-paying Californians half the cost could be recovered straight away, through the Federal and State tax credits, and the balance could be set against their tax liabilities over the next three to five years. The net result was that for a high-rate taxpayer (paying tax at the top rate of 50%) more than half the investment would be returned in the first year and all would be returned within five years, almost regardless of how much energy the wind turbines produced.[64] Given good, reliable turbines and a suitably windy location the potential return was very attractive.

The Altamont Pass through the Californian Coastal Range was favoured by U.S. Windpower and others because on summer days the hot air rising from the Central Valley, to the east, sucks in cooler air from the coastal zone, to the west, and this airflow funnels through the Pass and gives strong winds through the afternoon and evening. Electricity demand in California peaks during the summer months, as people use their air-conditioners to keep cool, and the value of the output from wind farms is enhanced by the fact that it also peaks during the summer months. Conditions similar to those in the Altamont Pass are also to be found in two southern Californian locations, in the San Gorgonio Pass just to the west of Palm Springs (and about 160 km east of Los Angeles) and in

Figure 5.17 Flowind, 1980s, 17 m dia., 150 kW, Altamont Pass, California, US.

the Tehachapi Mountains about 60 km south-east from Bakersfield. All three locations are sparsely populated, and up to the 1980s the land was mainly used for cattle grazing; all three featured prominently in the Californian wind boom.

In 1981, the year the Californian wind boom started, about 150 wind turbines were installed; these were mostly the U.S. Windpower 50 kW machines in the Altamont Pass, and the total installed capacity at the end of the year was approximately 10 MW. By the following year many other developers had had time to lease land, secure zoning approval and a grid connection, raise finance and obtain and install wind turbines; 1982 consequently saw a much increased level of activity and 1150 turbines were installed, with a total capacity of 65 MW. The pace of development quickened further in 1983 with the installation of 2500 turbines, whose capacity totalled 170 MW. And the boom continued through to 1985, with 4700 turbines totalling 380 MW in capacity installed in 1984 and a further 4300 turbines totalling 400 MW capacity added in 1985. Overall in just five years more than 12 000 wind turbines were installed, with a combined capacity of over 1000 MW. Almost all were horizontal axis machines of more than thirty different designs,[65] some with two blades, most with three and just a few dozen with four blades. However the total included about 500 vertical axis wind turbines, some of which are shown in Figure 5.17.

Danish imports

Though the installation of over 12 000 wind turbines in only five years was a remarkable achievement it was tarnished by the poor performance of many of the machines. Unlike their Danish counterparts the American designers of small wind turbines in the late 1970s had sought to design lightweight machines, often with innovative features that would – they hoped – lead ultimately to low cost when in volume production. Unfortunately none of them, with the possible exception of the U.S. Windpower design, had been developed to the stage where they were ready for volume production when the Californian demand for turbines suddenly developed in 1981. However market pressures were such that, ready or not, turbines were put into production. Very predictably the result was widespread failures, and since major highways run through the Altamont and San Gorgonio Passes these failures were very visible. The situation was aggravated when the developers responsible for some of the worst projects disappeared, leaving – sometimes for years – broken and derelict wind turbines on the hillsides and in full view of the highway.

Much more positive was the performance of U.S. Windpower, who moved the manufacture of their turbines from New England to the Altamont Pass and who, by 1985, had installed nearly 3000 of their 50 kW and 100 kW machines (see Figure 5.16) on their extensive Altamont Pass wind farm. Though their design was also lightweight (barely half the weight[66] of comparable Danish turbines) they were very professional in the way they organised maintenance and repairs, and in this they were helped by the proximity of their factory; they were consequently able to ensure that their turbines were available for generation most of the time. However all their turbine production went to supply their own wind farms; they did not sell them to other wind farm developers, who had to look elsewhere for turbines that could survive the strong winds in the Californian Passes and operate reliably. Increasingly, from 1983, Californian developers turned to Danish manufacturers for the turbines they needed.

One of the first to commit to buying Danish was Zond Systems, a company started by Jim Dehlsen in 1980 which over the next decade established itself as one of the most successful producers of wind-generated electricity in California. Zond installed most of its turbines in the Tehachapi Mountains but, like most wind farm developers, they did not manufacture their own machines. For their first project in late 1981 they bought and installed fifteen Storm Master turbines, which were made in southern California and had a three-bladed, 12 m diameter, rotor. They were stall regulated, with a rated power output of 40 kW, and their slender and very flexible fibreglass blades were downwind of the tower. Zond installed many more of these machines in 1982 but the Storm Master turbines – like so many of the lightweight American machines – were not robust and there were frequent failures. Dehlsen therefore decided to look elsewhere, visited Europe to see what was available and contracted with Vestas to supply 150 of their 15 m diameter, 65 kW rated, wind turbines for installation in late 1983.[67]

Other Californian wind farm developers bought a further 210 machines from other Danish manufacturers; and the 360 Danish turbines with a combined capacity of 20 MW accounted for 12% of the capacity installed in California that year. The Californian financial incentives were so generous that turbines which in Denmark would be sold for about 310 000 kroner, equivalent to about $31 000, could be sold on to Californian investors for prices in the range $100 000 to $180 000.[68]

The more robustly engineered Danish wind turbines performed well in California and this, as well as the attractive price, encouraged Zond to buy a further 550 Vestas machines in 1984, and other developers bought more than one thousand turbines from other Danish manufacturers. Danish turbine sales to California totalled 110 MW that year and accounted for 29% of the capacity installed in 1984.[69] As noted previously production of wind turbines for the Danish home market in the early 1980s had been fairly steady at about 140 machines per year, corresponding to a capacity of about 5 MW per year. By 1984 sales to California were – by capacity – more than twenty times larger and the four leading Danish manufacturers, Vestas, Bonus, Nordtank and Micon,[70] had all had to substantially expand their manufacturing capability. They were put under further pressure in 1985 when Zond ordered a further 1200 turbines from Vestas, and other developers bought an additional 2300 machines from the other Danish manufacturers. That year Danish turbine sales to California totalled 220 MW and accounted for 55% of all the capacity installed (with half the balance produced by U.S. Windpower) and Figure 5.18 shows just a

Figure 5.18 Micon, 1980s, 16 m dia., 65 kW, Altamont Pass, California, US.

few of the many thousand Danish machines that were then winning a reputation for their performance and reliability.

The market peaks

1985 marked the peak of the Californian wind boom. The Federal tax incentives expired at the end of the year and were not renewed; investors' enthusiasm for wind power was soured by the poor performance of some of the early projects and the bad publicity that had resulted; and a sharp decline in the value of the dollar made imports from Denmark 25% more expensive. The capacity installed in 1986 consequently dropped to 275 MW, down 30% from the previous year's 400 MW high, with imports from Denmark reduced from 220 MW to 180 MW. Though the Danish market share had increased to 65%, the reduction in actual turbine sales caused financial difficulties to companies that had expanded too fast; and non-payment from some developers, who were having their own financial problems, made the problem worse. The economic attractiveness of wind farm investments declined further when the Californian State tax incentives expired at the end of 1986, and was made worse by the continuing fall in the value of the dollar.[71] The capacity installed in 1987 consequently slumped to 155 MW, with imports from Denmark down to 55 MW. This massive downturn led to the bankruptcy or major financial re-structuring of all but one of the major Danish manufacturers (the exception being Bonus) though the re-structured companies – having shed their liabilities – were able to continue in business.

The expiry of the tax credits and the worsening dollar-to-kroner exchange rate made investment in wind farms by wealthy Californian residents much less attractive, but the best of the turbines installed – which were mostly Danish – had by then operated reliably for several years, long enough to give institutional investors the confidence to start investing in wind projects. And though the tax credits had gone the 1978 PURPA legislation remained; this, it may be recalled, required utilities to buy the output from renewable energy projects at the utilities' full avoided costs, and in California the Public Utilities Commission had decreed in the early 1980s that a rate of about 7 cents/kWh was appropriate; they further declared that renewable energy projects should be offered so-called Standard Offer (SO) contracts in which power purchase prices were specified for the first 10 years, after which the price would be set at the avoided cost determined in the light of circumstances at the time.[72] The two Californian utilities, Pacific Gas and Electric (PG&E) in the north and Southern California Edison (SCE) in the south, had subsequently signed these SO contracts with wind farm developers for more than two thousand megawatts. Though oil prices slumped in 1986, and were to stay low for the next 17 years,[73] wind farm developers still had SO contracts for several hundred megawatts of projects that had yet to be installed, and price escalation provisions within these contracts meant that by the late 1980s the starting price was close to 10 cents/kWh. This was sufficient to underpin continuing institutional investment in Californian wind farms and between 1988 and 1990 turbines with a capacity totalling a further 370 MW were installed, with Danish machines accounting for nearly half;[74] and this brought the total wind power capacity installed in California through the decade 1981–1990 up to 1820 MW.

Environmental concerns

In the early years of the decade, when oil prices were high and energy supply security was a matter for serious public concern, the first Californian wind farms had been generally welcomed. However much of this goodwill was dissipated when many of the turbines installed in the early 1980s proved to be inadequately engineered, and were seen from the highways that passed close by to be standing broken or idle for much of the time. This problem diminished as the decade progressed, and as more robust and reliable turbines imported from Denmark were increasingly used, but the fall in oil prices in 1986 then reduced the public perception of the need for wind farms. Projects proposed in the later 1980s consequently found themselves subject to much more detailed environmental scrutiny and an increased proportion was refused zoning approval. In many cases the issues hotly debated were local and site specific, but one concern which was raised frequently was the visual change that wind farms cause. Wind turbines must be sited in exposed locations and will consequently be clearly visible; and for some people the visual change is undesirable. However studies of public attitudes, both in the Unites States and in Europe, have shown repeatedly that wind farms are visually acceptable to a large majority.[75]

One other significant environmental issue which emerged in the Altamont Pass in the late 1980s was that a larger than expected number of birds had been killed by collision with the wind turbines or their power lines, and these included raptors such as the golden eagle. All structures cause some bird deaths and collisions with the glass facades of buildings kill nearly 100 million birds per year in the United States.[76] In the Altamont Pass a two year study that commenced in 1989 found 182 dead birds, two-thirds of which were raptors;[77] bird fatalities in the San Gorgonio Pass and in Tehachapi were much lower. Later studies indicated that the Altamont Pass is unusual in that it has a high raptor population, plus many thousands of wind turbines in a relatively small area. There is also evidence that the widespread use of lattice towers, see Figure 5.16, on which raptors like to perch contributed to the larger than expected number of bird fatalities. Subsequent experience both in the United States and in Europe has shown that 'there is minimal risk to birds from the operation of properly sited wind turbines'.[78]

Design evolution

As might be expected the strong demand for wind turbines to meet the needs of Californian wind farms through the 1980s stimulated substantial progress in the technology. There was continuing pressure on manufacturers to improve the cost-effectiveness of the turbines they were producing, and the manufacturers' response was to steadily increase the size of the turbines. For example the Danish machines installed in California in 1982 had an average rated output of just 55 kW and a diameter of about 15 m. Through the later 1980s Danish manufacturers brought out a succession of new models, each time incrementally increasing the rotor diameter and the rated power output, and by 1990 turbines such as the 225 kW rated, 27 m diameter, Vestas V27 shown in Figure 5.19 and the 450 kW rated, 35 m diameter, Bonus 450 shown in Figure 5.20 were in production.

By contrast U.S. Windpower, which produced more wind turbines per year than any other manufacturer, chose to continue with the production of the same 100 kW rated, 17 m diameter, machine throughout the 1980s and only commenced the development of a successor in 1987.[79] And while Danish manufacturers had followed a path of gradual design evolution U.S. Windpower, to be competitive, then had to develop a wind turbine that was a step change from its predecessor. U.S. Windpower changed its name to Kenetech in 1988 and the new machine, designated the KVS-33, had a 33 m diameter rotor with a rated power output of 300 kW. Kenetech also departed from their previous experience by having the three variable-pitch blades upwind of the tower. A gearbox with two parallel output shafts was then used to drive two 150 kW induction generators, and the power output was then delivered to the grid through two parallel AC/DC/AC advanced electronic power converters, an arrangement which allowed the rotor speed to be varied and which – at least in theory – would increase the annual energy output by several per cent.[80] The first prototypes were installed in the Altamont Pass in 1989, and were followed by 22 pre-production machines in 1991. However commercial pressures led to the KVS-33 being put into volume production before it was fully developed and field tested, and ongoing engineering problems with this machine were a significant factor in Kenetech's bankruptcy in 1996. Kenetech's assets were sold, but no-one chose to continue with production of their turbines.

Figure 5.19 Vestas V27, late 1980s, 27 m dia., 225 kW, Denmark. *Source*: Image courtesy of Vestas.

Though a total of 1820 MW of wind turbine capacity had been installed in California by 1990 several thousand of the small machines erected in the early 1980s had proved to be inadequately engineered, and had to be removed. However about 15 000 wind turbines with a total capacity of about 1500 MW continued in operation. Danish turbines – mostly made by Vestas, Bonus, Nordtank and Micon – accounted for nearly half this operational capacity and U.S. Windpower turbines accounted for about

Figure 5.20 Bonus 450, late 1980s, 35 m dia., 450 kW, Denmark. *Source*: Image courtesy of Siemens Wind Power.

one quarter; machines made by a wide variety of American manufacturers, plus a few non-Danish European ones, made up the balance. American wind turbine designers had often made disparaging remarks about the much heavier and generally less complex Danish machines, but experience in the Californian wind farms had shown that Danish wind turbines were more robust and reliable, as well as being very competitively priced. Output from the 1500 MW of operational capacity in 1990 totalled about 2.4 billion kWh.[81] Utilities generate power on a very large scale and they consequently tend to use large units such as the terawatt-hour (TWh), which is equal to one billion kWh. In utility terminology Californian wind farms had in 1990 produced 2.4 TWh of electricity, and had clearly demonstrated that wind power could now be seen as a utility scale electricity producer.

Danish success factors

It is interesting and somewhat sobering to note that the several hundred million pounds which had been spent in America and Europe on centrally directed wind turbine development programmes had contributed little, if anything, to the successful evolution of the modern wind turbine, typified by Danish machines such as those shown in Figures 5.19 and 5.20. This failure on the part of the centrally directed programmes, and the eminent aerospace engineering companies that were the main recipients of Government funding, can in part be explained by the assumption – made by many in the early 1970s – that companies which could make successful large aeroplanes would find the development and manufacture of successful large wind turbines relatively easy. The reality was very different. Aircraft only have to fly in the turbulent air close to the ground for a few minutes each flight, as they take off and land. Wind turbines have to operate in this very challenging environment for many thousands of hours per year, and for upwards of twenty years, with relatively little maintenance; and they have to be built very much less expensively.

It is perhaps fortuitous that in Denmark in the late 1970s circumstances, not least the legacy of pioneers such as la Cour and Juul, allowed the development by enthusiasts of small grid-connected wind turbines. Step-by-step these were improved to the stage where Danish agricultural machinery manufacturers saw the opportunity to get involved and to diversify their product ranges. Their tradition of engineering robust and reliable agricultural equipment helped to ensure that Danish wind turbines were as robust and reliable as was possible for a completely new product. Competition in Denmark to supply the small home market led to the steady development of larger and more cost-effective wind turbines, and Danish manufacturers were therefore well placed to benefit when the market for wind turbines in California suddenly emerged. Though the Californian boom of the early 1980s – followed by the market decline in the later 1980s – caused some acute growing pains Danish wind turbine manufacturers ended the decade as very clear market leaders; and with reliable and reasonably cost-effective products they were well placed for further growth in the 1990s, as and when the market for wind turbines might allow.

6

Progress and economics in Europe, 1973 to 1990

Denmark

Denmark's leading role in the development of the modern wind turbine, as described in the previous chapter, was greatly aided by the small home market in the early 1980s, and as the decade progressed this domestic market grew substantially. The wind turbines sold in the early 1980s had been seen to perform reasonably well and to be a sound investment; and this positive feedback reinforced the desire of many Danes to support a clean and renewable means of meeting, at least in part, their electricity needs. Turbine sales, which up to 1984 had averaged about 140 units per year, more than doubled in the second half of the decade; but as the turbines grew steadily larger the installed capacity grew much faster. At the end of 1984 Denmark had 640 grid-connected wind turbines with a total installed capacity of just 27 MW; by the end of 1990 more than 2880 turbines had been installed, and the installed capacity had grown more than tenfold to 343 MW. In 1990 these wind turbines had an average capacity factor of 23% and generated 0.61 TWh, just under 2% of Denmark's total electricity generation.[1]

Throughout the 1980s most turbine purchases were made by individuals and co-operatives, and wind energy's steadily improving economics more than compensated for the progressive reduction of the Government subsidy that when introduced in 1979 was worth 30% of the turbine cost (the subsidy was removed completely in 1989). Utility involvement was at first minimal but the Government specified in 1985 that they should install 100 MW, and by 1990 just under one-quarter of the 343 MW total was utility owned. Though most wind turbines were installed singly, or in small groups, a number of wind farms were also built and Figure 6.1 shows a small part of one of the largest, the Taendpipe/Velling Maersk wind farm which is about 75 km north of Esbjerg, near Ringkøbing and close to the west coast of Jutland.[2] The location is also

Figure 6.1 Taendpipe/Velling Maersk wind farm, late 1980s, Ringkøbing, Denmark.

close to the Vestas factory and offices at Lem, and the wind farm contains a total of 100 Vestas wind turbines with an aggregate capacity of 12.6 MW. The Taendpipe part of the wind farm was built first, in 1986, using 35 Vestas V17–75 machines and is owned through a co-operative by over 500 local families; as the designation indicates these turbines have a rotor diameter of 17 m and a rated output of 75 kW. The Velling Maersk part of the wind farm is owned by the local utility – Vestkraft – and was built in two phases, in 1987 and 1989; it contains 34 Vestas V20–90's, 2 Vestas V25–200's and 29 Vestas V27–225's. Although the wind farm uses turbines of four different sizes this is not at all obvious, and the overall appearance is visually pleasing.

Stall control or pitch control?

Danish wind turbine manufacturers were known through the early 1980s for the general similarity of their designs; they almost all used the same stall-controlled rotor configuration pioneered by Juul on the Gedser mill in 1957, with three fixed-pitch blades located upwind of the tower, grid-connected through an induction generator and with overspeed protection provided by rotatable blade tip brakes. Vestas was the first to depart from this pattern when they commenced production of their V25–200 machine in 1986; this had variable-pitch blades and all sub-sequent Vestas wind turbines have had variable-pitch blades. With fixed-pitch blades as the wind speed increases the angle between the airflow and the blades steadily increases until the flow over the blades stalls,[3] and at higher wind speeds the flow over the blades remains stalled; the way that a fixed-pitch, stall-controlled, wind turbine's power output then varies with wind speed is indicated in Figure 6.2(a). However with variable-pitch blades as the wind speed increases the blade pitch angle can be controlled so that the angle between the airflow and the blades never gets high enough to cause stall, and by suitable adjustment of the pitch angle the wind turbine's power output can be held constant above the rated wind speed, as is indicated in Figure 6.2(b). With stall control the power output peaks, just prior to stalling, and then drops down as the wind speed increases further and the flow over the blades goes more deeply into stall. Fluctuations in the wind speed – due to turbulence – have little effect on the fully stalled flow so the power output in high wind speeds is relatively steady. With pitch control the average power output in high winds can, as noted, be kept nearly constant, but the short-period fluctuations in the wind speed that result from turbulence can lead to substantial short-period power fluctuations; though these are of no consequence to the grid[4] they do have consequences for the design of the gearbox and other turbine components.[5]

The choice between pitch control and stall control makes very little difference to a wind turbine's efficiency; by the late 1980s both types of machine could achieve[6] peak power coefficients in the range $C_P \approx 0.40$ to 0.45, which as discussed in Appendix A means that nearly two-thirds of the kinetic energy in the wind flowing through the rotor is removed and converted to an electrical power output. Stall-controlled turbines are mechanically simpler, as their blades are fixed to the hub, whereas in variable-pitch machines each blade has to be supported by a large-diameter bearing at its root end and the hub must contain a mechanism to actuate the blades. However safety requirements dictate that turbines must have aerodynamic brakes; with pitch-controlled machines this is achieved simply by turning the blades through 90°, but stall-controlled machines require rotatable blade tips. Making the blade tip so that it can be rotated adds considerably to the complexity and cost of the blade, and becomes more difficult with larger blades. For the same rotor diameter and the same location there is no significant difference between the energy output of a stall-controlled turbine and the energy output of a pitch-controlled turbine and the purchaser's choice would usually be determined by price (and other important factors such as the warranty, the time to delivery and the manufacturer's reputation for reliability and after-sales service).[7] At small turbine sizes, such as the 10 to 15 m rotor diameters that were typical of the early 1980s, variable-pitch machines would have been appreciably more expensive; for larger turbines the cost of providing blade variable pitch becomes proportionately much smaller, and the difficulty and cost of engineering rotatable blade tips becomes greater, and Vestas with the 25 m diameter V25–200 and its successors was able to offer competitively priced variable-pitch machines.[8]

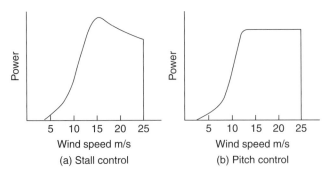

Figure 6.2 Typical wind turbine power curves.

Shut-down in high wind speeds

As is indicated in Figure 6.2 modern wind turbines are designed to be shut down when the wind speed exceeds about 25 m/s. This corresponds to storm force 10 on the Beaufort scale (see Table 3.1) and wind speeds of this magnitude are described as causing 'considerable structural damage' over land. However these very high wind speeds are very infrequent, and even for a location where the average wind speed is as high as 8 m/s (such as an elevated upland site in Wales or Scotland) wind speeds above 25 m/s would be expected for less than 5 hours per year.[9] Wind turbines could be designed to continue in operation at these very high wind speeds but loads on the blades and the rest of the turbine structure would then be higher and the turbine would cost more. And because these very high wind speeds occur so infrequently the resultant increase in the turbine's annual energy output would be minimal; for example a doubling of the shut-down wind speed to 50 m/s would increase the annual energy output by less than 1%. Designing for a shut-down wind speed much above 25 m/s would, in consequence, reduce a turbine's cost-effectiveness.

At the other end of the operational range of wind speeds it matters little, in terms of the annual energy output, whether a turbine starts to generate at 4 m/s or 5 m/s. Less than 10% of the total energy in the wind comes at wind speeds below the annual average, and the turbine's output in light winds is further diminished by the low efficiency of both the gearbox and the generator at the corresponding low power levels.[10] The main benefit of a low starting wind speed is aesthetic; people much prefer to see wind turbines turning and a machine that has a 4 m/s starting speed will be seen to be in operation for many more days per year than one whose starting speed is 5 m/s. For example on a good site where the annual average wind speed (at the turbine hub height) is 7 m/s a turbine with a 5 m/s starting speed will operate for 67% of the time, which means it will be idle for one day in three; however a turbine with a starting wind speed of 4 m/s will operate for 77% of the time and it will be idle for less than one day in four.

Though wind turbines are designed to operate in wind speeds ranging from 4 or 5 m/s up to about 25 m/s they must also be able to survive without damage the maximum wind speed that is likely to be encountered – just for a few seconds – once in fifty years, and this survival wind speed is typically about 60 m/s.[11] Aerodynamic forces go up as the square of the air speed, and withstanding the high forces associated with wind speeds of 60 m/s might be expected to add substantially to a turbine's cost.[12] However the solidity of a modern wind turbine is only about 5% and with the rotor stopped – as it is in wind speeds above about 25 m/s – the aerodynamic force on the blades in a wind speed of 60 m/s is no greater than the force on the blades when the turbine is operating. There is therefore no need to spend substantial sums of money to provide strength that would only be required once in fifty years.

Costs relative to the annual energy output

It was stated earlier that wind turbine economics in Denmark steadily improved through the 1980s, and that this more than compensated for the removal by 1989 of the 30% turbine subsidy that had been introduced in 1979. As has been noted the 15 m diameter, 55 kW rated, turbines which in 1982 accounted for about half the turbine sales cost 310 000 kroner, equivalent to a cost of 5640 kroner per rated kilowatt or to 1750 kroner per unit of rotor swept area (kroner/m^2). Though the cost per kilowatt is frequently used when comparing different means of generating electricity it must be used with care when applied to wind turbines. One manufacturer might give his 25 m diameter turbine a rated power output of 250 kW, while another would give a turbine the same size – and otherwise very similar – a rated power output of only 200 kW. If they were sold for the same price a would-be purchaser might well be tempted to buy the one with the higher rating, as its cost per kilowatt would be a substantial 20% lower. In fact if they were located side by side their energy outputs would be very similar, since this is primarily a function of the rotor swept area and the site's windiness. And in less windy locations the turbine with the lower power rating would probably have a somewhat higher annual energy output, as gearbox and generator losses would be smaller during the lengthy periods of operation at relatively low power levels. It is often more helpful to compare wind turbines on the basis of their cost per unit of rotor swept area; information on the annual energy output per unit of rotor swept area can then be used to calculate the cost per kilowatt-hour, which is normally the parameter that matters most.

In Denmark through the 1980s there were four successive generations of wind turbine designs.[13] The first generation machines with 10 m diameter rotors and power ratings in the range 22 to 30 kW, see for example Figures 5.12 and 5.13, were dominant at the start of the decade and still had a significant market share in 1982; their cost then averaged 2450 kroner/m².[14] They were succeeded by the second-generation machines with rotors about 15 m diameter and power ratings in the range 55 to 75 kW, such as the turbine shown in Figure 5.14, whose market share peaked in 1985; as indicated above their cost in 1982 was 1750 kroner/m². The third-generation machines with rotor diameters of about 20 m and power ratings in the range 90 to 100 kW, see for example Figure 5.15, reached the peak of their market share in 1987; but by 1990 the market was dominated by fourth-generation machines, such as that shown in Figure 5.19, with rotor diameters of about 25 m and power ratings in the range 150 to 250 kW.

Costs in 1990 for these fourth-generation machines, as given in one of the status reports that accompanied the European Wind Energy Association's 1991 strategy document *Wind Energy in Europe*,[15] averaged 2285 kroner/m². Inflation in Denmark through the 1980s was such that 1 krone in 1982 was worth 1.46 kroner by 1990;[16] at 1990 price levels the cost of the first-generation wind turbines was consequently 3580 kroner/m². So in progressing from the first-generation machines of the early 1980s to the fourth-generation machines of 1990 Danish manufacturers had been able to reduce the price per unit of rotor swept area by 36%, enough by itself to compensate for the removal of the 30% subsidy. Moreover, and as is shown in Figure 6.3, the productivity of the

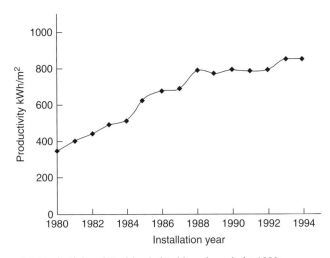

Figure 6.3 Productivity of Danish wind turbines through the 1980s.

turbines – that is to say their annual energy output per unit of rotor swept area – more than doubled, increasing from 340 kWh/m^2 in 1980 to 780 kWh/m^2 by 1990.[17] This very substantial improvement was partly the result of the improved reliability of the later turbines, but it was helped by many detailed design improvements as well as by the fact that the larger machines of 1990 had taller towers and so gained advantage from the fact that the wind speed increases with the height above the ground.[18]

Combining these figures for the productivity with the previous cost figures gives the turbine cost per *annual* kilowatt-hour of output. For the first-generation machines of the early 1980s this was 3580/340 = 10.5 kroner per *annual* kWh; however by 1990 the fourth-generation machines – with their 150 to 250 kW power ratings and rotor diameters of about 25 m – had reduced the turbine cost to 2285/780 = 2.9 kroner per *annual* kWh, a more than threefold reduction. For a complete wind farm (or even for a single wind turbine) there are additional costs which are often referred to as the *balance of plant* costs. These will include the costs of access tracks, turbine foundations, erection and commissioning, cabling between turbines and the grid connection, as well as costs for environmental impact assessment, insurance and finance. In Denmark through the 1980s these balance of plant costs would typically add about 25% to the turbine cost, and gave a capital cost in 1990 for the complete wind power installation of 3.6 kroner per *annual* kWh of electricity generated.[19]

Cost per unit energy output

To facilitate comparison with the cost of generating electricity from other sources we need to calculate wind power's generation cost per unit energy output; and this has two main elements, the capital recovery cost and the operations and maintenance cost. The capital recovery cost is the charge that has to be made every year to repay the original capital expenditure and to provide an agreed rate of financial return on the money used (rather like the mortgage repayments that have to be made on the purchase price of a house). As such the capital recovery cost is proportional to the capital cost, but it also depends on the period over which the cost is to be recovered as well as the required rate of financial return. For utility investors the period is usually specified to be twenty years, consistent with an expected lifetime of at least this length for a well-designed wind farm; the rate of return is usually specified to be either 5% or 8%, net of inflation.[20] (This net-of-inflation required rate of return is sometimes referred to as a *real* rate of return; and in some cost-of-energy calculations it is instead referred to as a test discount rate.) The annual charge required to repay a given capital cost over 20 years *and* to provide a 5% rate of return on the investment is equal to 8.0% of that capital cost; while to repay the cost over 20 years and provide an 8% return requires an annual charge equal to 10.2% of the capital cost.[21] As noted above the capital cost of Danish wind farms in 1990 was 3.6 kroner per *annual* kWh; the capital recovery cost for a 5% required rate of return, with repayment over 20 years, was therefore 0.080 × 3.6 = 0.29 kroner/kWh. Similarly the capital recovery cost for an 8% required rate of return, and with repayment over 20 years, was 0.102 × 3.6 = 0.37 kroner/kWh.

To determine the overall cost of energy from a 1990 Danish wind farm we must now add the cost of operations and maintenance. This heading includes both scheduled and unscheduled maintenance, consumables such as oil for the gearbox and grease for bearings, insurance, land lease payments, administration and an allowance for repairs, etc. The Risø National Laboratory in Denmark surveyed and analysed costs actually incurred through the 1980s and the 1990s and developed a model which quantified the year by year annual cost of operation and maintenance for turbines of several different sizes (and included allowance for major refurbishment after the 10th year of operation).[22] For the fourth-generation 150 to 250 kW machines that accounted for most sales in 1990 their figures indicate that the average annual cost of operations and maintenance was 4.6% of the turbine cost; and as the 1990 turbine cost was 2.9 kroner per *annual* kWh this gives a cost for operations and maintenance of 0.13 kroner/kWh. Adding this to the previously calculated capital recovery cost gives – for a 5% required rate of return – an overall cost of energy from a 1990 Danish wind farm of 0.42 kroner/kWh; for an 8% rate of return this would increase to 0.50 kroner/kWh. *At the mid 1990 exchange rate of £1 = 11.3 kroner these Danish wind energy costs become 3.7 p/kWh for a 5% rate of return and 4.4 p/kWh for an 8% return.*[23]

Cost comparisons

In its strategy document *Wind Energy in Europe* (published in 1991 but mostly using data compiled in 1990) the EWEA compared the cost of wind energy with the cost of electricity generated from other sources, and these costs are given in Table 6.1.[24] The European Currency Unit, or ECU, preceded the euro and in 1990 its value was equal to £0.70. The indicated costs assume a required rate of return of 8%, and the range given for wind corresponds to a range of average wind speeds; the low-cost end of this range would require sites with wind speeds that are substantially higher than in typical Danish locations. As might be expected the cheapest option for utilities is to generate electricity by burning

Table 6.1 *Energy costs in 1990*

	Energy costs	
Power source	ECU/kWh	p/kWh
Wind	0.041–0.074	2.9–5.2
Coal	0.042–0.056	2.9–3.9
Gas (CCGT)	0.028–0.042	2.0–2.9
Nuclear (PWR)	0.085–0.100	6.0–7.0

coal or gas, but the costs for electricity from these sources do not include the very real – but hard to quantify – consequential costs of the pollution and greenhouse gases that they produce. Wind energy costs are seen to be close to competitive with electricity from fossil fuels even when these external costs are ignored, and are very competitive with the costs quoted by the EWEA for nuclear power. These nuclear power costs come from the 1990 UK report *The Cost of Nuclear Power*, produced by the House of Commons Energy Committee,[25] which reviewed the cost of electricity from Pressurised Water Reactors (PWRs) – and in particular of the Sizewell B PWR whose construction commenced in 1988 – in the light of information disclosed in the run-up to the privatisation of the UK electricity supply industry in 1990.

The Netherlands and Germany

By 1990, in a little over a decade, Danish manufacturers had developed wind turbines that were robust, reliable and reasonably cost-effective; and nearly 2900 turbines with a combined capacity of 340 MW were providing almost 2% of Danish electricity needs. It was a remarkable achievement, and despite Denmark's small size and population[26] its installed wind power capacity was more than double the total in the rest of Europe. Next highest was the Netherlands, a country closely associated with traditional windmills, which had initiated a national wind energy research and development programme in 1976. By the early 1980s several companies – including Polenko, Lagerwey and Bouma – were making small wind turbines that were similar in diameter and power rating to those that were being produced in Denmark, and their prices were comparable.[27] However strict siting controls and a purchase price for surplus electricity that was lower than in Denmark led to a substantially lower volume of sales. And though some Dutch turbines were sold into California in the early 1980s the total numbers were small compared with the sales of Danish

Figure 6.4 Ijsselmeer dyke wind farm, 1987, Urk, Netherlands.

wind turbines.[28] A market stimulation programme was introduced in 1986, and led to the construction of projects such as the 7.5 MW wind farm along the Ijsselmeer dyke shown in Figure 6.4 (which used twenty-five 300 kW rated, 25 m diameter, wind turbines made by WindMaster Nederland). Despite this the total installed wind power capacity[29] in the Netherlands by the end of 1990 was barely 40 MW.

After the Netherlands the next highest installed wind power capacity was in Germany. As has been noted German wind power activity centred initially on the government-funded development of large wind turbines, such as the Growian and Monopteros machines shown in Figures 5.8 and 5.9, but with no more success than the similarly premature large wind turbine projects in the United States. MAN, who had built the 100 m diameter Growian, also developed a small wind turbine with a diameter of about 12 m which had two, upwind, variable-pitch blades and was grid-connected using an induction generator. It had a rated power output in the range 20 to 40 kW, depending on the windiness of the site, and several hundred were sold into the Californian market in the early 1980s. However it did not perform as well as was expected, and MAN discontinued its manufacture. In the late 1980s a growing number of small wind turbines were bought by individuals from Danish and Dutch manufacturers, and several German manufacturers of small wind turbines also emerged, most notably Enercon and Tacke. By 1989 about 150 of these small machines had been installed, with a total capacity of about 12 MW. The introduction in March 1989 of a generously funded '100 MW' subsidy programme – designed to encourage the installation of 100 MW within five years – led to a sharply increased rate of turbine sales, and by the end of 1990 the total installed capacity was approximately 35 MW.[30]

No other country in Europe[31] had an installed wind power capacity in excess of 10 MW at the end of 1990, and this included the UK. Given that its wind resource is acknowledged to be the best in Europe, and its long tradition of using wind power, this lack of progress in the UK seems somewhat surprising. Wind energy activities in the UK through the 1970s and 1980s are reviewed in the next chapter, and the explanation for this lack of progress given.

7

UK progress, 1973 to 1990

Wind power: the least promising renewables option?

In the UK, as elsewhere, the 1973 oil crisis led the Government to review the prospects for using renewable energy sources; and by 1976 its judgement was that the two technologies which showed most promise of making a useful contribution to UK energy needs before the end of the century were wave energy and solar heat.[1] Tidal power, via a barrage across the Severn Estuary, was also judged capable of making a significant contribution to our energy needs, but studies had led to disappointingly high estimated costs as well as to serious technical uncertainties and major environmental concerns.

Enthusiasm for wave energy derived from the realisation, in the early 1970s, that there was a large amount of power in the waves approaching the UK coast from the Atlantic Ocean. Measurements made at Ocean Weather Ship India, 700 km west of the Hebrides, indicated that the *average* wave power flux was 77 kW/m. This led to the assessment, by the Department of Energy and others, that 'about half the present total requirement for electricity in the UK could in principle be met in a stretch of ocean *as short as 600 miles*',[2] that is to say 970 km; the italics are mine as deploying wave energy conversion devices along a distance of this magnitude would be challenging in the extreme. In the mid 1970s the total UK electricity consumption was about 215 TWh/year,[3] so this forecast implied a wave energy contribution of about 100 TWh/year. Wind energy, by contrast, was seen to have only limited potential. The Electrical Research Association (ERA) had, as part of the UK's modest wind energy programme in the early post-war years (see Chapter 4), made wind speed measurements at a large number of locations (mostly hill tops); and based on this legacy of data they estimated that up to ten thousand 1 MW wind turbines could be installed and could make a contribution of up to 16 TWh/year towards our energy needs.[4] Most of this would come from the 3100 turbines that the

ERA proposed for hill-top sites, but there were doubts whether more than a small proportion would be visually acceptable. The CEGB in particular doubted whether sites could be found for wind turbines to make more than a 1% contribution to their electricity demand.[5]

The Department of Energy was somewhat more optimistic and estimated that wind energy might make a contribution of 12 TWh/year by the year 2000 (equivalent to about 3% of forecast electricity consumption),[6] though its 'ultimate potential for growth beyond 2000' was judged to be 'small'; this compared unfavourably with their estimate that wave energy could – by the year 2000 – make a contribution of 30 TWh/year and had an 'ultimate potential for growth beyond 2000' that was judged to be 'very large'. These assessments of the ultimate potential contributions to our energy needs had led to wave energy being classified as the most promising renewable energy option, with wind energy classified as the least promising. And as might be expected funding for research and development reflected these priorities, with £400 000 spent on wave energy over the period 1973 to 1977, compared with just £40 000 spent on wind energy.[7]

The offshore wind potential

This pessimistic assessment of the potential contribution that wind energy could make to meeting UK electricity needs was, however, flawed in that it completely overlooked the option of siting wind turbines offshore in the shallow waters that surround most of our southern coasts. From 1976 onwards I argued the case for building offshore wind farms each large enough to give an energy output equivalent to a conventional power station on land, and pointed out that the southern North Sea to the north and east of the Wash, see Figure 7.1, was a particularly suitable area.[8] Meteorological Office data for this sea area indicated that the average wind speed was about 7.2 m/s at the standard height of 10 m above the surface. American studies were then indicating that the most economic size for a wind turbine would be about 75 m diameter, with a hub height of about 70 m and a rated power output of about 2.5 MW. With turbines such as these deployed offshore the hub height average wind speed would be about 9 m/s and the average power in the wind would therefore be approximately 860 W/m²; up to the turbines' tip height the average power in the wind is then 92 kW/m, fully comparable with the power available in the waves approaching the north-west coast of Scotland.[9]

Heronemus had in the early 1970s proposed using large floating structures, each with a steel lattice superstructure supporting between three and seventy

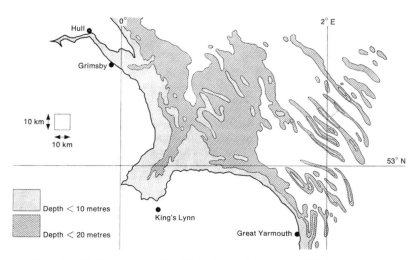

Figure 7.1 Shallow waters off the Wash. *Source*: Musgrove (1978a).

wind turbine rotors, to harness the wind resource off the coast of the United States.[10] I preferred the simpler option of extending the towers of individual wind turbines down to the seabed. Shallow water areas, such as those shown in Figure 7.1, are then favoured so as to keep the tower and foundation costs as low as practicable. In the figure the stippled area shows where the water depth is less than 10 m, and the hatched area shows where the water depth is less than 20 m. As can be seen the 20 m depth contour extends more than 70 km eastwards from the coast of Lincolnshire, and one can in fact cross the southern North Sea to the Netherlands – a distance of 300 km – without exceeding a depth of 30 m. Britain's first offshore gas field was discovered at West Sole, about 90 km east of Grimsby, in 1965; and by the mid 1970s there were many gas platforms in the area and the gas brought to the shore by seabed pipelines was starting to make a significant contribution to UK energy needs.[11] In consequence there was by then a good body of knowledge about sea and seabed conditions and wave loads on structures, and a good understanding of how to fix structures to the seabed. This meant that there would have been little or no risk associated with the foundations for offshore wind turbines.

I therefore proposed the construction of a series of 1000 MW offshore wind farms, each one containing four hundred 2.5 MW wind turbines. These – as noted – would each have a rotor diameter of about 75 m, and an average spacing between them of 500 m (equal to about 7 rotor diameters) seemed appropriate. The total area required for the 1000 MW wind farm was therefore 100 km^2 and for simplicity I assumed a 20 by 20 turbine array so that the wind farm envelope was a 10 km by 10 km square; a square of this size is shown to scale in

Figure 7.1. With an estimated capacity factor of 40% one such wind farm would provide an energy output of 3.5 TWh/year, equal (then) to 1.6% of the total UK electricity consumption. The area off the Wash could clearly accommodate several 1000 MW wind farms, and other shallow water locations around the coast of England and Wales[12] could provide many additional suitable sites, such that the wind power potential in waters less than 20 m deep would be at least 20% of UK electricity needs; the deployment of wind turbines into slightly deeper waters would more than double this potential. Wind speeds over the sea are higher than over coastal areas (because the sea surface is smoother than the land) but one has to go about 10 km offshore to get the full benefit. Most offshore wind farms would therefore be sited at least this distance from the coast and though – weather permitting – they would still be visible they would not be visually prominent.

Though initially sceptical the Department of Energy in due course accepted that the offshore wind resource was substantial,[13] and funding for wind power research and development then steadily increased. The CEGB also came to accept the large potential contribution from offshore wind and in July 1978 the Chairman – Glyn England – acknowledged that offshore wind systems with a capacity of 20 000 MW might be deployed in the shallow waters around the coast and could provide about a quarter of the CEGB's output.[14] Using forecasts from the United States for the cost in volume production of multi-megawatt wind turbines, and making allowance for the extra cost of the longer offshore towers and their foundations as well as the cost of the cable connection back to the shore, I estimated that the capital cost of a 1000 MW offshore wind farm would be approximately £410/kW at 1976 price levels. This cost, which recent experience has shown to be reasonably accurate,[15] was an order of magnitude lower than the capital cost estimates made in 1979 for the wave power devices that were then being considered.[16] My corresponding energy cost estimate for the output from a 1000 MW offshore wind farm was then 1.2 p/kWh, just slightly higher than the 1.1 p/kWh average cost of the fuel burnt in the CEGB's oil- or coal-fired power stations.[17]

Managing wind power's variability

There was some initial concern over how best to cope with the fact that the output from wind farms is variable, and dependent on the weather. One seemingly obvious solution would be to provide large-scale energy storage; on a utility scale the most practicable option is hydro-electric pumped storage[18] but it is an expensive way to store energy. Far cheaper, and almost always the preferred option, is to store energy in the form of unburnt coal, oil or gas. In other words the output from multiple offshore wind farms would be used to shut down – in whole or in part – a number of conventional power stations so that many millions of tonnes of coal (or oil, or gas) could be left unburnt for years or decades longer. On windless days the power stations would, of course, operate as they do in the absence of wind farms to meet consumer demand, but when the wind blows fuel is saved. And a single 1000 MW offshore wind farm – with its output of 3.5 TWh/year – would reduce the UK coal burn by approximately 1.5 million tonnes per year and save the emissions of about 4 million tonnes per year of the greenhouse gas carbon dioxide,[19] though the importance of the latter would not be fully appreciated until well into the 1990s.

The issue of how best to operate a typical utility mix of power stations when thousands of megawatts of wind farms are added was, from the late 1970s, the subject of many detailed system integration studies.[20] These showed that in the worst case scenario, corresponding to the output from wind farms being completely unpredictable (so that conventional power stations have to be kept operating on part-load at all times to provide 100% back-up), the fuel savings are about 15% smaller than when the output from the wind farms is completely predictable.[21] In these circumstances a single 1000 MW offshore wind farm would reduce the UK coal burn by 1.3 million tonnes per year, instead of the 1.5 million tonnes per year indicated previously. In practice the output from wind farms can be predicted fairly accurately for several hours ahead and the fuel savings are only about 5% smaller than when the output from the wind farms is completely predictable.

The winter months are, on average, markedly windier than summer months and wind turbines consequently produce most of their output during the winter. This enhances the value of wind-generated electricity as electricity demand is also at its peak in winter. The winter peak in the output from wind farms also has the consequence that the fossil-fuelled power stations that can be shut down, in whole or in part, when the wind farms are generating will mostly be those that have lower than average efficiency. As Taylor and Rockingham noted the average fuel cost in 1979–80 for the CEGB's fossil-fuelled power stations was 1.4 p/kWh but for the least efficient power stations, which would be off-loaded first, the fuel cost was 2.0 p/kWh.[22] System integration studies also showed that the value of adding wind generation to a utility system was greater than just the value of the fuel saved; there was also some capacity credit. In other words if you added 10 000 MW of wind farms to the system[23] you could avoid the need to replace a few thousand megawatts of conventional power stations as they reached the end of their life without any reduction in the security of the electricity supply to customers. The exact magnitude of this capacity credit depends on the specifics of the utility system, on the geographical diversity of the wind farms and on their overall installed capacity, but for low levels of wind contribution the capacity credit is about equal to the average output from the wind farms. As the wind penetration increases the capacity credit becomes a smaller percentage of the installed capacity but it still adds considerably to the value of wind energy to a utility system.

The Department of Energy wind programme

The Orkney multi-megawatt wind turbine

Following their recognition that the offshore wind resource was substantial the Department of Energy identified and funded three main strands of wind energy activity.[24] The first of these was to design, and in due course build, a 60 m diameter multi-megawatt wind turbine intended for use on exposed hill-top sites. As first conceived in 1977 it was to be 3.7 MW rated with two fixed-pitch steel blades located upwind of a concrete tower. The main participants, Taylor Woodrow Construction, British Aerospace and GEC Energy Systems, formed themselves into a consortium called the Wind Energy Group (WEG) to take forward the design, and as it evolved the power rating was reduced to 3 MW and the decision was made to use a teetering rotor with variable-pitch blade tips. Designated LS-1, and shown in Figure 7.2, it was eventually completed and first generated electricity in February 1988 on Burgar Hill, in Orkney.[25] The rotor configuration was very similar to the much larger Boeing Mod-2 turbines, see Figure 5.3, which were installed at several locations in the United States between 1980 and 1982, though the higher rating of the smaller LS-1 is a consequence of its optimisation for a very windy hilltop location; the forecast annual average wind speed on Burgar Hill at the turbine's 45 m hub height was a quite exceptional 12.5 m/s.

  The leisurely pace of the UK wind programme unfortunately meant that the LS-1 was obsolete by the time it was built, and a succession of problems then led to the commissioning phase lasting nearly four years. Regular operation commenced in January 1992, and through 1992 the turbine generated 5.4 million kWh of electricity, corresponding to a capacity factor of 20%. However in December 1992 large cracks were observed near the hub ends of the steel blades and the turbine was shut down. The owner, Scottish Hydro-Electric, (the utility then serving the north of Scotland) decided it was not worth repairing but a group of enthusiasts banded together, leased it for a nominal sum, and arranged for it to be repaired and put back into operation. It ran for a further two years but the income generated was insufficient to

Figure 7.2 LS-1, 1987, 60 m dia., 3 MW, Orkney, Scotland. *Source*: Image courtesy of Steve Macken.

pay for its ongoing operating costs; it was finally shut down in 1997 and demolished in November 2000.

Offshore wind farm assessment

The second strand in the Department of Energy's wind programme was to assess in some detail the engineering aspects of siting and installing large numbers of offshore wind turbines. This was primarily a desk-based study, led by Taylor Woodrow, and when completed in December 1979 it concluded that there were no insuperable technical problems.[26] The study assumed that the turbines would be similar to the two-bladed 60 m diameter LS-1 then being developed by WEG and that 196 would be deployed in a 14 by 14 array. The estimated capital cost of £1475/kW was substantially lower than contemporary estimates for wave power devices[16] but was more than double my estimate (of £410/kW in 1976) which uplifted for inflation through to January 1980 became £650/kW. Taylor

Woodrow's high estimated capital cost led to a forecast cost of energy that was substantially larger than the potential fuel-saving benefit; and though they noted that their costs erred on the side of caution (they assumed for example that the wind farm would take 8 years to build and that the turbine rotors would need to be replaced after only 13 years) there is no doubt that their high cost estimates weakened interest in offshore wind.

1979 was an important year politically in that the General Election in May had seen the defeat of Jim Callaghan's Labour Government and the election of a Conservative Government, which was to remain in office for the next 18 years. The new Prime Minister, Margaret Thatcher, was a strong supporter of nuclear power and in December 1979 the Government announced its programme to build 15 000 MW of nuclear power capacity over a ten-year period that was to commence in 1982.[27] In the event this programme led to the construction of just one nuclear power station, the 1200 MW Sizewell B, which was approved in 1987 after a lengthy public inquiry. It was eventually completed in 1995 at a cost that was 40% higher than estimated, and plans for several successors were dropped in 1989 as the Government moved towards privatisation of the electricity supply industry. The Conservative Government's early enthusiasm for nuclear power unfortunately led to reduced interest in renewables, which was soon followed by reduced funding. And though the CEGB had[14] taken a very positive interest in offshore wind when Glyn England was Chairman he was replaced in 1982 by the less supportive Walter Marshall, previously Chairman of the United Kingdom Atomic Energy Authority.

By this time it was clear from experience in the United States and Germany, where the development of multi-megawatt wind turbines had been the prime focus of their more liberally funded wind programmes, that building reliable and cost-effective large wind turbines was more challenging than had been anticipated, and there was a good deal more to be learnt before these large machines could be deployed commercially. With offshore wind farms there are high fixed costs associated with constructing foundations and turbine support structures that will withstand the once-in-a-lifetime maximum wave loads that may be experienced, and the cost of such structures increases only slowly with turbine size. Offshore wind farms can therefore only be competitive if they are able to use reliable and competitively priced multi-megawatt wind turbines, and the disappointing experience with experimental large wind turbines in the early 1980s meant that progress offshore would have to wait. Despite the UK's quite exceptional offshore wind resource the Department of Energy was content to classify offshore wind as a *long-shot* technology[28] (a description it used for technologies that in its judgement had little prospect of ever being deployed cost-effectively) and leave progress to others.[29] In the event it would be the late

1990s before reliable and competitively priced large wind turbines became available (from Danish manufacturers) and made commercial offshore wind farms possible, as is discussed in the next chapter.

Vertical-axis wind turbines

The third main strand in the Department of Energy's wind programme was the development of straight-bladed vertical-axis wind turbines. As discussed previously the modern vertical-axis wind turbine was invented in France by Georges Darrieus, but only came to prominence in the early 1970s with the work undertaken at the National Research Council in Canada, and subsequently at Sandia Laboratories in the United States. American vertical-axis wind turbine developments focussed on the so-called 'egg-beater' configuration with curved blades, as shown in Figures 5.5 and 5.17. However the complexity of this shape led to the use of blades made from extruded aluminium, despite the poor fatigue properties of this material.

In my judgement a straight-bladed H-configuration was preferable. Straight blades are much easier to manufacture and my first experimental machine, shown in Figure 7.3,[30] simply had solid wooden blades. Though this small turbine had a guyed support tower I proposed that larger machines would, like most horizontal-axis wind turbines, have cantilever towers. H-configuration vertical-axis wind turbines were – in my view – particularly well-suited for use in the offshore wind farms I was concurrently proposing, and in this application guy cables would be quite impracticable. In the 3 m machine shown in the figure the blades were hinged where they attached to the cross-arm so that as the wind speed increased, and the rotor turned faster, the angle between the blades and the vertical steadily increased. Blade inclination on a vertical-axis wind turbine has an effect similar to changing the pitch of the blades on a conventional horizontal-axis machine,[31] and so provides a means of limiting the power output in strong winds.

If you take a given design of small wind turbine and scale up the rotor from, for example, 10 m diameter to 100 m diameter then you will find that the blade bending stress that results from aerodynamic forces is unchanged. However the blade-bending stress that results from a blade's own weight goes up linearly with size, and so would be ten times greater in the larger machine.[32] With horizontal-axis wind turbines as the blades turn from the 3 o'clock position to the 9 o'clock position the weight-related blade-bending stress goes from a maximum in one direction through zero to a maximum in the opposite direction, and the continual rotation of the blades means that this stress is continually reversing, and so is fatigue inducing. For small wind turbines this weight-related

Figure 7.3 Vertical-axis wind turbine, 1977, 3 m dia., Reading, England. *Source*:
Image © Crown copyright 1977.

bending stress is insignificant but it becomes increasingly significant as the
turbine size increases to 100 m diameter, and potentially much more. With
vertical-axis wind turbines the effect of scaling is the same; the blade-bending
stress that results from aerodynamic forces is unchanged, and the blade-bending
stress that results from the blade's own weight goes up linearly with size. The
vital difference is that with an H-configuration vertical-axis rotor the weight-
related blade-bending stress is constant; and, as with bridges, when self-weight
loads and bending stresses are constant structures can be made very large.

   The above reasoning led me to conclude that H-configuration vertical-axis
wind turbines could be made much larger than conventional horizontal-axis
wind turbines, and therefore had the potential to be more cost-effective in
offshore wind farms. The Department of Energy was persuaded that the development of such machines should be supported, and Sir Robert McAlpine &
Sons Ltd and Northern Engineering Industries plc formed a company – Vertical
Axis Wind Turbines Ltd – to take the development forward. The result was a

Figure 7.4 VAWT 25, 1986, 25 m dia., 130 kW, Carmarthen Bay, Wales. *Source*: Image courtesy of npower renewables and International Power.

succession of design studies followed by the construction of two experimental wind turbines. The first of these, the VAWT 25 shown in Figure 7.4, was completed in 1986 and installed on a coastal test site provided by the CEGB at Carmarthen Bay, South Wales. The rotor diameter was 25 m and steel was used both for the main structure of the blades and for the cross-arm; the 25 m high tower was made from pre-stressed concrete. In most tests the turbine was grid-connected through an induction generator and the rotor turned at 21 rpm. Given the low site wind speed (averaging just over 6 m/s) the rated power output was 130 kW.

The VAWT 25 operated reliably[33] and though the blades could be inclined – into an arrowhead configuration – through more than 60° the experimental programme demonstrated that power limitation in strong winds could also be achieved very satisfactorily by keeping the blades upright and allowing the flow to stall.[34] This allowed the design of the successor machine to be simplified, with the blades attached rigidly to the cross-arm. This wind turbine, the VAWT 35 shown in Figure 7.5, had a diameter of 35 m and was completed in 1990.

Figure 7.5 VAWT 35, 1990, 35 m dia., 500 kW, Carmarthen Bay, Wales. *Source*:
Image courtesy of npower renewables and International Power.

Though for convenience it too was tested[35] at the CEGB's Carmarthen Bay test
site it was designed for use in windier locations, where average wind speeds of
about 8 m/s could be expected; it therefore had a rated power output of 500 kW.
Steel was again used for the main structure of the cross-arm but the two 24.3 m
long blades were made from fibreglass.

   With vertical-axis wind turbines the blades only interact efficiently with
the wind for about half of each revolution; the optimum solidity is consequently about 10%, approximately double the optimum solidity of horizontal-
axis wind turbines, and each tapered blade of the 35 m test machine consequently had an average width of 1.75 m. The turbine was grid-connected
through an induction generator and though the gearbox and generator could
have been located at the base of the tower (as is sometimes proposed for
vertical-axis wind turbines) this would have required the use of a long and
heavy torque tube and would not have been cost-effective; these components
were therefore located at the top of the 30 m high tower, just below the cross-
arm. The relatively high solidity of vertical-axis wind turbines means that

their optimum rotor speed is somewhat lower than would be the case for a horizontal-axis wind turbine of similar size, and the VAWT 35 rotor turned at just over 20 rpm; as a result the blade circumferential speed of 37 m/s was substantially less than the blade-tip speed of about 60 to 80 m/s that is typical for modern horizontal-axis wind turbines.[36]

Following the completion and commissioning of the VAWT 35 in mid 1990 the turbine was comprehensively tested through the following winter, but the programme was terminated prematurely by a blade failure in late February 1991.[37] By the time the cause had been investigated, and a replacement blade designed and certified, National Power (the new owner of the Carmarthen Bay test site following the 1990 privatisation of the CEGB) had decided that the loss-making test site had fulfilled its purpose and should be closed; the VAWT 25 and the VAWT 35, together with the other wind turbines that were on test there, were subsequently dismantled. Operation of the VAWT 35 through the winter of 1990–91 had given enough data to validate the turbine's performance but the economic assessment of H-configuration vertical-axis wind turbines through the late 1980s had led to the conclusion that in production they would be somewhat more expensive than contemporary horizontal-axis wind turbines.[38] It was recognised that vertical-axis wind turbines had the potential to be made very large, and that in offshore applications this could result in a smaller number of turbines being required to provide a given wind farm capacity; the savings in the total cost of the smaller number of offshore foundations and support structures might then be sufficient for the overall wind farm economics to favour the use of vertical-axis wind turbines. However by the late 1980s oil was cheap, the threat posed by global warming and climate change was not widely recognised and the Department of Energy had classified offshore wind as a long-shot technology; in its view offshore wind had little prospect of being deployed cost-effectively regardless of the type of turbine used, and the UK vertical-axis wind turbine programme came to an end.

Other UK wind turbine developments

Lawson-Tancred

Outside the three main strands of work supported by the Department of Energy, and described above, wind energy activity in the UK was very limited. Several University groups were researching various aspects of small wind turbines through the 1970s and 1980s,[39] and Sir Henry Lawson-Tancred built the 17 m diameter wind turbine shown in Figure 7.6 at Aldborough, Yorkshire, in 1977.

Figure 7.6 Aldborough, 1977, 17 m dia., 100 kW, England.

With three fixed-pitch blades located upwind of the tower, and a fantail for yaw orientation, its design was strongly influenced by the 100 kW machine installed on the Isle of Man in 1959 (see Figure 4.15).[40] Though initially rated at 30 kW its power rating was soon increased to 100 kW. It was used for a variety of experimental studies over the next decade but plans to put it into production, and to develop a megawatt-scale successor, did not materialise.

Fair Isle

Through the 1980s a very small number of Danish wind turbines were installed; and most notably these included a 60 kW rated, 14 m diameter, modified Windmatic machine which was erected and commissioned on Fair Isle in 1982. Fair Isle is a small and very windy island located to the north of Scotland, about halfway between the island groups of Orkney and Shetland, and is home to about 70 people. A diesel-powered electricity system was installed in 1962 but the cost of electricity was such that a supply could only be provided for a few hours per day. The installation of the wind turbine allowed the diesel use to be greatly reduced, but more importantly it also allowed electricity to be supplied for many more hours per day.[41] The turbine operated very successfully through to 1996 when it was refurbished and a second wind turbine, rated at 100 kW, was added to help meet the increased electricity demand; the original 60 kW machine was then in 1997 given a B listing (usually only given to buildings of regional importance) for being of Special Architectural or Historic Interest.

Carmarthen Bay wind turbine test centre

1982 was also the year that the first wind turbine was installed at the Carmarthen Bay wind turbine test centre. The CEGB had declared in 1980 that it wished to purchase a 'commercially proven' wind turbine and it subsequently contracted with the Glasgow-based James Howden and Company to supply a 24 m diameter, 200 kW rated, machine made by WTG Energy Systems of Buffalo, New York. The turbine was installed in November 1982 at the test site which the CEGB provided on reclaimed land adjoining their obsolete and soon-to-be-closed Burry Port coal-fired power station. Road access to the site was good and the coastal location, just 6 km west of Llanelli and overlooking Carmarthen Bay, gave good exposure to the prevailing winds from the south and west. The WTG machine had three fixed-pitch blades located upwind of a lattice tower, and rotatable tips for safe shutdown in high winds.[42] Unfortunately it was less of a proven design than the CEGB and Howden had hoped and a succession of problems meant that it only ran in total for 400 hours. Howden eventually replaced it in 1986 with a wind turbine they had made, the 28 m diameter 300 kW rated machine shown in Figure 7.7, which had a steel tubular tower and three upwind wood-epoxy blades with variable-pitch tips. The small wind turbine visible in the background of the figure, on the lower right-hand side, was an experimental shrouded vertical-axis machine installed by Balfour Beatty in 1985.[43]

Howden

The long-established and publicly quoted Howden Group of engineering companies had expertise in the design and manufacture of large industrial fans, and they installed the first wind turbine of their own design on Burgar Hill, Orkney in 1983. It was somewhat smaller than the machine shown in Figure 7.7, with a diameter of 22 m, but the rotor configuration was essentially the same; and given the windiness of the Burgar Hill location this first prototype Howden turbine was also rated at 300 kW. It performed well and in early 1985 Howden installed a 26 m diameter 330 kW turbine for Southern California Edison (SCE)

Figure 7.7 Howden, 1986, 28 m dia., 300 kW, Carmarthen Bay, Wales. *Source*:
Image courtesy of npower renewables and International Power.

at a Palm Springs test site. This was followed later in the year by the installation
in the Altamont Pass, California, of a seventy-five turbine wind farm using 31 m
diameter, 330 kW rated, machines; a further ten 15 m diameter 60 kW rated
machines and a 45 m diameter 750 kW machine were also installed at the same
location.[44]

Howden received several additional orders, including one from the CEGB for
a 55 m diameter, 1 MW rated, wind turbine to be installed at Richborough, Kent,
next to an oil-fired power station. However what seemed to be the emergence of
a competitive UK wind turbine manufacturer received a set-back in February
1986 when the blades failed on the Palm Springs machine. Further blade
failures later that year at the Altamont Pass wind farm led to its prolonged
shut-down and the consequent loss of revenue from electricity sales, plus the
cost of re-designing and replacing all the blades, totalled more than £13
million.[45] As a result of this loss in their wind division the Howden Group as
a whole had to report a loss at the end of the 1986/87 financial year, and ongoing
losses in their wind division led to the Howden Group's announcement – in
April 1989 – that they were withdrawing from the wind energy business. They

continued to meet their contractual commitments and their 1 MW Richborough wind turbine was completed in August 1989.[46]

Wind Energy Group

Prior to building the 3 MW rated LS-1, shown in Figure 7.2, the Wind Energy Group had built an experimental two-bladed, 20 m diameter, wind turbine. This machine, designated MS-1, had a rated power output of 250 kW and was installed on Burgar Hill, Orkney, in 1983 on a site adjoining Howden's 300 kW experimental turbine.[47] The MS-1 was a versatile machine that could be tested in a variety of operating modes, and its primary purpose was to help with the development of the design of the LS-1. However the Wind Energy Group (WEG) made use of the experience gained with it to design a 25 m diameter, 250 kW rated, three-bladed machine, designated MS-2, and the proto-type was installed on a coastal site near Ilfracombe, Devon, in December 1984. The MS-2 had full-span variable-pitch blades made from wood-epoxy, and the blades were upwind of the steel tubular tower; grid connection was, as usual, through a speed-increasing gearbox and an induction generator.

In co-operation with U.S. Windpower twenty MS-2 wind turbines were installed in the Altamont Pass before the end of 1986, where they performed well.[48] However WEG recognised the need to further reduce the energy cost so they then designed a two-bladed, 33 m diameter successor, designated MS-3. Two prototypes were built in 1988; one was installed in the Altamont Pass, close to WEG's MS-2 wind farm, and the other – shown in Figure 7.8 – was installed at the Carmarthen Bay test centre. The 300 kW rated MS-3 had a teetering rotor with full-span variable-pitch blades made from wood-epoxy, and the blades were upwind of the steel tubular tower.[49] Though the two prototypes performed well the MS-3 – handicapped by the absence of a UK home market for wind turbines – just could not compete with the Danish manufacturers, who by then had been producing wind turbines in volume for almost a decade. WEG would have to wait until 1992 before they would have the opportunity to supply their turbines to some of the first UK wind farms.

Institutional barriers

The UK, exposed as it is to weather systems sweeping in from the Atlantic, has a very good wind resource;[50] it is in fact better than Denmark's, and the UK land area is also about six times larger.[51] The absence of a home market for wind turbines in the 1980s therefore requires explanation. There were in fact two

Figure 7.8 WEG MS-3, 1988, 33 m dia., 300 kW, Carmarthen Bay, Wales.

problems. The first was that until the Energy Act was passed in 1983 individuals had no right to connect wind turbines to the electricity grid. The Energy Act gave them this right and required the regional electricity area boards to publish the prices they would pay for electricity supplied to them. After allowing for the various fixed and other charges that the Area Boards made the net result was equivalent to the payment of at best about 1.8 p/kWh for the wind turbine's output,[52] which was only about two-thirds the price paid in Denmark.[53] The second and bigger problem for anyone in the UK who was minded to buy a wind turbine was the magnitude of the property taxes – which in the UK are some-what confusingly called *rates*. For wind turbines owned by anyone other than an electricity utility the standard method of assessment was the so-called 'contrac-tor's test'; this was based on the cost of the installation and led to the property taxes – the rates – being set at a level equivalent to about 1.6 p/kWh.[54] At this level the rates absorbed almost all the income from electricity generation, and what was left was insufficient to pay the costs of operation and maintenance;[55] there would certainly be no net income to repay the capital cost of the wind turbine. With no financial return whatsoever there was in consequence no UK market for wind turbines through the 1980s.

Electricity utilities also had to pay rates but their payments were determined by a formula specific to the publicly owned utilities and the rates they paid were very much lower, typically only about 0.1 p/kWh.[56] The British Wind Energy Association (BWEA) repeatedly lobbied Government through the 1980s, seeking to get the rates paid on privately owned wind turbines reduced to a similar level, but without success. It was only when the Government decided in the late 1980s to privatise the electricity industry, and sought to encourage competition both in generation and supply, that it recognised the need to ensure that all generators had to be assessed for rates on a similar basis and took the necessary action. From 1990 generators – whether existing power stations that post-1990 were in private sector ownership, or new-build gas-fired power stations that the Government hoped to encourage, or wind turbines – would all be assessed for rates based on their maximum power output and would all pay rates at about the same level. For wind turbines this meant that from 1990 they too would only have to pay about 0.1 p/kWh in rates.[57]

Prior to 1990 only electricity utilities – with their formula rating – could own wind turbines and avoid the punitive 1.6 p/kWh level of rate payments. UK wind enthusiasts at the BWEA's tenth annual conference, held in London in March 1988, were therefore pleased to hear the CEGB Chairman Walter Marshall (who was by then Lord Marshall of Goring) announce that the CEGB would co-fund with the Department of Energy a £30 million programme to build Britain's first three wind farms, each to have a capacity of about 8 MW provided by about 25 wind turbines.[58] The proposed timescale was fairly leisurely – Lord Marshall expressed the hope that if all went well they could be operational in 1993 – and though precise locations were not given one would be in Cornwall, one in West Wales and one in the Northern Pennines. In the event little progress was made on any by the time the CEGB was privatised two years later and only one, at Cold Northcott in Cornwall, would subsequently get built.

Privatisation and the Non-Fossil Fuel Obligation (NFFO)

In February 1988, just one month before Lord Marshall's wind farm announcement, the Government outlined its plans for electricity privatisation. The CEGB's power stations would be allocated to one or other of two generation companies, National Power or Powergen, and the twelve regional electricity area boards would become twelve independent companies who between them would have the responsibility to supply customers with electricity bought from National Power or Powergen or from any other suitable source (such as power stations that they or others might choose to build).[59] In recognition of the higher cost of nuclear power the electricity suppliers would between them be obliged to purchase at a premium price the output from all UK nuclear power stations, and the means for achieving this was to create the so-called Non-Fossil Fuel Obligation or – for short – NFFO (pronounced 'noffo'). This would allow the approximately £1 billion per year cost of the premium price paid for nuclear power to be recovered via a levy imposed on the electricity produced by fossil fuel power stations.[60] It was at first thought that NFFO would be set at a level that just covered the output from nuclear power stations but the wind energy community – together with the proponents of other renewable energy sources – lobbied to get support for renewables through NFFO; and in May 1989 the Government announced that the NFFO legislation would include some provision for the support of renewable energy projects. Through the latter part of 1989 the process for submitting renewable energy projects under NFFO was defined, and those wishing to secure NFFO contracts for the output from projects, such as wind farms, that they were hoping to build had to supply a considerable amount of detail relating to the proposed sites and the proposed turbines; and based on their forecast project costs they had to state the price per kWh they needed for their projects to be able to proceed.

In September 1989 the Energy Minister stated that nuclear power stations would be offered life-of-station power purchase contracts, and leaks indicated that the price would be in excess of 7 p/kWh. Even at this price, and with long-term power purchase contracts, potential investors were not happy with the financial return relative to their perception of the risks; and in November 1989 the decision was made to retain all nuclear power stations in public ownership. Also, in pricing the output from their fossil fuel power stations it had been CEGB practice, as a public sector utility, to calculate the repayment of the capital cost over a power station's lifetime (typically 30 to 40 years) and to provide a financial return – as specified by the Government – of 5%.[61] On this basis the CEGB had in 1988/89 supplied the area boards with electricity at an average price of 3.6 p/kWh. Post-privatisation the private sector would expect a substantially higher rate of return, typically about 11%, and would expect to see the capital cost of power stations repaid over a much shorter period – 20 years at most. The CEGB had indicated, in their 1988/89 annual report, that their fossil fuel power stations had a value of £12.6 billion (with a replacement cost of over £22 billion). If the Government had sought full recovery of this £12.6 billion value from the private sector purchasers the price of electricity would have increased very substantially. As this was politically unacceptable the power stations were instead sold off at a very much reduced price, totalling only about £4 billion. In this massively subsidised electricity generation environment the only new power stations that could hope to compete without price support would be the low capital cost gas-fired power stations with, at least temporarily, low-cost gas.

Those bidding wind energy and other renewable energy projects for support through the 1990 NFFO arrangements would also require private-sector funding. In the expectation that the power purchase contracts awarded under NFFO would provide a guaranteed premium price for a period of 20 years the initial wind bids averaged about 6 p/kWh. However, following a decision made by the European Commission[62] bidders were told in April 1990 that the NFFO contracts would only be allowed to pay a premium price until December 1998. Bid prices consequently had to be revised upwards (to about 9 p/kWh) and when the results of the first renewables NFFO were announced in September 1990 they included nine wind projects.[63] Four of these, totalling 2 MW, were for individual wind turbines that were already operational, including the 1 MW Richborough machine; the remaining five projects, totalling 26 MW, were for wind farms that ranged in size from 1 MW to 9 MW though only one, a 4 MW project at Delabole in Cornwall, had planning consent. And by the end of 1990 the Department of Energy had already initiated the process that would lead to a second round of NFFO premium price contracts being awarded to successful bidders in late 1991. So although the 1980s had been a time of considerable frustration for those wishing to build wind projects in the UK the new decade was starting much more positively. The punitive level of property taxes (rates) had at last been dealt with and there was the expectation that continuing rounds of NFFO would ensure that the price paid for the output from wind farms (and other projects) would be sufficient to make them financially viable. UK wind proponents had high hopes for the 1990s.

8

Development and deployment, 1990 to 2008

Global overview

By 1990 reliable, robust and reasonably cost-effective wind turbines with diameters up to about 35 m and power outputs up to about 400 kW could be supplied by several manufacturers, see Figures 5.19 and 5.20; and the Danish rotor configuration with three blades upwind of a tubular tower was dominant. Though some aspects of what had happened in California in the early 1980s had been deeply disappointing the decade had ended with 15 000 wind turbines operating there, with an installed capacity of 1500 MW supplying electricity to the Californian grid on a utility scale. And by 1990 in Denmark 2900 wind turbines with a capacity of 340 MW were providing just under 2% of Danish electricity needs. Though the installed capacity in the rest of the world barely totalled 140 MW the modern wind turbine had evolved to the stage where large-scale deployment was a real option. The high oil prices that had prompted the development of renewable energy sources in the early 1970s had subsided in the mid 1980s, but there was growing evidence that greenhouse gas emissions – caused mostly by the combustion of fossil fuels – were altering the Earth's climate, with potentially very damaging consequences.[1] Wind energy had progressed to the stage where it was the most cost-effective of the high-resource electricity-generating renewable energy options, though its potential was not widely appreciated.

To help remedy this situation the European Wind Energy Association (EWEA) published in mid 1991 its 'Strategy for Europe to realise its enormous wind power potential'; after reviewing the resource and the progress that had been made with the technology it proposed what seemed at the time some very ambitious targets for growth.[2] In mid 1991 the installed wind capacity in Europe was just 510 MW,[3] but the EWEA argued that by the year 2000 this could and should be increased by almost a factor of eight to

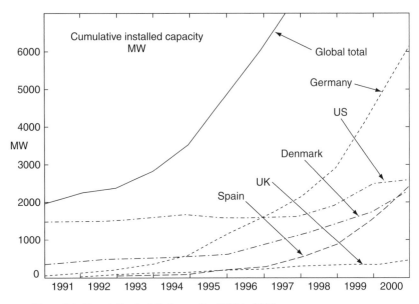

Figure 8.1 Cumulative installed capacity, 1990 to 2000.

4000 MW. In the event the European installed wind capacity exceeded 4000 MW by mid 1997, and by the end of the year 2000 the European installed wind capacity had reached 13 000 MW. As is shown in Figure 8.1 the main countries responsible for this rapid growth were Germany, and to a lesser extent Denmark and Spain. Through the 1990s most capacity growth was in Europe, but both the US and India saw significant activity in the latter half of the decade (with India – not shown in the figure – having over 1200 MW installed by the end of 2000). As a result the global cumulative capacity increased from 1980 MW at the end of 1990 to 17 700 MW at the end of 2000, corresponding to an average annual growth rate through the decade of 24%.

The exponential growth seen through the 1990s has continued since the turn of the century with the global cumulative capacity reaching 93 900 MW at the end of 2007, see Table 8.1.[4] The average annual growth rate over this period was 27%, slightly higher than in the preceding decade, and it would have been higher still but for the fact that demand for wind turbines has in recent years exceeded supply and resulted in many projects being delayed by one to two years. The quite exceptional rate of growth that has been sustained over the past seventeen years is illustrated by the

Table 8.1 *Installed wind power capacities, 1990 to 2007*

	1990	1995	2000	2005	2007
Europe – in year (MW)	140	810	3670	6240	8600
Global – in year (MW)	190	1290	4250	11 600	19 500
Europe – cumulative (MW)	440	2520	13 000	40 900	57 100
Global – cumulative (MW)	1,980	4780	17 700	59 200	93 900

fact that in 1990 the annual capacity increase was just under 200 MW, which was provided by about one thousand wind turbines each with an average capacity of approximately 200 kW: by 2007 the annual capacity increase was one hundred times larger, at just under 20 000 MW, provided by about ten thousand wind turbines with an average capacity of approximately 2 MW.

Up to 2003 the main driver for the increased use of renewable energy sources such as wind power was concern over global warming and its effects, and the consequent need to reduce emissions of carbon dioxide and the other greenhouse gases. However the price of oil, which averaged just $28/barrel from the mid 1980s through to the early 2000s,[5] has in recent years increased substantially and in mid 2008 reached a peak of almost $150/barrel. Though the financial crisis and the beginnings of global recession subsequently led to oil prices dropping to below $50/barrel by the end of 2008 few anticipate that low oil prices will be sustained, and the expectation is that future oil prices will average well above the levels seen through the 1990s.[6] These oil price expectations, combined with concern over energy supply security and continuing concern over the effects of global warming, make it likely that high rates of growth in the installed wind power capacity will continue. Though the exponential growth seen over the past seventeen years clearly cannot be sustained indefinitely there is – as yet – no sign that growth is slackening. The countries that have contributed most to the rapid growth in installed capacity in recent years can be seen in Figure 8.2 and the ten leading countries are listed in Table 8.2.[4]

Germany, which in the 1990s installed more than twice as many wind megawatts as any other country, has continued to lead the way and by the end of 2007 its installed wind power capacity was 22 200 MW. The installed capacity has also grown rapidly in the US, Spain and India, and the very recent upsurge of activity in China is noteworthy. However, relative to its size it is Denmark which makes the greatest use of wind power.

Table 8.2 *Top ten countries for installed wind power capacity 2007*

Country	Installed capacity (MW)		Land area km^2	Population density people/km^2	Power density kW/km^2
	Cumulative	In 2007			
Germany	22 200	1630	349 000	236	64
US	17 000	5270	9 160 000	33	1.9
Spain	15 100	3530	500 000	81	30
India	7800	1570	2 970 000	386	2.6
China	5900	3310	9 330 000	143	0.6
Denmark	3120	0	42 400	129	74
Italy	2730	600	294 000	198	9.3
UK	2490	530	242 000	252	10
France	2370	900	546 000	117	4.3
Portugal	2150	430	92 000	116	23

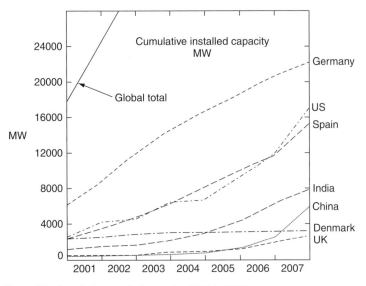

Figure 8.2 Cumulative installed capacity, 2000 to 2007.

Denmark

The 1990s, a decade of growth

By the end of 1990 Denmark had already installed 340 MW of wind power capacity, which as noted was sufficient to provide nearly 2% of Danish electricity. In that same year the Government committed to an action plan, *Energy*

2000, which was followed and expanded in 1996 by *Energy 21*; both were designed to encourage more sustainable development in the energy sector and they set a goal for Denmark to have 1500 MW of wind power capacity by 2005, sufficient to provide 10% of Danish electricity.[7] The investment subsidy programme that had been successfully used to encourage wind turbine purchases through the 1980s had been phased out as turbine economics improved, and had ended in 1989. However Denmark continued to support wind energy by requiring that utilities purchase the output from turbines at 85% of the pre-tax retail price, equivalent to about 0.33 kroner/kWh; added to this was a carbon dioxide tax rebate of 0.10 kroner/kWh and a wind energy support payment of 0.17 kroner/kWh to give a total of about 0.60 kroner/kWh.[8] At the 1990 exchange rate of £1 ≈ 11 kroner this was equivalent to about 5.5 p/kWh, and was sufficient to support a growing home market. In the 1990s, as in the 1980s, most turbine purchasers were individuals or co-operatives, and the widespread ownership or co-ownership that resulted helped with the acceptance of wind turbines by local communities.

Through the early 1990s Danish wind turbine installations averaged about 50 MW per year, using at first turbines with rotor diameters of about 25 m and with rated power outputs in the range 200 to 250 kW. However continuing strong competition within Denmark encouraged manufacturers to develop progressively larger machines that offered better economics;[9] this led to increased turbine sales and by the late 1990s the Danish installed wind power capacity was growing by about 300 MW per year, mostly provided by turbines with diameters of about 50 m and with rated power outputs in the range 600 to 1000 kW. The Danish wind turbine manufacturers which emerged as the market leaders during the 1980s – Vestas, Bonus, Micon and Nordtank – continued to dominate turbine supply in Denmark through the 1990s, and although the Danish home market was crucial to their development they all exported about 70 to 80% of their total production.[10] Though the turbines they manufactured grew steadily larger through the 1990s their appearance, apart from their size, changed very little. Figure 8.3 for example shows part of Carno wind farm in mid Wales, where 56 Bonus 600 kW wind turbines were installed in 1996, making it for a while the largest wind farm in Europe. Like their smaller predecessors these turbines with their 44 m diameter rotors are stall controlled and use rotatable blade tips to stop the rotor when required. The arrangement within the nacelle is also much the same as with earlier Danish machines with the low-speed shaft (to which the blades are attached) driving a speed-increasing gearbox, and with the high-speed output shaft from this driving an induction generator. The power output from each turbine is then taken by cables[11] down to ground level within the tower where a transformer is used to

Figure 8.3 Carno wind farm, 1996, Wales. *Source*: Image courtesy of npower renewables.

increase the voltage from 690 V to an appropriate higher voltage (which at Carno is 33 000 V). As is usual in Europe buried cables then take the power output from each turbine to an electricity sub-station located near the edge of the wind farm, and pole-mounted overhead power lines then deliver the wind farm's overall power output to the local utility's distribution network, where it is used to help meet local electricity needs. Each turbine's reinforced concrete foundation is also buried, and covered with one to two metres of soil, so that the pre-existing vegetation can be restored right up to the base of each tower.[12]

By 1996 Vestas was established as the largest wind turbine manufacturer, and was supplying just over 20% of the rapidly growing world market; its main competitors – Bonus, Micon and Nordtank – each had about 10% market share.[13] Selling and maintaining wind turbines in a growing number of countries around the world was expensive and Nordtank and Micon saw benefit in a merger, which took place in 1997; the new company, called NEG Micon, was then comparable in size to Vestas. In addition to buoyant export markets the Danish home market was also thriving through the late 1990s, with the result that the Energy 21 target of 1500 MW by 2005 was achieved more than six years early, in the first half of 1999. The Government decided that a reasonable limit for turbine installations on land would be about 2500 MW, and therefore amended the arrangements for purchasing the output from wind turbines so that machines bought on or after 1 January 2000 would receive substantially less than the 0.60 kroner/kWh that had previously been paid. The result was a surge of turbine purchases in late 1999, which qualified for the higher pre-2000 price for their output even though most were installed in 2000. Danish wind turbine installations in 2000 were consequently at a record level, with 560 MW installed, to give a year-end total of 2300 MW; 84% of this was owned either by private individuals or by co-operatives in which more than 100 000 families had shares, and wind power in 2000 provided 13% of the total Danish electricity consumption.[14]

Offshore wind makes progress

Denmark had recognised in the 1980s that offshore wind farms had the potential to provide substantially more electricity than wind turbines sited on land, and in the early 1990s local utilities were encouraged to build two offshore demonstration wind farms.[15] The first of these was completed in 1991 using 11 Bonus 450 kW wind turbines (similar to those shown in Figure 5.20) and is located 1.5 to 3 km offshore from Vindeby, which is on the north-west coast of the island of Lolland and about 125 km south-west from Copenhagen. The water depth varies across the site from 2.5 to 5 m, shallow enough to allow the use of broad-based, tapered-cylindrical, reinforced concrete foundations which sit on the sea bed; with added ballast each weighs about 1000 tonnes. The Vindeby wind farm was followed in 1995 by the Tunø Knob wind farm, located about 3 km west of the small island of Tunø and about 25 km south-south-east from Århus, which used 10 Vestas V39 500 kW wind turbines. The water depth here is about 5 m so again it was possible to use concrete gravity foundations. Operational experience from these two demonstration wind farms was encouraging and as the decade progressed, and larger turbines became available, the prospects for building commercial offshore wind farms improved.

Offshore wind farm foundation costs are substantially higher than on land, as are turbine installation costs, and as access is more difficult the costs of operations and maintenance are increased. Offshore wind speeds are however higher than on land so the turbines give about 50% more energy.[16] With relatively small machines, such as those used at Vindeby and at Tunø Knob, the extra costs more than outweigh the benefit of the extra energy output, but with larger wind turbines the foundation costs become a smaller proportion of the total and the premium that has to be paid for the output from offshore wind farms diminishes. Anticipating that by about 2000 multi-megawatt wind turbines would be commercially available, and that their use offshore would provide electricity at a cost comparable with the cost of electricity from wind farms on land, Denmark made plans in 1997 for the construction of a series of large offshore wind farms which would by 2030 provide an additional 4000 MW of wind power capacity.[17]

After a comprehensive environmental assessment and consenting process the first major project was built in 2002 by the local utility Elsam at Horns Rev, in the North Sea to the west of Jutland and 14 km off the coast from Blåvandshuk. This 160 MW wind farm extends over an area of 20 km^2 and uses 80 Vestas 2 MW turbines, each with a rotor diameter of 80 m;[18] the same machines were used a year later for the first UK offshore wind farm and are shown in Figure 1.3. The 39 m long blades are made from fibreglass-epoxy and weigh about 7 tonnes each. As with all Vestas turbines made since the mid 1980s the power output is controlled in high wind speeds by varying the blade pitch, and a wound rotor induction generator is used so that the rotor speed can be varied between 9 and 19 rpm.[19] The water depth at Horns Rev ranges from 6 to 14 m, and monopile foundations were used. These are simply thick-walled steel tubes which are typically 35 to 45 m long and weigh approximately 200 tonnes; their diameter is 4 m, about the same as the base of the tower. Each one is hammered about 25 m into the seabed, so that its upper end remains above sea level, and an offshore crane is then used to attach the tower, which weighs 160 tonnes, followed by the 80 tonne nacelle and the blades. Buried cables take the power output from each turbine to an offshore sub-station, which sits on its own monopile foundation; within this sub-station a transformer increases the output voltage (from 36 kV to 150 kV) so as to minimise power losses in the buried cable that takes the wind farm's power output to the shore.

A second large offshore wind farm was built in 2003 at Rødsand, in the Baltic Sea to the south of Lolland, and about 12 km off the coast from Nysted. Financed and owned by the utility Energi E2 in partnership with the Danish company Dong Energy and the Swedish power company Sydkraft this 166 MW project – shown in Figure 8.4 – uses 72 Bonus 2.3 MW wind turbines, each with a rotor diameter of 82 m and spaced an average of 640 m apart.[20] The water depth at Nysted/Rødsand ranges from 6 to 9 m and allowed the use of reinforced concrete gravity foundations. As at Horns Rev buried cables take the power output from each turbine to an offshore sub-station, from where a higher voltage buried cable takes the total wind farm power output to the shore.

The layout of the machinery within the nacelle of the Bonus 2.3 MW wind turbine is representative of many current multi-megawatt designs and is shown in Figure 8.5 (which makes an interesting contrast with the older and much smaller machine shown in Figure 5.15). The fibreglass-epoxy blades (3), each 40 m long and weighing about 9 tonnes, are attached to the nodular cast iron hub (5) using the large-diameter pitch bearings (4). Although these wind turbines,

Figure 8.4 Nysted/Rødsand wind farm, 2003, Denmark. *Source*: Image courtesy of Siemens Wind Power.

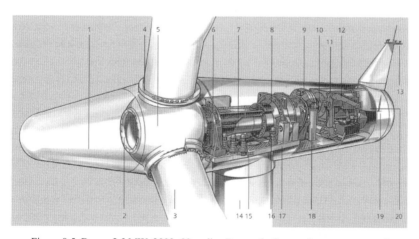

Figure 8.5 Bonus 2.3 MW, 2002, 82 m dia., Denmark. *Source*: Image courtesy of Siemens Wind Power.

like all previous Bonus machines, are stall controlled it is more cost-effective at megawatt scale to provide aerodynamic braking (when required) by turning the whole blade through almost 90° than to engineer the blades to incorporate rotatable blade tips.[21] The main bearing (6) supports the front end of the low-speed shaft (7) which transmits the power from the blades to the gearbox (8) and these primary mechanical components are all mounted on the steel bedplate (17). The high-speed shaft from the gearbox to the two-speed induction generator (11) cannot be seen but the brake disc (9) for the parking brake is mounted on it, and the coupling (10) connects it to the generator's input shaft; use of a two-speed generator allows the rotor to turn at 11 rpm in light winds and at 17 rpm in stronger winds, and gives a small increase in the energy output. The generator cooling fan is at (20). Pendant cables (not shown) take the 3-phase 690 V power output from the generator to the base of the tower where it is transformed to 33 kV before going by cable to the offshore sub-station. The wind direction and speed sensors at (13) provide inputs to an electronic controller which supervises and controls all aspects of the turbine's operation; when there is a sustained change in the wind direction the controller activates the multiple yaw motors (16) which mesh with the yaw ring (15) that is fixed to the top of the tower (14) and slowly turn the nacelle until the rotor is again facing into the wind. The service crane (12) is to assist with maintenance; (18) is the gearbox oil filter, (19) is the nacelle cover, (1) is the spinner – which helps to reduce the nacelle drag – and (2) is the bracket to which it is attached.

These first two large commercial offshore wind farms have on the whole both performed quite well.[22] Though Horns Rev had a poor first year, with low availability which was mainly due to transformer problems, all the turbines were taken ashore for remedial work in 2004 and since then the availability has been high. As the average wind speed at the 70 m hub height is 9.7 m/s the wind farm has consequently achieved a high capacity factor, typically in excess of 40%. Nysted/Rødsand is somewhat less exposed than Horns Rev but the wind farm availability has been over 95% and the capacity factor has been close to 40%. The environmental assessment of both wind farms has been comprehensive and thorough, with measurements made both before and after construction. The results were reported at a specially convened conference in November 2006 and generally indicated 'that impacts on the environment are extremely small'.[23]

Both the Horns Rev and the Nysted/ Rødsand wind farms were financed on the basis that the utility owners would be paid 0.45 kroner/kWh, equivalent in 2002 to 3.8 p/kWh.[24] This payment of a specified price per kWh generated is often referred to as a feed-in tariff, and has been the preferred method for encouraging wind power (and other renewables) in several European countries. Danish wind projects on land through the 1990s were supported through what was effectively a feed-in tariff of about 0.60 kroner/kWh. Though this was just adequate in the early 1990s the economics of wind energy improved substantially through the decade, and by the late 1990s it had become too generous, as is illustrated by the fact that Horns Rev and Nysted were able to proceed with a feed-in tariff of just 0.45 kroner/kWh despite the higher cost of offshore wind farms.[25]

Danish overview

The fast-growing world market for wind turbines has made it necessary for leading manufacturers to rapidly expand their production. The resulting financial pressures led to the merger in early 2004 of Vestas and NEG Micon, with the enlarged company retaining just the Vestas name. And in late 2004 Bonus, which had been privately owned, was bought by Europe's largest engineering group – Siemens – and re-named Siemens Wind Power. Vestas is by a substantial margin the world's largest wind turbine manufacturer with an annual turnover of nearly £4 billion, 15 000 employees and a market share in 2007 of 20%; Siemens, with a market share of 6%, is the sixth-largest wind turbine manufacturer.[26] In Denmark, as elsewhere, the trend to larger turbines is continuing. The Vestas product range now extends up to the 3 MW rated V90, which has a rotor diameter of 90 m, and a prototype 4.5 MW machine with a rotor diameter of 120 m is being tested. The Siemens product range extends up to the SWT-3.6–107, which is 3.6 MW rated and has a rotor diameter of 107 m. At this size Siemens found that it was cost-effective to move away from stall control and the power output in high wind speeds is controlled, as in most other multi-megawatt wind turbines, by varying the blade pitch.[27] The power quality of pitch-controlled turbines is poor unless the rotor speed is allowed to vary; the rotor inertia can then be used to smooth the power output and the loads transmitted through the drive train. Vestas provide a variable speed capability by using a wound rotor induction generator in an arrangement they call OptiSpeed.[28] Siemens use a normal induction generator together with a power electronics system that allows a wide range of rotor operating speeds and gives compliance with relevant grid codes.[29]

The surge of turbine installations in 2000 took the total installed capacity in Denmark to 2300 MW at the end of that year, with just 10 MW offshore. The capacity on land increased by 400 MW over the next two years, encouraged by financial incentives to replace large numbers of small and ageing turbines with smaller numbers of modern machines, but outside this scheme the price offered for the output from new wind turbines was too low to stimulate further investment. The land-based wind power capacity was consequently almost static – at about 2700 MW – between 2002 and the end of 2007, though the construction of Horns Rev in 2002 and Nysted in 2003, plus some smaller offshore projects, raised the offshore wind power capacity to 400 MW. The overall Danish installed capacity at the end of 2007 totalled 3120 MW, provided by 5210 wind turbines; these generated 7.7 billion kWh (= 7.7 TWh) in 2007, at an average capacity factor of 26%, and provided 20% of the electricity consumed in Denmark.[30]

In February 2008 the Government stated that it would be targeting to provide 20% of Danish *energy* from renewable sources by 2012; this will require that an additional 1300 MW of wind power capacity be built and wind turbines will then provide about 30% of Danish electricity. About two-thirds of this additional capacity will be offshore, and the first 400 MW will be provided by extensions to the existing Horns Rev and Nysted wind farms that have already been contracted. The Horns Rev extension will be built by Dong Energy using 91 Siemens 2.3 MW wind turbines, and this 209 MW wind farm is due for completion in late 2009. The contract to build was signed in 2005 and specified a feed-in tariff that will pay 0.52 kroner/kWh, equivalent then to 4.8 p/kWh.[31] The contract to build the 200 MW Nysted/Rødsand extension was won in April 2008 by E.on Sweden with a feed-in tariff that will pay 0.63 kroner/kWh, equivalent to 6.5 p/kWh, and this wind farm is due for completion in late 2011.[32]

Wind turbine prices reduced substantially through the 1990s, so that from a utility investment perspective (capital cost repayment over 20 years, and a 5% required rate of return) the cost of energy from wind farms on land reduced from about 0.48 kroner/kWh in 1990 to about 0.28 kroner/kWh in 1999[33]. The 0.45 kroner/kWh cost of the output from Horns Rev and Nysted, the first large commercial offshore wind farms, therefore represented a premium of about 50% on the cost of energy from land-based wind farms. It was hoped that this would reduce with experience, and that future projects would give energy cost reductions. However prices for commodities such as steel and copper rose sharply on world markets for several years prior to 2008, leading to substantial turbine price increases; and the continuing boom in turbine sales has led to shortages and consequent further price increases. The effect of these price increases is reflected in the 2008 cost of the feed-in tariff required by the Nysted/Rødsand wind farm extension, which at 0.63 kroner/kWh is 40% higher than the 2002 cost of Horns Rev; inflation in Denmark over this period totalled approximately 10% so in real money terms (net of inflation) the cost increased by about 30%.

Though 20% of Denmark's electricity comes from wind turbines the cost to the consumer is low; and as the Danish Energy Authority has noted the direct cost of the premium paid for the output from wind turbines is partially offset by the downward pressure that this output puts on the market price for electricity. The value to consumers of this downward pressure on electricity prices has been quantified by the Danish national laboratory at Risø, and in 2006 they found that it reduced the additional cost paid for wind generated electricity from 0.022 kroner/kWh to 0.0037 kroner/kWh; at the prevailing exchange rate this corresponds to a reduction in the additional cost of wind-generated electricity from 0.2 p/kWh to 0.03 p/kWh.[34]

Germany

1991: a feed-in tariff established

Though Denmark has led the world in the utilisation of wind power, with wind turbines providing 20% of its electricity in 2007, its relatively small size has meant that several other countries now have much larger installed wind power capacities. As is evident from Figures 8.1 and 8.2 Germany was the first major market to develop in the 1990s, and has been the world leader in installed capacity since 1996. The initial stimulus was provided by a package of financial incentives designed to encourage the installation of 100 MW of wind capacity by 1994; this was soon oversubscribed and was expanded to 250 MW in 1990.[35] Though this federal 250 MW wind programme was helpful in building momentum, much more important was the approval in autumn 1990 by the Federal Parliament of the Electricity Feed-in Law (EFL), or *Stromeinspeisungsgesetz*. Effective from January 1991, this required utilities to connect renewable energy projects and to buy their output at a price that for wind installations was equal to 90% of the average sales price to retail consumers. This relatively high percentage of the retail electricity price was justified on the grounds that it helped to create a 'level playing field' for clean sources of energy, given that the price paid by consumers for the electricity generated by conventional power stations did not reflect the costs imposed on society by the pollution they produced.[36]

Though individual provinces could – and did – provide some additional financial support, and purchasers of wind turbines could access low interest rate loans, the main driver for the rapid growth in the installed wind power capacity in Germany through the 1990s was the feed-in tariff mandated by the EFL.[37] In 1991, for example, this paid DM 0.167/kWh for the output from wind turbines, equivalent then to 5.7 p/kWh and similar to the price paid in Denmark. This stimulated a growing number of wind turbine purchases and the annual installed wind power capacity increased rapidly. In 1991, the first year of the feed-in tariff, 50 MW was installed but by 1995 this had risen to 500 MW; and in 2000 a total of 1670 MW was installed, bringing the cumulative installed capacity in Germany to the 6110 MW shown in Figure 8.1.

Larger turbines provide better economics so – as in Denmark – the average size used for new installations increased steadily through the decade; in 1991 for example new turbines had on average a rated power capacity of just under 200 kW, but by 1995 this had risen to 500 kW and by 2000 the average size had increased to just over 1.1 MW.[38] As might be expected most of the capacity was installed in the windier, northerly, provinces bordering on the North Sea and the Baltic Sea, led by Lower Saxony where by the end of 2000 the cumulative capacity was 1750 MW. Through the early 1990s most projects were small, typically just two or three turbines, and were locally owned either by individual farmers or by small groups of local people; as such they rarely encountered opposition. In the later 1990s projects became larger, and were often developed by people without local connections; they would therefore more frequently meet with opposition. Most of the larger projects that were built were financed through project-specific companies co-owned by many individual investors, both local and non-local, and as a result many tens of thousands of Germans have a wind farm ownership interest.[39]

Domestic manufacture

Though many of the turbines required by the growing German market were imported from Denmark the market was large enough to encourage the growth of domestic wind turbine manufacturers and one of these, Enercon, grew to be the largest supplier of wind turbines in Germany and a major supplier world-wide. Enercon was started in Aurich, Lower Saxony, by Aloys Wobben in 1984 and its first turbine was the 55 kW rated E-16, which had a rotor diameter of 16 m and 3 fixed-pitch Danish-made blades located upwind from a lattice tower. It was stall controlled and in most respects very similar to the Danish machines then being made. However Enercon also made power electronic variable-speed drive systems for industrial applications, and these were used on the E-16 turbines so that the rotor speed could be varied over a three-to-one range. The E-16 was succeeded in 1989 by the 32 m diameter, 300 kW rated, E-32 which had variable-pitch blades as well as variable rotor speed, and this was followed in 1993 by the innovative E-40. With a 40 m rotor diameter and a 500 kW power rating this was the first turbine in the Enercon range to eliminate the need for a gearbox by using a large-diameter ring generator directly driven by the rotor, whose speed is allowed to vary in the range 18 to 38 rpm; power electronics is then used to give an electrical output at the correct voltage and frequency for grid connection.[40]

The E-40 was extremely successful and many thousands were sold, including the one shown in Figure 8.6 which was installed by Ecotricity in 1996 at Lynch

Figure 8.6 Enercon E-40, 1996, 40 m dia., 500 kW, Gloucestershire, England.
Source: Image courtesy of Ecotricity.

Knoll, in Gloucestershire; the housing for the 4.8 m diameter ring generator is clearly visible behind the hub of the rotor. All subsequent Enercon wind turbines have had the same configuration as the E-40, using directly driven large-diameter ring generators instead of the more usual gearbox and generator combination. And through the 1990s Enercon were the market leaders in Germany, typically supplying about 25% to 30% of all the turbines installed.

Germany's second-largest wind turbine manufacturer through the early and mid 1990s was Tacke Windtechnik, based in Salzbergen in Lower Saxony, which was formed in 1990 as an offshoot from a long-established manufacturer of industrial and marine gearboxes.[41] Tacke produced a succession of conventional three-bladed, fixed-pitch, stall-controlled wind turbines in the 1990s, ranging up to a 43 m diameter 600 kW rated machine, and gained a market share in Germany of about 20%. In 1996 they took the major step of developing an advanced 1.5 MW rated, 65 m diameter, variable-speed machine with variable-pitch blades, but financial problems led to their bankruptcy in 1997. Tacke were then bought by Enron Wind, a subsidiary of the large American energy company, who continued to make Tacke machines; and when Enron collapsed in 2002 their wind assets were bought by General Electric, which has continued to make machines derived from the Tacke 1.5 MW.

The feed-in tariff for the output from wind turbines remained substantially constant through the 1990s at about DM 0.17/kWh; inflation over this period averaged about 2% per year so in real money terms (net of inflation) the value of the feed-in tariff declined by about 20%. But as has been noted wind energy costs reduced through the decade by about 40%, so the overall effect was to improve the economics of wind power.[33] This helped to make some less windy, inland, locations financially viable, as did the use of towers that were substantially taller than in Denmark. However moves to liberalise the German electricity supply industry in the late 1990s made it necessary to change the way that support was given to renewable energy technologies, including wind.

From 2000: the Renewable Energy Law

Despite continuing strong opposition from the big electricity utilities a Renewable Energy Law, the *Erneuerbare-Energien-Gesetz* or EEG, was passed in February 2000, and crucially this continued to provide a feed-in tariff for the output from renewables. For wind power the new feed-in tariff was designed so that development of less windy inland sites would be encouraged, without allowing excessive returns to be made on new projects in windy areas. To achieve this the EEG stipulated that the feed-in tariff would pay a specified price for the output from new wind projects for a period of 20 years; for a minimum of 5 years the price paid would be DM 0.178/kWh and it would then drop to DM 0.121/kWh. At the average exchange rate for 2000 these prices were equivalent to 5.5 p/kWh and 3.8 p/kWh. New projects in less windy locations would have the higher price paid for more than the 5 year minimum, with the length of this high price period calculated in accordance with rules specified in the EEG legislation.[42] Moreover, and in the expectation that the economics of wind energy would continue to improve, the EEG stated that the specified feed-in tariff prices would only apply to projects installed up to the end of 2001; for projects built in 2002 the prices would be 1.5% lower and for subsequent years the prices would reduce by a further 1.5% per year. In a departure from previous practice the EEG also stated that utility-owned wind projects would be allowed to benefit from the feed-in tariff.

The new feed-in tariffs gave investors confidence, after a period of uncertainty, and as can be seen in Figure 8.2 the German installed wind power capacity grew rapidly. In the peak year of 2002 more than 2300 wind turbines with a capacity totalling 3200 MW were installed. The installation rate subsequently declined primarily because suitable sites were becoming increasingly scarce; but the situation was aggravated by a revision of the EEG in 2004 which slightly reduced the price levels for the output from wind projects built in the latter part of that year and prescribed more rapid reductions for future years.[43] As noted previously, and contrary to expectations, turbine prices have increased in the past few years, partly as a result of the sharply rising price of commodities such as steel and copper but also as a consequence of turbine supply shortages. The combination of increased turbine prices and a year-by-year reducing feed-in tariff made financing new projects in Germany increasingly difficult.

Despite these problems the total installed capacity continued to grow – though at a reducing rate – and 2007 saw 883 wind turbines installed, adding 1670 MW and bringing the cumulative total wind power capacity to 22 200 MW by the end of the year.[44] As the German market contracted after 2002 Enercon's market share increased and they supplied 50% of the capacity installed in 2007, with Vestas supplying 24% and a smaller German manufacturer – Repower – supplying a further 11%; the balance was shared between many other manufacturers. The buoyant German market underpinned Enercon's growth over the years, but though the domestic market remains important it now exports 70% of its production and is the world's fourth-largest manufacturer (after Vestas, GE Wind and Gamesa) with a 13% share of the global market for wind turbines.[26]

The 2300 turbines installed in 2002 had, on average, a rated power output of 1.4 MW; however by 2007 the average for new installations had risen to 1.9 MW, provided by turbines such as the 2 MW rated, 70 m diameter, Enercon E-70 shown in Figure 8.7. The basic design is similar to the Enercon E-40, shown in Figure 8.6, and the E-70 has three variable-pitch blades made from fibreglass-epoxy and located upwind of the steel tubular tower. The rotor drives directly a large-diameter ring generator, and the power electronics used to give an electrical output at the correct voltage and frequency for grid connection also gives compliance with grid codes[29] as well as allowing the rotor to operate at speeds in the range 6 to 22 rpm. Reducing the rotor speed when the wind speed is low gives a small improvement in energy capture,[45] but what is frequently more important is that the consequent lower blade tip speed leads to substantially lower turbine noise levels. In noise-sensitive locations compliance with specified noise limits is usually most demanding when wind speeds are low (because ambient noise levels rise rapidly at higher wind speeds).[46] Reducing the rotor speed in these light wind conditions helps to ensure that turbine noise levels meet the conditions necessary to prevent noise nuisance.

Figure 8.7 Enercon E-70, 2008, 70 m dia., 2 MW, Somerset, England. *Source*: Image courtesy of Ecotricity.

German overview

As a consequence of the stimulus given by the EEG renewable energy sources provided 12% of the electricity consumed in Germany in 2006, with wind contributing nearly half. The BMU (the Federal Ministry for the Environment, Nature Conservation and Nuclear Safety) calculated that in 2006 the direct cost to consumers of the EEG's feed-in tariffs was €3.3 billion.[47] However they also calculated that the electricity provided by renewables had saved consumers about €5 billion by displacing electricity that would otherwise have been generated by the least efficient fossil-fuelled power stations, which had led to reductions in wholesale electricity prices. They further calculated that the savings in emissions of carbon dioxide and air pollutants were worth €3.4 billion and that savings in coal and gas imports should be valued at €1 billion. The €3.3 billion cost of the feed-in tariffs in 2006 was therefore offset by €9.4 billion of benefits. The BMU also stated that the renewable energy sector had been responsible for the creation of 230 000 jobs, with wind energy contributing 82 000 to this total.

In 2007 wind turbines generated 39.5 billion kWh and provided 7.2% of the electricity consumed in Germany;[44] the average capacity factor was 21%, significantly lower than Denmark's 26% simply because many of the German turbines are located well inland, in less windy locations. As Table 8.2 indicates Germany's wind power density of 64 kW/km^2 – equivalent to a 1 MW wind turbine for every $16\,km^2$ of land area – is only slightly smaller than Denmark's $74\,kW/km^2$;[48] it is therefore at first somewhat surprising that in Denmark in 2007 wind provided 20% of the electricity consumed, while in Germany the contribution was just 7.2%. The explanation for this apparent anomaly is that Germany's population density is almost double Denmark's, and its electricity consumption per square kilometre of land area is therefore also approximately double.[49] The lower capacity factor of German wind turbines also contributes to the difference.

Renewable energy sources provided 14% of the electricity consumed in Germany in 2007 with wind, as noted, contributing 7.2%. The Government plans to increase substantially the use of renewables and has stated that it 'wants at least 27% of electricity consumption to originate from renewables by 2020 and 45% by 2030'.[50] Much of this increase will need to be provided by wind turbines, both on land and offshore. New installations on land will continue, but at a diminishing rate because of the growing scarcity of suitable sites. There is however considerable potential to increase the wind power capacity by re-powering, that is by replacing many of the relatively small machines that were installed in the 1990s with larger and more powerful wind turbines. For example the 5200 turbines installed up to the end of 1997 – many on very good sites – contributed just 2080 MW and re-powering most of these older installations would add several thousand megawatts of capacity.

Germany's offshore wind potential is limited by its relatively short coastline; it has about 300 km of North Sea coast and about 500 km of coast facing the Baltic Sea but tourism and nature conservation constraints are such that almost all the offshore wind projects now being planned are between 25 and 100 km offshore. The Government anticipates that by 2030 offshore wind turbines with a total rated capacity of about 25 000 MW will be operational, and will provide about 80 billion kWh/year, but no fully offshore wind farm has yet been built. Nineteen projects with a capacity totalling over 6000 MW are being progressed through the consenting process,[51] and sixteen of these are in the North Sea.

One issue that has delayed construction has been the magnitude of the feed-in tariff for offshore wind projects. The EEG in 2004 specified that this should be 9.1 cents/kWh for a minimum of 12 years; for the remainder of the 20-year period of the feed-in tariff the price paid for the wind farm's output would then reduce to 6.2 cents/kWh. At the average exchange rate in 2008 of £1 = €1.26 these prices correspond to 7.2 p/kWh and 4.9 p/kWh respectively. However wind turbine prices have increased substantially since 2004, and the cost of offshore wind projects has gone up even more than the cost of wind projects on land. The EEG was therefore amended in June 2008 such that all offshore wind farms commissioned by 2015 will be paid 15 cents/kWh for the first 12 years, reducing to 3.5 cents/kWh for the remainder of the 20-year tariff period.[52] At the £1 = €1.26 exchange rate these prices correspond to 11.9 p/kWh and 2.8 p/kWh respectively, and this new and much improved tariff is expected to lead to the start of construction for several projects in 2009. Offshore wind farms completed after 2015 will receive a somewhat lower price for their output, reflecting the expectation that project costs will by then have reduced. In Germany, as in Denmark, the Transmission System Operator (TSO) has the responsibility for providing and paying for the cable connections from offshore wind farms to the grid system on land,[53] and transmission system constraints are likely to limit the capacity installed offshore by the end of 2011 to between 2000 MW and 3000 MW.

Spain

The 1994 feed-in tariff

Spain is the only other European country in which wind power development has been on a scale comparable with Germany, as can be seen from Figures 8.1 and 8.2. A relatively low level of activity in the early 1990s was transformed by legislation – passed by Royal Decree in December 1994 – which was designed to encourage the greater use of renewable sources of energy.[54] It stipulated that the power output from renewables projects should be bought at a premium price that would be fixed for a five year period; and for wind power the price was set at 11.6 pesetas/kWh, equivalent then to 5.6 p/kWh. The relatively short duration of the guaranteed premium price proved to be no barrier to securing finance for projects and the installed wind power capacity grew rapidly. Targets were set and the permitting process was controlled by the individual provinces, and to maximise local benefits – and in particular to help create local jobs – regional governments would grant concessions to manufacturers in return for commitments to set up factories in their areas.

The feed-in tariff arrangements have been amended on several occasions, most recently by the Royal Decree of March 2004.[55] This gives the owners of wind farms two options for the sale of their power; they can either accept a feed-in tariff price for all their output or they can sell their output on the open market and receive an additional premium payment for choosing this option. The choice has to be made in advance for a period of one year after which the electricity producer can, if it wishes, change to the alternative. The level of the feed-in tariff and the alternative price premium are both indexed on an annual basis to the average price of electricity, and for 2008 the feed-in tariff was set at

7.32 cents/kWh with the alternative price premium set at 2.93 cents/kWh; at the average exchange rate for 2008 of £1 = €1.26 these prices correspond to 5.8 p/kWh and 2.3 p/kWh respectively. Though there is less certainty than with the German 20-year fixed-price feed-in tariff the Spanish tariff arrangements are sufficient to give investors confidence in the return on wind farm investments: and as is shown in Figure 8.2 there has been continuing substantial growth in recent years in the installed wind power capacity.

The size of the Spanish market was large enough, even in the 1990s, to encourage the development of local wind turbine manufacturers and two of these, Gamesa and Acciona, are among the ten largest manufacturers world-wide. Gamesa's wind energy activities commenced in 1993 through a technology partnership arrangement with Vestas, which ended in 2001. It has for many years been the dominant manufacturer in Spain and in 2007 it supplied 47% of the 3500 MW installed, with Vestas providing 20% and Acciona 19% and with the remaining 14% shared between many other suppliers. Gamesa is, unusually, a wind farm developer as well as a wind turbine manufacturer and Figure 8.8 shows part of its Serra de Meira wind farm in Galicia. Wind farms have been installed in most regions of Spain but Galicia, in the north-west and facing the Atlantic Ocean, has a better than average wind regime and has been one of the most active provinces.

Figure 8.8 Serra de Meira wind farm, 2005, Galicia, Spain. *Source*: Image courtesy of Gamesa, © Gamesa Corporación Tecnológica.

The Serra de Meira wind farm shown in the figure has a total of 58 Gamesa G52–850 wind turbines, which as their model number indicates are rated at 850 kW and have a 52 m rotor diameter. As is now usual the variable-pitch blades, which are made from fibreglass-epoxy, are located upwind from the steel tubular towers. And each turbine uses a wound rotor induction generator[56] in conjunction with power electronics so that the rotor speed can be varied in the range 15 to 31 rpm. A similar 2 MW machine is also made by Gamesa and this has a rotor diameter in the range 80 m to 90 m, depending on the site windiness. Gamesa has greatly expanded its international activities in recent years and in 2007 it ranked as the third-largest manufacturer (after Vestas and GE Wind) with sales totalling just over 3000 MW, which corresponds to a 14% share of the global market.[26] Spain's second largest manufacturer, Acciona, is a large engineering and infrastructure group and the main product of its wind division is a conventional three-bladed wind turbine rated at 1.5 MW and with a diameter of 77 m. Acciona's sales totalled 870 MW in 2007, giving it a 4% share of the global market, and it ranked as the seventh-largest manufacturer.

Spanish overview

Spain is one of the largest countries in Europe, see Table 8.2, and there is a good wind resource in many regions. In 2002, when the installed wind power capacity was about 4000 MW, the Government set a target to have 13 000 MW installed by 2011; however it was apparent by 2005 that this target would be surpassed well before 2011 and it was therefore increased to 20 000 MW by 2010. With 3530 MW installed in 2007, and a total installed capacity of 15 100 MW by the end of that year, it seems probable that this 20 000 MW target will also be exceeded before 2010; a revised target of 29 000 MW by 2016 is now being considered. Though Spain is 40% larger than Germany its population is only about one half as large and the 11 000 MW that was operational in 2006 was sufficient to provide 9% of Spain's total electricity consumption;[57] the present target of 20 000 MW corresponds to wind turbines providing about 16% of Spain's electricity. Spain's wind power density of 30 kW/ km^2, see Table 8.2, is less than half that already achieved by Denmark and Germany and there is potential on land for a wind power capacity well in excess of the 20 000 MW already planned. Spain also has a substantial offshore wind resource, though areas where the water depth is less than 35 m are relatively limited. The Government has decided that all suitable sites should be open to public tender and does not expect to issue the first offshore wind farm licence until 2012; it is therefore unlikely that any Spanish offshore wind farm will be operational before 2014.[58]

United Kingdom

The disappointing 1990s

Germany, Denmark and Spain all successfully used feed-in tariffs to encourage investment in wind projects through, and since, the 1990s. However when the UK Government decided to encourage renewable energy projects it chose to use a very different mechanism, the Non-Fossil Fuel Obligation, or NFFO. As discussed in Chapter 7 the main purpose of NFFO was to provide a means for paying a premium price for the output from nuclear power stations, with the approximately £1 billion per year cost recovered by a levy of about 10% on all electricity sales. It was then decided that the levy should be increased to about 11% so as to provide additional funds that could be used to support a wide range of renewable energy projects. The Government selected projects for support by issuing – from time to time – calls for proposals from those wishing to build renewable energy schemes. In their responses proposers had to provide technical and financial details of the projects they wished to build and bid competitively for power purchase contracts. For wind projects the proposers had to provide details of the site that had been selected, including evidence of its average wind speed, plus details of the type and number of turbines that would be used as well as details of the grid connection and the overall project costs, plus evidence that the project could be financed; they then had to bid the price they required for the electrical output from their project. In due course the Government would decide how many of the lowest-priced projects from each of the various renewables technologies would be supported, and would then issue a so-called NFFO Order; this effectively required the companies supplying electricity to consumers to buy the output from the selected projects, if and when they were successfully consented and built, at the specified NFFO contract prices.

Between 1990 and 1998 there were in total five NFFO Orders and the first of these, in September 1990, gave NFFO contracts to just nine wind projects. Four of these totalling 2 MW were for existing operational wind turbines, including two of those on test at Carmarthen Bay; the remaining five – totalling 26 MW – were for wind farms. As discussed in Chapter 7 these NFFO1 contracts were only able to pay a premium price through to the end of December 1998 (for reasons related to the nuclear origin[59] of NFFO) and the premium price for the five wind farms averaged 9 p/kWh. Only one wind farm, Delabole, had planning permission when the NFFO1 Order was made and it then took many months to complete the financing arrangements. Construction commenced in June 1991 and the wind farm – Britain's first – was operational by the end of

December, using ten Vestas wind turbines each with a 34 m diameter rotor and a rated power output of 400 kW.[60] Though none of the other four proposed wind farms had planning consent by September 1990 their planning applications had in each case been approved by their local planning authority, which would usually be responsible for giving consent.[61] However the Government, which has the right to intervene if it considers that there are planning issues of more than local importance, instructed the local planning authorities not to issue consents; in due course it then decided that each of the three largest projects should have a Public Inquiry, a time-consuming and expensive process. The smallest project, the four turbine Chelker wind farm, was after many months released for the local planning authority to determine, and was then consented and built.

First to go to Inquiry in April 1991 was Cemmaes, a proposed 24 turbine 7.2 MW wind farm in North Wales about 12 km north-east from Machynlleth and close to the southern boundary of Snowdonia National Park. The visual change caused by the wind farm – and its acceptability – was the principal issue at the Inquiry though many other issues, including turbine noise, were also covered.[62] Approval for the project was given in September 1991 and the developer, National Wind Power,[63] could then conclude the turbine selection process. They chose to use the only available British-made machine, the Wind Energy Group's two-bladed 300 kW rated MS-3, which had been tested at Carmarthen Bay and is shown in Figure 7.8. Cemmaes wind farm was then constructed in 1992 on the 400 m high Mynydd y Cemais ridge and was operational by the end of November; Figure 8.9 shows the completed wind farm with Snowdonia's mountains in the background. Visual change was the principal issue at the subsequent Kirkby Moor and Ovenden wind farm Public Inquiries but both these projects were also approved, and both were built in 1993 using 400 kW Vestas turbines similar to those used for Delabole in 1991.[64]

The second round of NFFO was initiated in late 1990 and concluded in November 1991 with the award of NFFO2 contracts to forty-nine wind projects, whose capacity totalled 196 MW. As with NFFO1 these contracts paid a premium price only until the end of December 1998. Allowing time for consenting and construction this would typically leave NFFO2 projects with about five years of premium price operation, compared with about six years for NFFO1 projects. The prices bid into the NFFO2 process were therefore substantially higher than for NFFO1 and all the most competitively priced projects were given contracts at the same price, which was 11 p/kWh. Twenty-two NFFO2 projects, with a capacity totalling 121 MW, were in due course consented and constructed. Most used Danish wind turbines, with Vestas and Bonus predominating, but two used the Dutch WindMaster machines, two

Figure 8.9 Cemmaes wind farm, 1993, North Wales.

used Japanese machines made by Mitsubishi and one used the American-made Carter wind turbines; two NFFO2 wind farms, Cold Northcott in Cornwall and Llangwyryfon in mid Wales, used British-made two-bladed WEG turbines identical with those used at Cemmaes wind farm and shown in Figure 8.9.

The Government had given some financial support to the three WEG wind farms, so as to help WEG compete with overseas manufacturers who had been producing wind turbines in volume for several years. Following the supply of 65 of their machines to these wind farms it was hoped that WEG would be able to compete without further support and secure a share of future wind farm supply contracts. However these plans for growth received a major setback on the night of 8 December 1993 when a severe storm struck North Wales. On the very exposed Mynydd y Cemais ridge the measured wind speeds were of hurricane strength, and though wind turbines are designed to withstand wind speeds of this magnitude it was evident next morning that several Cemmaes machines had been damaged, and two had lost their blades.[65] As a precautionary measure all the turbines at this and the other WEG wind farms were then shut down. Careful investigation revealed that the failures had been due to a weakness in the mechanism used to vary the blade pitch, and the turbines were in due course modified and put back into service. WEG continued with the development of wind turbines designed to be more cost-effective than the 33 m diameter MS-3 but were unable to secure further volume orders either in the UK or in the rapidly expanding European market. In 1998 WEG and its associated blade-making company were bought by NEG Micon and the UK manufacture of wind turbines ended.[66]

Though the NFFO2 Order was made in November 1991, just one year after NFFO1, there was a very unwelcome interval of over three years before the NFFO3 Order was made in December 1994. NFFO4 followed, after another lengthy interval, in February 1997 and the final NFFO5 Order was made in September 1998. The NFFO Orders only applied in England and Wales; separate legislation was required in Scotland and no arrangements were made in the early 1990s to encourage renewables in Scotland. However from 1994 onwards there were three Scottish Renewables Obligation (SRO) Orders that ran in parallel with the NFFO3, NFFO4 and NFFO5 Orders. Arrangements for these three later NFFO Orders and for the three SRO Orders were similar to those for NFFO1 and NFFO2, though project proposers had to provide even more information in support of their site-specific bids. The minority whose bids were successful were given 15-year index-linked premium price power purchase contracts at their bid price.[67]

The volume of contracts awarded was always very much smaller than the volume of bids submitted. For example 198 wind projects were bid into NFFO3, representing 1542 MW of proposed wind capacity, but contracts were only awarded to the 55 that were bid most competitively, corresponding to a capacity of just 386 MW. The most competitive bids were generally for projects in the windiest locations and these high site wind speeds, together with the 15-year length of the power purchase contracts, led to NFFO3 contract prices that averaged 4.3 p/kWh – very much lower than the 11 p/kWh paid by the short duration NFFO2 contracts; they were also substantially lower than the $5\frac{1}{2}$ to 6 p/kWh then being paid by the feed-in tariffs in Denmark, Germany and Spain. Many project developers had failed to win any NFFO3 contracts and when they next had the opportunity to do so, in 1997, they bid more aggressively with the result that NFFO4 contract prices were nearly 20% lower, and averaged 3.5 p/kWh. And the vital importance – to developers – of winning power purchase contracts led to a further reduction in 1998 when NFFO5 prices averaged just 2.9 p/kWh; see Table 8.3.[68]

In total the three later NFFO Orders, plus the corresponding three SRO Orders, led to contracts being signed for 244 wind projects with a total capacity of 2458 MW. As noted the contract prices were substantially lower than the feed-in tariff prices in Denmark, Germany and Spain; and UK politicians claimed that this demonstrated the superiority of the very competitive NFFO/SRO process. However it proved impossible to secure planning consent for all but a small proportion of the projects contracted and the growth of wind power capacity in the UK through the 1990s was very much smaller than in Denmark, Germany and Spain. By the end of 2000 the installed capacity in

Table 8.3 *NFFO Summary*

NFFO Order	Date made	Contracted price, p/kWh	Contracted capacity, MW	Capacity built, MW	Percentage built
NFFO1	Sept. 1990	≈ 9	28	28	100%
NFFO2	Nov. 1991	11	196	121	62%
NFFO3	Dec. 1994	4.3	339	117	35%
NFFO4	Feb. 1997	3.5	768	90	12%
NFFO5	Sept. 1998	2.9	791	0	0%

the UK was just 410 MW, see Figure 8.1, compared with 2300 MW in very much smaller Denmark, 6110 MW in less windy Germany, and 2400 MW in Spain (which had introduced its feed-in tariff several years later than Denmark and Germany).[69]

Why NFFO failed

Why was the NFFO/SRO process so unsuccessful? Though a wide variety of issues[70] was always raised by those who objected to wind farms the two most significant concerns were usually noise and visual change. Most people have heard the noise produced by helicopter rotors or by aircraft propellers and they anticipate that the larger rotors of modern wind turbines will be even more noisy. However the noise made by a blade is strongly dependent on its speed through the air, and the relatively low tip speed of modern wind turbines results in a low level of noise emissions.[71] First-time visitors to an operational wind farm almost always comment that the noise level is very much lower than they had expected; and they are surprised to find that they can stand by the tower of a wind turbine working at full power and converse without raising their voices.

Although modern wind turbines are relatively quiet the background noise levels in rural areas can be very low, and care must be taken to ensure that there is no noise nuisance at the nearest dwellings. This typically requires that the distance between the turbines and the nearest dwellings should be several hundred metres, though the precise figure depends on the number of turbines and their size and rotational speed; it also depends on the local topography. To ensure a uniform high standard of noise protection for wind farm neigh-bours the DTI convened a Noise Working Group in the early 1990s and in

1996 this produced a report which gives explicit guidance both to wind farm developers and to planning officers; all wind farms built subsequently have been required to comply with the stringent noise conditions that this report specifies.[72]

Visual change is a much more subjective issue but as the Department for Business, Enterprise and Regulatory Reform comments, in its note *Wind Power: 10 Myths Explained*, 'the consistent conclusion of ... surveys is that the large majority of people living near wind farms like them'.[70] Repeated public attitude surveys have also shown a continuing high level of public support for wind energy and for wind farms (typically 70% to 80% of those polled were in favour, 5% to 10% were opposed and the remainder were indifferent).[73] Unfortunately this support was not reflected in wind farm planning decisions through the 1990s. The minority who were opposed were usually very vocal in their opposition and as Gipe comments, based on his observations both in the US and the UK, 'following a textbook strategy seen time and time again opponents elevated the value of the status quo, denigrated the project's benefits, denied its need, challenged the breadth of support, amplified its impacts, and criticized farmers for their greed in endorsing the project'.[74] The UK planning system requires elected local councillors to approve – or reject – most wind farm planning applications;[75] given opposition from a vocal minority, and the absence of clear planning guidance from Government on the need for wind energy, these local councillors would all too often take the easy option and reject a proposed wind farm.

The bureaucratic and very competitive nature of the NFFO/SRO process was also less than helpful. No wind farm was financially viable without a NFFO/SRO contract but to win one, especially in the three later NFFO/SRO rounds, required that a very windy location be identified and proposed; such locations are often in very scenic areas and any proposed wind farm was likely to prompt opposition from landscape protection organisations. In addition, and in contrast with the simplicity of the Danish and German feed-in tariffs, the complexity of the NFFO/SRO bidding process made it difficult for local people to take their own projects forward.[76] Specialist wind farm developers therefore emerged who would only rarely be locally based; and the irregular and unpredictable NFFO/SRO process led to a situation where they could not justify the substantial expense involved in taking a project forward through the planning process, which would need to include the preparation and submission of a comprehensive and costly Environmental Impact Assessment,[77] unless and until they had secured a NFFO/SRO contract. The contract conditions then gave them very little flexibility in making changes in response to local consultation.

The Renewables Obligation

The 1997 General Election returned a Labour Government for the first time since Margaret Thatcher's success in 1979. Prior to the election Labour, led by Tony Blair, had stated that they would do more to encourage the use of renewable sources of energy, and after a lengthy period of consultation the decision was made to replace the seriously flawed NFFO/SRO process. The opportunity to introduce a feed-in tariff was missed and the UK instead adopted a relatively complex Renewables Obligation (RO). In its first year, which commenced in April 2002, this required each electricity supplier to ensure that 3.0% of all the electricity it sold to its customers came from qualifying renewable energy sources.[78] The specified percentage would then increase year-by-year until it reached 10.4% in the year commencing April 2010.[79] And to encourage compliance the RO legislation specified that suppliers who failed to meet their obligation in full should pay a penalty of 3p/kWh for any shortfall.

Electricity sales in 2002/3 totalled just over 300 billion kWh and the RO consequently required suppliers to include just over 9 billion kWh from qualifying renewable energy sources in the electricity they sold to their customers; and to do this they either had to generate the renewable electricity themselves or buy it from other sources. In 2002/3 the total volume of renewable electricity available was only about 6 billion kWh,[80] so suppliers had to pay penalties of about £90 million on the 3 billion kWh shortfall. A peculiarity of the RO is that the annual penalty payments paid by suppliers are then repaid to them; however the repayment received by each supplier is directly proportional to the total amount of renewable electricity that it provides to its customers. Suppliers have to demonstrate the extent of their compliance with the RO by submitting Renewable Obligation Certificates (ROCs) to the system administrator[81] and they must buy these from qualifying electricity generators, who are issued with 1 ROC for every MWh of renewable electricity they produce. So at the end of 2002/3 each supplier had to submit its ROCs to the system administrator, and every ROC submitted reduced the penalty it had to pay by £30 (corresponding to the 3p/kWh penalty level). In addition every ROC submitted entitled the supplier to a share of the penalty repayments, which that year was worth about £15 per ROC.[82] Every ROC issued in 2002/3 was therefore worth about £45 to an electricity supplier, equivalent to about 4.5 p/kWh. And every kWh generated by a qualifying renewable electricity facility, such as a wind farm, therefore had its value increased by a similar amount.

Investors and banks lending money for the construction of a wind farm need to have confidence that they will in due course recover their investment, and earn a return at least comparable with what they might achieve by lending their money elsewhere. They will therefore want reassurance that the chosen site is sufficiently windy, and that the turbines to be used are reliable. They will also want to be certain that the output from the wind farm can be sold over an extended period at a price that will give them the financial return that they require. In 2001/2, before the introduction of the RO, the output from wind farms could be sold through an open auction process for about 2.5 p/kWh; in 2002/3, with the RO in force, the average selling price increased to about 6.5 p/kWh.[83] The 2002/3 value was underpinned by the fact that ROCs were worth about 4.5 p/kWh that year, but there was a great deal of uncertainty about what they might be worth five or ten years later; it was even suggested by some that if the RO succeeded in encouraging the construction of a large number of qualifying renewable electricity projects then the ROC price could collapse.[84] As a consequence of this uncertainty electricity suppliers would only commit to long-term power purchase agreements for the output from proposed new wind farms at a price well below the 2002/3 auction price of 6.5 p/kWh; and typically the price at which a 15-year power purchase agreement would be offered was in the range 4.5 to 5 p/kWh. However despite being at a large discount to the auction price this range was well above the level of the NFFO4 and NFFO5 contract prices, see Table 8.3, and was sufficient to make wind farms on much less windy sites financially viable.

At NFFO4 and NFFO5 price levels sites for wind farms had to have average wind speeds at hub height in excess of 8 m/s, and such sites are typically exposed hill-top locations like the Mynydd y Cemais ridge shown in Figure 8.9. With long-term power purchase contracts available at prices in the

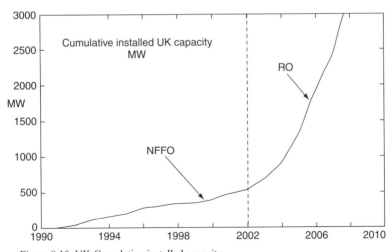

Figure 8.10 UK Cumulative installed capacity.

range 4.5 to 5 p/kWh wind farm developers could consider sites with average wind speeds at hub height down to about 6.5 m/s and then, as in Denmark and Germany, large areas of relatively flat open countryside were potentially suitable.[85] The result was an upsurge of development activity which began as soon as the detailed RO arrangements were known. And despite continuing planning difficulties the UK installed wind power capacity began to grow much more rapidly, as Figure 8.10 clearly shows. With NFFO it took 13 years, from 1990 through to the end of 2002, for the UK to install just 550 MW of wind power capacity, an average of just over 40 MW per year; in the 3-year period prior to the end of 2007, with the RO, the increase in wind power capacity averaged over 500 MW per year.[86]

Offshore wind farms

As noted previously the UK has a very large offshore wind resource,[87] but the generation of electricity from offshore wind farms at an affordable cost requires the use of megawatt-scale wind turbines. Denmark had built two demonstration offshore wind farms in the early 1990s, at Vindeby and Tunø Knob, using sub-megawatt turbines and they had performed well.[15] By the late 1990s megawatt-scale wind turbines had been developed by several leading manufacturers and Denmark was planning to use them in the construction of a series of large offshore wind farms.[17] Danish progress rekindled UK interest and the first UK offshore wind turbines were installed and commissioned in December 2000, just 1 km off the Northumberland coast and close to the port of Blyth. Developed by Borderwind, and built with the aid of an EU grant, this project used two Vestas 66 m diameter machines each with a rated power output of 2 MW.[88] By then the level of interest in developing offshore wind farms in the shallow waters around the UK was such that the Crown Estate, which owns the seabed, decided to initiate a formal process for allocating future sites.[89] Interested developers were therefore invited to apply for a lease for a site which, in their judgement, was suitable for an offshore wind farm. In this first round of offshore wind development the Crown Estate specified that individual projects should have no more than 30 turbines, and developers were limited to one site each. Eighteen lease applications were approved in April 2001, for sites that were mostly between 5 and 10 km from the coast, and developers could then initiate environmental monitoring and seabed investigations prior to submitting their applications for the necessary consents.[90]

The first of the Round 1 offshore wind farms to be consented and built was North Hoyle, off the North Wales coast, which was commissioned in

November 2003, see Figure 1.3. Earlier in the same year the Crown Estate had initiated a second round of bidding for leases, which this time was confined to three specified large areas for which Strategic Environmental Assessment had been undertaken.[91] And in December 2003 site licences were awarded for 15 projects, whose total proposed capacity was approximately 6000 MW. These Round 2 projects were generally much larger than Round 1 projects and included two, the London Array and Triton Knoll, which were each about 1000 MW. The locations of all the Round 1 and Round 2 sites are shown in Figure 8.11, with Round 1 sites indicated by their smaller size and lighter shading.[92] The dashed line around the coast shows the 12 nautical mile (about 22 km) territorial limit and as can be seen some of the Round 2 sites are outside this, though all are well within the UK's Exclusive Economic Zone.

By the end of 2007 five of the Round 1 projects were operational and the latest of these, Burbo Bank, was completed in mid 2007 off the north-west coast. This project used twenty-five Siemens 3.6 MW rated wind turbines – each with a diameter of 107 m – and Figure 8.12 shows the wind farm nearing completion, with the jack-up barge used for erecting the turbines in the fore-ground. The commissioning of Burbo Bank gave the UK a total installed offshore wind power capacity of 404 MW,[93] just slightly in excess of Denmark's total, but the differential is expected to grow considerably over the next few years. The UK added about 160 MW in 2008, and will add about 400 MW in 2009, with more projects due for completion in 2010. Several of the Round 2 projects have also now been consented, including the 1000 MW London Array and the 500 MW Greater Gabbard, and will soon commence construction.[94]

As indicated previously the cost of offshore wind farms is substantially greater than the cost of wind farms on land,[95] and to be financially viable each Round 1 project required a Government grant that in most cases was £10 million. The much larger Round 2 projects would have required much larger grants, which the Government was unwilling to provide, and their financial position was made worse both by the rising price of wind turbines and by concern over the future value of ROCs. The Government was able to deal with the latter by first increasing the RO target from 10.4% in 2010 to 15.4% in 2015, and then by stating that it would further increase the RO target – as and when necessary – to 20%.[96] The rising cost of fossil fuels after 2003 also led to a substantial increase in the wholesale price of electricity, which increased the value of the output from wind farms and partially offset the consequences of the turbine price increases.[97] And as part of their overall review of the Renewables Obligation the Government recognised the need to further enhance the value

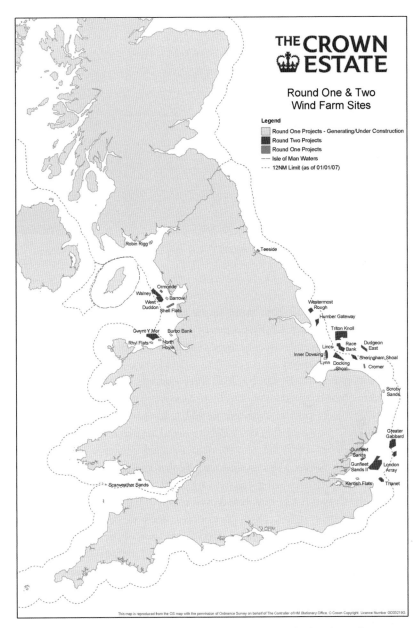

Figure 8.11 Round 1 and Round 2 offshore wind farm site locations. *Source*: Image courtesy of The Crown Estate, © Crown copyright 2008.

Figure 8.12 Burbo Bank wind farm, 2007, UK. *Source*: Image courtesy of Siemens Wind Power.

of the output from offshore wind farms and made the decision that such projects should receive 1.5 ROCs per MWh, instead of the previous 1 ROC per MWh.[98] The overall effect of these changes was sufficient to allow Round 2 projects to be financed.

It is anticipated that by 2011 at least three-quarters of the Round 1 offshore wind farms will be operational and their capacity will total about 1200 MW; Round 2 projects can be expected to add a further 5000 to 7000 MW by about 2015.[99] However the UK offshore wind resource is extremely large and in December 2007 the Government initiated Strategic Environmental Assessment for further areas around the UK that could provide an additional 25 000 MW of offshore wind capacity by 2020. This was followed in May 2008 by the Crown Estate initiating a third round of bidding for offshore wind farm development rights. Nine zones have been identified as potentially suitable for multiple large offshore wind farms; two of these are off the east coast of Scotland, two are off the coast of Wales and the west of England, and the remaining five are off the east (3) and south (2) coasts of England. The largest individual site is about 200 km east from the north-east coast of England and has an area of about 8000 km^2; for comparison the Round 1 sites each have an area of about 10 km^2. Each zone will have a single developer (or consortium) and the Crown Estate expects to award zones to developers by the end of 2009. This will, it is hoped, allow time for consenting and a start to construction for the first Round 3 sites in 2014, with the full 25 000 MW able to be operational by 2020.[100]

UK overview

The Renewables Obligation was introduced in 2002 to encourage the greater use of renewables for electricity generation, so helping to meet UK targets for reducing greenhouse gas emissions. The initial target of 10% electricity from renewables by 2010 was, as noted, subsequently increased to 15% by 2015; and by early 2008 there was – in addition – 'an aspiration' for 20% of UK electricity to be provided from renewables by 2020. However EU Member States agreed in Spring 2007 that by 2020 renewable energy sources should provide 20% of the total *energy* that they used; and in January 2008 the European Commission proposed that the UK share of this target should be for renewables by 2020 to provide 15% of the total *energy* it used.[101] The lower than average target for the UK reflects the fact that in 2007 renewables supplied only 1.8% of its energy, well below the EU average. Existing policies are expected to lead to renewables providing just 5% of UK *energy* by 2020, so some major policy changes will be required if 15% is to be achieved.

The UK Government proposals for providing 15% of UK *energy* from renewables by 2020 were subsequently set out – in June 2008 – in a Consultation Paper,[102] and following a period of consultation a revised UK Renewable Energy Strategy will be published in Spring 2009. The Consultation Paper suggests that electricity-generating renewables will need to provide almost half the total renewables contribution, with most of this electricity provided by wind farms, both on land and offshore. More specifically it suggests that by 2020 about 32% of UK electricity will need to be provided by renewables, with wind power supplying two-thirds. For offshore wind it suggests that the UK should by 2020 target to have an installed capacity of about 14 000 MW; and it equates this to the addition of about 3000 wind turbines, each with a rated power output of about 5 MW. On land it suggests that the target for 2020 should also be about 14 000 MW; and it equates this to the addition of about 4000 wind turbines, each with a rated power output of about 3 MW.

At the end of 2007 the UK had an installed wind power capacity of 2490 MW, see Table 8.2, and about 400 MW of this was offshore. Wind farms on land and offshore together gave an energy output of 5.3 billion kWh, equal to 1.3% of UK electricity generation and equivalent to the electricity used by 1.2 million UK homes.[103] The average capacity factor was 27%. To achieve the targets suggested by the Government for 2020 would require the installation on land of about 1000 MW per year, plus about 1100 MW per year offshore. The on-land target is well below what Germany has achieved every year since 1999; and reaching 14 000 MW by 2020 would take the UK to a power density of just

58 kW/km^2, lower than the levels already achieved by both Germany and Denmark, see Table 8.2. The worldwide turbine production and installation rate reached almost 20 000 MW in 2007 and is doubling every three to four years. By 2020 the annual production and installation rate can be expected to be well in excess of 100 000 MW. By comparison with these figures the turbine volumes required to meet the suggested UK wind power targets, both on land and offshore, are relatively minor. The continuing installation of wind turbines offshore is a relatively recent development, but the necessary supply chain of the vessels required for turbine installation, cable laying and other related activities is rapidly being put in place. Given continuing political will the suggested UK wind power targets for 2020 would seem very achievable, and the offshore target in particular could prove to be very conservative.[104]

United States

Low growth beginnings

Though the US had – through the 1980s – installed much more wind power capacity than any other country the level of activity through the early and mid 1990s was minimal, as Figure 8.1 clearly shows. Between 1990 and 1997 the overall increase in the installed wind power capacity was just 90 MW,[105] even smaller than in the UK. As is discussed in Chapter 5 most of the financial incentives that had stimulated the installation of many thousands of wind turbines in California in the early 1980s had ended by 1986. By then the era of high oil prices that had prompted the introduction of these incentives had also ended. A small number of legacy projects continued to be built for a few more years, with the benefit of premium price power purchase contracts that had been signed prior to 1986;[106] and these included Zond's 77 MW Sky River project in Tehachapi, which was completed in November 1991 using 342 Vestas 225 kW machines.

Zond was one of the pioneer developers in California in the early 1980s, and by the early 1990s it was operating and maintaining a fleet of well over two thousand wind turbines. Based on this experience Zond felt able to initiate the manufacture of its own machines and the first prototype of what became the Z-40 wind turbine was completed in 1994. The Z-40 had a rotor diameter of 40 m and a rated power output of 500 kW, and was a conventional design with three variable-pitch fibreglass blades located upwind of the tower. Like most machines at the time it was grid-connected using an induction generator, and the rotor speed of 29 rpm gave a blade-tip speed of 61 m/s – similar to most Danish machines.

Zond's Z-40 was used for a number of small projects in the mid 1990s including some overseas.[107] However the biggest wind project in the mid 1990s was in Minnesota. The State legislature there had approved a request from Northern States Power to expand its nuclear waste storage facilities, but with the condition that the utility should add 425 MW of wind power to its supply. Kenetech had won the contract to supply the first phase of 25 MW in 1994 with its 330 kW rated 33 m diameter KVS-33 wind turbines, and was expected to win the bidding for the next phase of 100 MW in 1995. However the contract was won by Zond with its 750 kW rated Z-46, a turbine which at the time existed only on paper. The Z-46 was an up-scaled version of the Z-40, and as the model number indicates it had a rotor diameter of 46 m. The first prototype Z-46 was built in 1996 and the 100 MW Minnesota wind farm was completed on Buffalo Ridge, near Lake Benton, in September 1998.[108] By then Zond had been bought by Enron, one of the largest gas and electricity companies in the US, and re-named Enron Wind; and Kenetech, who through the 1980s and well into the 1990s had been the world's largest manufacturer of wind turbines, had – as discussed in Chapter 5 – collapsed into bankruptcy.[109]

The Production Tax Credit (PTC) and 'green power'

Zond had won the Minnesota 100 MW contract in 1995 by offering to supply the wind farm's output to Northern States Power at just 3 cents/kWh, and this very low price was only possible with the benefit of the Production Tax Credit (PTC). The PTC was introduced in 1992 to encourage the use of a range of renewable energy technologies;[110] but to avoid some of the problems that had been encountered in California in the early 1980s the financial support it provided was directly linked to a project's energy output over an extended period, *not* to its capital cost. The PTC applied to wind projects that commenced operation between 1 January 1994 and 30 June 1999, and gave a tax credit of 1.5 cents/kWh (adjusted annually for inflation) for the first ten years of operation. Provided that the company owning the project had US tax liabilities the PTC was worth a full 1.5 cents/kWh for 10 years, and allowed Zond/Enron Wind – and others – to reduce the price at which they sold a wind farm's power output by a corresponding amount.

Through the 1990s oil prices remained low,[5] and the support provided by the PTC was insufficient by itself to stimulate any significant level of wind power activity. However the Energy Policy Act of 1992 which had brought in the PTC also started the slow process of deregulating the electricity industry and opening it up to competition (a process that was by then already well under way in several European countries, including the UK). Growing concern in the US

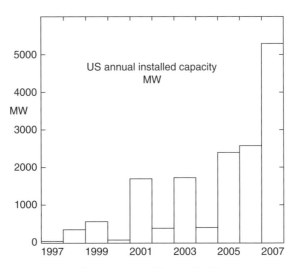

Figure 8.13 Annual installed capacity, 1997 to 2007, US.

about greenhouse gas emissions and climate change then led to the introduction of 'green power' products, which allowed consumers to pay a small premium and buy at least part of their electricity from sources such as wind farms. And in the later 1990s a small number of States passed legislation which required utilities to contract for specified levels of renewables capacity.[111] Wind turbine economics had, as noted earlier in this chapter, steadily improved through the decade and wind power was consequently well placed to benefit from such legislation. The result, as indicated in Figure 8.13, was an upsurge of activity in 1998 which saw 340 MW of wind power capacity installed, followed in the first half of 1999 by more than 500 MW. And though most of the wind turbines were Danish-made Enron Wind were able to secure a market share of about 30%.

Despite sustained lobbying from the American Wind Energy Association (AWEA) the PTC lapsed at the end of June 1999, and almost all the capacity added in 1999 was completed prior to this deadline. Legislation in December 1999 extended the PTC through to the end of December 2001, but it was too late to allow projects to be financed and built in 2000 and the capacity installed that year slumped to just 60 MW. A slowly growing number of States were introducing some form of Renewables Portfolio Standard (RPS), requiring all sellers of electricity to source a proportion of their supply from qualifying renewables sources.[112] And in 2001 with the benefit of the PTC, and with pent-up demand carried forward from 2000, a total of almost 1700 MW of wind power capacity was installed. This gave the US a cumulative installed capacity of 4250 MW by the end of 2001, see Figure 8.2, a total that was second only to Germany's.

The PTC lapsed again at the end of December 2001, and although it was renewed in March 2002 this was too late to benefit all but a small proportion of the projects planned for the year. The capacity installed in 2002 consequently slumped to just 400 MW, as Figure 8.13 indicates. Enron Wind's parent company, the Enron Corporation, had experienced a very high profile collapse into bankruptcy in late 2001 and one consequence was the subsequent sale of Enron Wind. The General Electric Company judged that the time was right for it to acquire a wind turbine manufacturing capability and in March 2002 it bought the manufacturing assets of Enron Wind, which then became GE Wind Energy.[113]

Momentum builds

The PTC renewal in March 2002 extended the availability of the tax credit, which by then was worth 1.8 cents/kWh, through to the end of 2003. And 2003, with a backlog of projects carried forward from 2002 and with an RPS in place in fourteen States, became another record year. Wind power projects with a total capacity of just over 1700 MW were completed, with more than half coming from just four States – California, Minnesota, New Mexico and Texas – which each added over 200 MW; eleven other States provided the balance. By the end of the year California, with its legacy of wind farms built in the 1980s, had a total installed capacity of 2020 MW and Texas had 1300 MW. And GE Wind, helped by a strong performance in its home market, jumped to second position in the ranking of the world's leading wind turbine manufacturers with an 18% share of worldwide sales, which that year totalled 8200 MW.[114]

GE Wind had focussed their resources on a 71 m diameter 1.5 MW wind turbine derived from the machine Tacke were developing when they collapsed and were bought by Enron;[109] and for several years this was the only model that GE Wind sold. Some of the 1.5 MW machines they installed in the Klondike wind farm in Oregon in 2002 are shown in Figure 8.14, and as can be seen the configuration is conventional with three blades upwind of a steel tubular tower. The power output is controlled – as in almost all megawatt-plus machines – by varying the blade pitch, and goes through a speed-increasing gearbox to an induction generator. Power electronics is then used to allow the rotor speed to vary through the range 11 to 22 rpm, and to give a grid-compliant electrical output.[115]

The PTC lapsed for a third time at the end of December 2003 and was only renewed, through to the end of 2005, in October 2004. Few projects could go forward without its benefit so 2004 was another relatively poor year for wind power installations, as Figure 8.13 clearly shows, with just 400 MW installed.

Figure 8.14 Klondike wind farm, 2002, Oregon, US. *Source*: Image courtesy of GE Energy.

But as might be expected in 2005, with the benefit of the PTC and with a backlog of projects postponed from 2004, the capacity installed climbed to a new high of 2400 MW. Wind turbines were installed in a total of 22 States led by Texas (which added 700 MW), Oklahoma (which added 300MW) and Iowa (which added 200 MW).[116] And the market was dominated by GE Wind – whose 1.5 MW turbines provided 60% of the total installed capacity – and by Vestas, whose market share was 30%. Most of the balance was provided by the Japanese manufacturer Mitsubishi, who had first supplied wind turbines to the US in the late 1980s.

In mid 2005 the PTC was extended, for once before its expiry, through to the end of 2007; and well before the end of this period it was extended again, for a further year, through to the end of 2008.[117] This four-year period of PTC stability greatly helped those involved in the developing wind industry, and wind power's progress was further helped by the steadily increasing number of States with Renewable Portfolio Standards. Wind power's economics had also become close to competitive with electricity from gas-fired power stations; in 2005/6 well-sited wind farms could provide electricity at a long-term contract price (net of the PTC benefit) of about 3.5 to 4 cents/kWh, equivalent then to 1.9 to 2.2 p/kWh. This price subsequently increased, partly as a result of increased world steel and copper prices, but also as a consequence of the growing shortage of wind turbines caused by soaring world demand. Wind power's increasing cost was, however, more than compensated by continually increasing oil and gas prices. And utilities became increasingly aware of the benefits that committing to long-term contracts for the output from wind farms could bring, by reducing their exposure to potentially high and volatile future fuel prices.

The result, as shown in Figure 8.13, has been a period of sustained growth with the 2400 MW installed in 2005 followed by 2560 MW in 2006, and then by 5270 MW in 2007. In each of these three years more capacity was installed in the US than in any other country, and as both Table 8.2 and Figure 8.2 indicate the total wind power capacity of 17 000 MW at the end of 2007 was second only to Germany's 22 200 MW. Texas – with 4360 MW installed – was the most active state, followed by California with 2440 MW; and four other states (Minnesota, Iowa, Washington and Colorado) each installed more than 1000 MW.[118] GE remains the dominant wind turbine supplier in the US and, in addition to the well-established 1.5 MW machine shown in Figure 8.14, its product range now includes a 2.5 MW turbine with a rotor diameter of 100 m and – for offshore locations – a 3.6 MW machine with a rotor diameter of 111 m.

In 2007 GE supplied wind turbines with a total capacity of nearly 2300 MW to projects within the US, giving it a market share of 43%; and worldwide its wind turbine sales totalled nearly 3300 MW, giving it a 15% share of the global market where it was second only to Vestas.[26] In the US in 2007 Vestas was second to GE with 21% market share, and was followed by Siemens with 16%. The remaining 20% was mostly shared between the Spanish manufacturer Gamesa, the Japanese Mitsubishi and the Indian Suzlon, but an interesting American new entrant with 1% market share was Clipper Windpower. Clipper was formed in 2001 by Jim Dehlsen, founder of Zond and one of the pioneers of wind farming in California in the early 1980s. Following the installation of the first prototype of its 2.5 MW, 89 m diameter, wind turbine at Medicine Bow, Wyoming, in 2005 Clipper put the machine into production and supplied turbines to two wind farms in 2007, with many more planned for installation in 2008 and subsequently. Though conventional in appearance, with three variable-pitch blades located upwind of a tubular tower, the design features what Clipper describe as a distributed powertrain. In essence the speed increasing gearbox has four output shafts, each powering its own 660 kW generator, and power electronics is used to allow the rotor speed to vary in the range 10 to 16 rpm.[119]

US overview

The Government sponsored annual wind power report for 2007, compiled by Wiser and Bolinger, noted that this was the first year in which the output from wind turbines exceeded 1% of the total electricity generated in the US.[120] In some states the wind contribution was substantially higher, and both in Minnesota and Iowa wind contributed over 7% of the electricity produced. In the US, as elsewhere, there had been concerns about the effects of wind power's variability on the operation of electrical grid systems. However growing experience with significant levels of wind penetration, supported by a number of major wind integration studies that were completed in the period 2003 to 2007, have shown that there are no fundamental problems in managing the consequences of wind's variability and the costs of integration – up to penetration levels of about 30% – are well below 1 cent/kWh and typically below about 0.5 cents/kWh.

Wiser and Bolinger indicate that wind farms built in the US in 2007 had on average a capacity of 120 MW, provided by wind turbines whose power rating was just over 1.6 MW, and both these numbers had almost doubled since 2000. The average wind farm capacity factor had also increased over the same period, rising from about 30% (in 2007) for projects built in 2000 to nearly 34% (again in 2007) for projects built in 2006. It is noteworthy that even the low end of this range is substantially higher than in Europe, reflecting the fact that there are large areas in the US with a very good wind resource. Throughout the five-year period 2003 to 2007 the average cost of energy from wind projects was in the range 3.5 to 4 cents/kWh (net of PTC benefits), and was competitive with wholesale power prices (which typically were in the range 3 to 7 cents/kWh). Wind energy costs were a minimum in about 2003 when wind farm capital costs were about $1300/kW. Substantial wind turbine price increases subsequently led to increased project costs; these averaged about $1700/kW in 2007 and were expected to average about $1900/kW in 2008. Wind energy output costs consequently increased from under 3.5 cents/kWh for projects completed in 2002/3 to nearly 4.5 cents/kWh for projects completed in 2007; however wind remained competitive as wholesale power prices also increased after 2002/3. Excluding the PTC benefit, which in 2007 was worth 2.0 cents/kWh, the 2007 wind energy cost would have been 6.5 cents/kWh; at the then current exchange rate of £1 = $2.0 this was equal to just 3.3 p/kWh.

Preliminary data for 2008 indicates that it was another record year for wind power in the US, with approximately 8300 MW of capacity added, taking the cumulative total to just over 25 000 MW.[121] Germany added about 1700 MW in 2008, taking its cumulative capacity to just under 24 000 MW, so the United States regained its position as the global leader in installed wind power capacity. The renewal of the PTC in late 2008 for one further year – to the end of 2009 – will help to ensure continued substantial growth in the installed US wind power capacity through 2009, despite the effects of the recession that developed in the latter part of 2008. In the medium to longer term growing concerns over energy supply security and the cost of oil and gas imports, combined with increasing pressure to reduce greenhouse gas emissions, should ensure that the outlook for wind power in the US is very positive; and President Barack Obama's support for the more rapid deployment of renewable sources of energy is also very encouraging. One feasible scenario is examined in some detail in the report *20% Wind Energy by 2030*, published in mid-2008 by the Department of Energy (DOE). This notes that the on land economic wind resource in the US is more than 8 million MW and 'is sufficient to supply the electrical needs of the entire country several times over'. Given realistic constraints on the rate of deployment of wind turbines it indicates that by 2030 it would be possible to install just over 300 000 MW of wind power capacity, and this would be sufficient to supply 20% of the electricity consumed nationwide.[122]

The scenario proposed in the DOE report anticipates that the wind power installation rate would steadily increase to a maximum by 2018 of about 16 000 MW per year, which is only three times larger than the 2007 installation rate; it would then remain approximately constant. Over 80% of the capacity would be land based but the US also has a very large offshore wind resource[123] and 54 000 MW – out of the 304 000 MW total – would be deployed offshore, mostly in shallow waters off the east coast. Integrating the variable output from over 300 000 MW of wind power capacity into the US electricity network is considered in some detail, with the conclusion that the consequences can be managed and the cost of dealing with wind power's variability is modest; typically 10% or less of the wholesale value of wind energy. And the total cost of the programme to supply 20% of US electricity from wind power is calculated to be only 2% more than a base-case scenario in which no wind power capacity is added; the extra cost to the electricity consumer would consequently be substantially less than 0.1 cents/kWh, or approximately 50 cents per month per household.[124]

Set against these minimal additional costs would be a number of substantial benefits whose monetary value cannot readily be calculated. The fossil fuel savings that would result from the use of wind power on the scale proposed would reduce by more than half the number of new coal-fired power stations that would need to be built by 2030, and would also reduce – by an estimated 60% – the volume of liquefied natural gas (LNG) that would need to be imported. The fossil fuel savings would also reduce emissions of the greenhouse gas carbon dioxide by about 820 million tonnes per year, equal to 14% of the present US carbon dioxide emissions. Water supply is a growing problem in many parts of the US, and the power industry is a major water consumer; however the 20% wind programme would reduce consumption by about 450 billion gallons per year, equal to about 17% of the power industry's forecast 2030 consumption. The 20% wind programme would also be a major source of employment; it is calculated that by 2030 it would create over 170 000 direct jobs, and if indirect and induced jobs are included the total is over 500 000.[125]

As Table 8.2 indicates the 17 000 MW of wind power capacity installed in the US by the end of 2007 corresponds to a wind power density of 1.9 kW/km^2. Increasing the installed wind capacity to 304 000 MW by 2030 would increase the wind power density to just 33 kW/km^2, still well below the levels already achieved both by Denmark and Germany. That the US can supply 20% of its electricity with a wind power density less than half that required by Denmark, despite having a per capita electricity consumption that is twice as high,[126] is a consequence of the relatively low US population density of just 33/km^2; in Denmark – see Table 8.2 – the population density is nearly four times higher. One consequence of the low US population density is that electricity transmission distances are longer, and achieving 20% electricity from the wind by 2030 will require significant expenditure to upgrade the electricity transmission network – as is discussed in the DOE report.[127] And though 20% electricity from the wind by 2030 is a reasonably ambitious interim target the US wind power potential is so large that in the longer term the contribution that wind power could make to reducing the need for fuel imports – as well as to reducing greenhouse gas emissions – is very much larger.

Canada, Netherlands and Japan

Progress since 1990 in Denmark, Germany, Spain, the UK and the US has been reviewed in some detail, and these countries have used a variety of financial measures to encourage the increased use of renewable energy technologies in general, and wind energy in particular; the payment for an extended period of a specified price per unit of electricity produced, either for the whole of a wind turbine's output (as in Germany) or as an addition to the wholesale price of electricity (as in the US), has been seen to be particularly effective. Many other countries have introduced measures to encourage the increased use of renewable energy technologies, and as wind is the most cost-effective of the high resource renewable energy options it has been the principal beneficiary of these measures. The ten countries where the installed wind power capacity exceeded 2000 MW at the end of 2007 are listed in Table 8.2. And just off the list with 1850 MW installed was Canada, similar in size to the US and with a correspondingly large wind resource which it has been slow to utilise. At the end of 2004 Canada's total installed wind power capacity was just 440 MW, but

1400 MW was then added in the next three years and this increased level of activity seems likely to continue.[128]

The Netherlands – renowned for the use of traditional windmills in earlier centuries – had 1750 MW of modern wind turbines installed by the end of 2007. As noted in Chapter 6 only the US and Denmark had more wind power capacity installed than the Netherlands in 1990, and at that time there were several reasonably successful Dutch wind turbine manufacturers. Though the Netherlands is a small country, only one-seventh the size of the UK and smaller even than Denmark, its population of nearly 17 million is three times larger than Denmark's and its population density of 490/km^2 is one of the highest in Europe. The Netherlands' small size and high population density have both constrained subsequent wind power development but it is worth noting that the 1750 MW installed by the end of 2007 corresponds to a wind power density of 52 kW/km^2, which is surpassed only by Denmark and Germany. The Netherlands does have a substantial offshore wind resource and the first Dutch offshore wind farm was completed in autumn 2007 about 14 km off the coast from Egmond aan Zee, which is 30 km north-west from Amsterdam; 36 Vestas V90 3 MW wind turbines give this project a power capacity of 108 MW. A second offshore wind farm, the 120 MW Q7 project which is 23 km off the coast from IJmuiden, uses Vestas V80 2 MW turbines and was completed in mid 2008. Several additional offshore wind farms are under development but none is likely to be built before about 2012.[129]

Japan, with 1540 MW of wind turbines operational by the end of 2007, is the only other country that then had an installed wind power capacity of more than 1000 MW. Japan's land area is similar to Germany's and its 1540 MW corresponds to a wind power density of just 4 kW/km^2. One of its leading engineering companies – Mitsubishi – has been manufacturing wind turbines since 1987, but outside Japan its sales have been almost exclusively to the US.[130] Its 2008 product range includes a 57 m diameter 1 MW wind turbine and a 92 m diameter 2.4 MW machine; both models are conventional in their design, with three variable-pitch fibreglass blades located upwind of a steel tubular tower. Mitsubishi sales have averaged about 2% of the world market, and it is just outside the ranks of the top ten manufacturers.

India

India was one of the first developing countries to encourage the use of modern grid-connected wind turbines, and with financial support from the Danish overseas aid agency the first Danish machines were installed in the mid

1980s. By 1990 the total installed capacity was about 30 MW, comparable with what was then installed in Germany and the Netherlands; and only Denmark and the US had substantially more. Capacity growth was, however, very limited until the Government and individual States took steps in the mid 1990s to encourage private investment in electricity generation and provided some financial incentives. Though the detail varied from one State to another these incentives typically included 100% depreciation of the capital cost in the first year, subsidies towards the cost and a premium price for electricity sold to State electricity boards. Another significant incentive, peculiar to India, was to allow wind farm owners to 'bank' their output with the electricity board. Demand for electricity in India is growing rapidly and often exceeds supply, with the result that power cuts are common. Power banking means that a factory owner who buys a wind farm (anywhere in the same State) can feed the output into the grid, and can subsequently use the energy that has been 'banked' to provide power to the factory and avoid being required to have power cuts.[131]

The incentives were sufficient to encourage the installation of several thousand wind turbines in the mid 1990s and though the pace of development slowed in the late 1990s India had – by the end of 2000 – an installed wind power capacity of 1220 MW, a total exceeded only by Germany, Denmark, Spain and the US.[4] India's Electricity Act of 2003 required states to bring in measures to ensure that renewable energy sources provided a growing proportion of the electricity generated. Though the detailed arrangements were left to individual states several set challenging targets and provided feed-in tariffs to encourage developments; and as is shown in Figure 8.2 the installed wind power capacity subsequently increased rapidly.

Through the 1990s most of the wind turbines were imported, but the market was large enough to encourage several Indian companies to start making them; and one of these companies, Suzlon, has since grown to become one of the world's leading wind turbine manufacturers. Suzlon's background was in the textile industry, but having initially bought two imported wind turbines to help provide a more secure supply for his factory the owner, Tulsi Tanti, saw the opportunity for local manufacture. In collaboration with the German company, Südwind Energy, Suzlon built its first wind turbine – a 270 kW machine – in 1996 and by 2000 it had supplied turbines with a total capacity of 100 MW to projects in India. With continuing rapid growth Suzlon had by 2003 joined the ranks of the world's top ten wind turbine manufacturers, and by the end of 2007 it had become the fifth largest (after Vestas, GE Wind, Gamesa and Enercon) with a 9% share of the global market.[132] Suzlon wind turbines have, like most of their competitors, three variable-pitch blades located upwind of a steel tower and range in size from a 52 m diameter 600 kW machine to an 88 m diameter machine with a rated power output of 2.1 MW. Almost half of the 2007 production of just over 2000 MW was installed in India, where Suzlon had a massive 65% market share. In many Indian wind farms lattice towers are used, because of their lower cost; however the turbines that Suzlon exports to the US, Europe and elsewhere usually have the visually more pleasing tubular towers.[133]

India's wind power capacity growth has been relatively steady in recent years, see Figure 8.2, at about 1600 MW per year; and by the end of 2007 the total installed capacity was 7800 MW. Capacity factors are relatively low, typically about 15%, and the total wind-generated electricity output in 2007 was therefore about 9 billion kWh, equivalent to nearly 2% of India's total electricity consumption. As noted in Table 8.2 the installed capacity corresponds to a wind power density of only 2.6 kW/km^2, and though the population density is a relatively high 386/km^2 the per-capita electricity consumption of just 430 kWh/year is more than an order of magnitude smaller than in Europe or North America.[126] The economically viable wind resource on land has been estimated to be at least 45 000 MW, so wind has the potential to make a substantial contribution to meeting India's electricity needs.[134] There is in addition a very large offshore wind resource, though there are no plans yet to build wind farms offshore.

China

In the late 1980s a few tens of grid-connected wind turbines were installed in China, mostly Danish made and part-funded by Danida, the Danish overseas aid agency. By the end of 1990 the total wind power capacity in China was approximately 15 MW, though the modern grid-connected machines accounted for only about 5 MW; the balance was provided by more than 100 000 small battery-charging windmills. Though similar in their general design to the battery-charging windmills that were widely used in the US and elsewhere in the 1930s (and discussed in Chapter 4) these were made in China and used by nomadic herdsmen in Inner Mongolia to provide electricity for a variety of applications, the most important of which was to provide power in the family yurt for a television.[135]

The installed grid-connected wind power capacity slowly increased through the 1990s, reaching 45 MW in 1995 and 340 MW by the end of 2000, still with most of the wind turbines imported from Europe. Early assessments of the wind resource had shown that China's wind potential was very large, and even though the wind power capacity growth at that time was slow it was sufficient to encourage some local manufacture. And one of the first Chinese wind turbine manufacturers was Goldwind, who with German technical support produced their first machine – with a rated power output of 600 kW – in 1998. By 2002 Goldwind were able to supply 20% of the 70 MW installed that year, with almost all the balance provided by German and Danish manufacturers, though the cumulative installed capacity at the end of 2002 was a still very modest 470 MW.

Power shortages caused by China's rapid economic growth helped the development of policies to more actively encourage renewable energy sources, and in 2003 the Government set a target for 4000 MW of wind power capacity to be installed by 2010, with a further target of 20 000 MW installed by 2020. The Renewable Energy Law passed in 2005 gave further support, and its overall target for China to obtain 15% of its energy from renewables (including hydropower) by 2020 led to the targets for wind power being increased to 5000 MW by 2010, and to 30 000 MW by 2020. The Renewable Energy Law stipulated that the output from wind projects should be bought at a fixed premium price; it also specified that for projects approved after 2005 the Chinese content in the turbine supply should be at least 70%, a requirement which encouraged many leading wind turbine manufacturers to establish – or expand – factories in China.[136]

Active government support led to an upsurge of development activity in China. Projects up to 50 MW capacity can be approved by provincial governments, while larger projects additionally require central government approval, which inevitably takes longer. Most projects have therefore been below the 50 MW threshold. And as is shown in Figure 8.15 the rate of growth in the installed wind power capacity in the past four years has been quite remarkable. The 200 MW installed in 2004 was more than double the capacity added the previous year, and this was followed in 2005 by another record year which saw 500 MW installed; the cumulative capacity was then more than doubled in 2006 by the completion of a further 1330 MW. In 2007 the cumulative capacity was more than doubled for the second consecutive year by the installation of 3310 MW, corresponding – see Table 8.2 – to a level of activity exceeded only in the US and Spain. By the end of 2007 the total installed capacity in China had reached 5900 MW, and the 5000 MW target for 2010 had been passed more than three years early.[137] And preliminary data for 2008 indicates that China's remarkable growth is continuing, with 6300 MW installed in the year to give a cumulative total of just over 12 000 MW by the end of 2008.[138]

The rapid growth in the size of the market for wind turbines has encouraged many more companies to commence their manufacture, often with technical assistance from European consultancy companies such as Germany's Aerodyn,

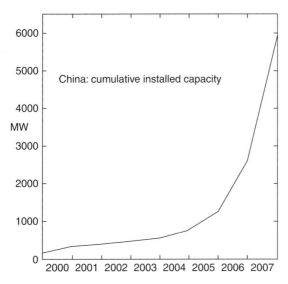

Figure 8.15 Wind power capacity growth in China.

Austria's Windtec and the UK's Garrad Hassan. Recent estimates indicate that there are now about 40 wind turbine manufacturers in China, though some are still at the prototype development stage. Goldwind remains the market leader. Its output of just 14 MW in 2002 was then sufficient to give it a 20% market share, but in 2007 it delivered 830 MW – all to Chinese projects – and achieved a 25% market share. And this level of production was sufficient to make it the world's eighth largest manufacturer.[26] Goldwind makes wind turbines that range in size from a 43 m diameter 600 kW rated machine up to a 70 m diameter machine which has a rated power output of 1.5 MW. All have the configuration that is now almost standard, with three blades upwind of a steel tower. On the smaller machines the blade pitch is fixed and the rotor speed is also constant, so that the power output in strong winds is limited by stall; rotatable blade tips then give the necessary aerodynamic braking capability. As in Europe and the US on the larger machines the power output in strong winds is controlled by varying the blade pitch and power quality is improved by allowing the rotor speed to vary.[139]

In 2007, for the first time, Chinese wind turbine manufacturers supplied more than half the total capacity installed. Goldwind's principal competitor, Sinovel, achieved a 21% market share, and other Chinese manufacturers – led by Dongfang and Windey – supplied a further 10%, bringing the Chinese manu-facturers' total to 56%; most of the balance was provided by Gamesa, Vestas and GE Wind. Sinovel's output in the year was 680 MW, which was sufficient to make it the world's tenth largest manufacturer. So far most of the capacity has been installed in the north, led by Inner Mongolia where more than 1500 MW has been installed. Though China's best wind resource is in the north there is also a significant wind resource along the east coast, and the total resource on land is estimated to be in excess of 250 000 MW. China also has a substantial offshore wind resource, which is estimated to be in excess of 700 000 MW.[140]

The output from wind farms is usually sold through long-term power purchase contracts that are awarded following a competitive tendering process. Winning bid prices in 2008 were typically about CNY 0.5/kWh, where CNY is the agreed abbreviation for the currency, the Chinese Yuan Renminbi; with £1 = CNY 12.9 this corresponds to about 3.9 p/kWh. Many projects qualify for support under the Kyoto Protocol's Clean Development Mechanism, which can then add up to about CNY 0.1/kWh ($\approx$ 0.8 p/kWh) to the income stream for the first few years.[141]

As has been noted the Renewable Energy Law passed in 2005 set two targets for wind power's growth and the first of these, 5000 MW by 2010, was passed in 2007. The second target was to have 30 000 MW installed by 2020 and it seems probable that this too will be passed several years early; the Chinese Renewable Energy Industry Association suggests that the cumulative installed capacity will exceed 50 000 MW by 2015 and BTM Consult – the long established Danish wind power market consultancy – forecasts that by 2012 the cumulative installed capacity will exceed 42 000MW, with this total including 5000 MW offshore.[142]

The output from China's rapidly growing wind turbine manufacturers has so far been absorbed by China's own rapidly growing market for wind turbines. There is, however, the expectation that manufacturers will soon be seeking to export some of their wind turbines. Given China's relatively low labour costs, and the export success it has achieved in recent years in many other manufacturing sectors, this development could have a significant impact on the global market for wind turbines within the next decade.

9

The future: from marginal to mainstream

Traditional windmills made a significant contribution to meeting people's energy needs in England, and in many other countries, for almost seven hundred years following their invention in the twelfth century. However they were superseded in the nineteenth century by steam engines, which were later supplemented by a variety of internal combustion engines. These could provide power on demand and countries – led by the UK – used them in rapidly growing numbers to develop their manufacturing and transportation capabilities. And as prosperity gradually increased these power sources made possible the supply of an ever-increasing range of consumer goods and services to growing populations. People in developed countries now enjoy a very much higher standard of living than their nineteenth-century predecessors, but this prosperity is underpinned by the combustion – on a massive scale – of coal, oil and gas, the so-called fossil fuels.

Through most of the twentieth century these fuels were abundant, and available at low cost, and most observers anticipated that this happy state of affairs would continue indefinitely.[1] The oil crisis of 1973 therefore came as a considerable shock, but though oil prices were high for the next twelve years they slumped down in 1986 and were to stay low throughout the 1990s.[2] Though people again became complacent about the price and availability of fossil fuels there was growing evidence that the carbon dioxide produced by their combustion was starting to affect the global climate. The potentially very damaging consequences of the emission into the atmosphere of large quantities of carbon dioxide and other greenhouse gases have been the focus of steadily increasing scientific and public concern since the 1980s, and there is now widespread acceptance of the need to take action to drastically reduce these emissions.[3] However it will not be easy to achieve the massive reductions required; and the problem is made even more challenging by the fact that developing countries will concurrently be seeking to greatly increase their own energy consumption, so that they too can enjoy a standard of living comparable with that already achieved by developed countries.

Low-carbon options for electricity generation

There is consensus that much more effort should be put into using energy more efficiently, through a wide range of measures which include improving the level of insulation in housing, increasing the efficiency of domestic electrical appliances and increasing the fuel efficiency of cars and other modes of transport. However there will still remain the need to supply energy on a very large scale, but without correspondingly large emissions of carbon dioxide. Worldwide the biggest single source of carbon dioxide emissions is electricity generation, and electricity is expected to provide a growing share of our future energy needs.[4] And there are, in practice, three main low-carbon routes to meeting an increasing proportion of our electricity needs; nuclear power, and/or power from renewable sources, and/or continued massive consumption of fossil fuels but with carbon capture and storage.

Carbon capture and storage

The last of these, the continued massive consumption of fossil fuels – and in particular coal – but with carbon capture and storage (CCS) has been much discussed in recent years.[5] Several alternative ways of removing the carbon dioxide from the fuel have been proposed, some pre-combustion and others from the post-combustion flue gases, but all involve considerable additional complexity and cost. Once removed the carbon dioxide has to be pumped to where it can be safely and inexpensively stored for centuries, and locations such as depleted oil or gas reservoirs or deep saline aquifers have been suggested as potentially suitable. Given that a single large (1000 MW) modern coal-fired power station produces about six million tonnes of carbon dioxide per year[6] the gas handling, pumping and storage problems are formidable, and add further to the cost.

Though the first mini CCS power plant commenced operation in September 2008 at Spremberg in eastern Germany with a power rating of just 5 MW, and the UK plans to build a pilot project at the 100 MW power level by about 2014, the use of CCS in any commercial 1000 MW power station is unlikely before 2020.[7] The equipment needed to provide CCS adds substantially to the total cost of a coal-fired power station, and the power required to separate the carbon dioxide and pump it into storage reduces the efficiency; it is therefore clear that the overall effect on the cost of electricity from coal-fired power stations will be substantial, but it cannot yet be quantified with any confidence.[8] As the House of Commons Environmental Audit Committee in its July 2008 report cautions 'it is not clear when CCS will be available, or whether it will ever be available'; and it emphasises that 'the possibility of CCS should not be used as a fig leaf to give unabated coal-fired power stations an appearance of environmental acceptability'.[9]

Nuclear power

With nuclear power the ability to generate electricity on a large scale is not in doubt, and the associated emissions of carbon dioxide are very much lower (per unit of electricity generated) than from coal-fired or gas-fired power stations.[10] Several aspects of the technology do, however, give cause for concern, and one of these is the issue of radioactive waste disposal.[11] Although nuclear power has been used commercially for over fifty years there is still no satisfactory solution to the problem of managing nuclear waste. And the Royal Commission on Environmental Pollution considered the issue of sufficient importance to rec-ommend that no new nuclear power stations should be built 'until the problem of managing nuclear waste has been solved both to the satisfaction of the scientific community and the general public'.

Safety and the risk of an accident comparable with the one at Chernobyl in 1986,[12] which led to the large-scale release of radioactive material, is another cause of public concern. Although modern reactors are designed to be inher-ently safer than their predecessors, the small risk of a catastrophic accident remains. There is also the risk of a successful terrorist attack on a vulnerable nuclear facility, such as one where high-level radioactive waste continues to be stored on the surface pending resolution of the waste disposal problem. Though the risks may be small the potential consequences are incalculable; this is recognised in international law by the way that the insurance liability of nuclear power station owners is capped, as well as by the nuclear-related exclusion clauses in domestic insurance policies.[13]

A third cause for serious concern is nuclear proliferation. If developed coun-tries, such as the UK, decide that the construction of nuclear power stations is an essential part of the measures that they need to take to reduce greenhouse gas emissions, so as to counter the threat posed by climate change, then their lead will be followed by many developing countries. Several, including India, Pakistan and

North Korea, have already used the expertise gained with nuclear power stations as a foundation on which to build a nuclear weapons capability, and there is concern that Iran is following their example. As the Sustainable Development Commission remarked 'the widespread adoption of nuclear power would greatly increase the chances of nuclear proliferation'.[14] And as more countries acquire nuclear weapons the probability that they will be used is greatly increased.

There is, in addition to the concerns noted above, considerable uncertainty with respect to the economics of nuclear power. For example the UK Government, in its January 2008 'White Paper on Nuclear Power', indicated that its central estimate for the cost of nuclear power was approximately 4 p/kWh, a figure based on an estimated construction cost of £2.8 billion for a 1600 MW power station.[15] However within a matter of months the chairman and chief executive of E.ON, the German energy company which owns one of the major UK electricity generators and suppliers, stated that the cost of such a power station was in fact expected to be in the range €5 billion to €6 billion, corresponding then to between £4 billion and £4.8 billion. This would imply an energy cost of about 6 p/kWh; and recent cost estimates from utilities in the United States that are proposing to build nuclear power stations suggest that the cost could be substantially higher.[16]

The uncertainty as to the cost of nuclear power is largely a consequence of the fact that there is no recent, relevant experience of their construction, either in Europe or North America. And though the third-generation nuclear reactor designs that are being proposed for construction in the UK and elsewhere promise to be a significant step forward in terms of their safety and reliability – and possibly also in their economics – none has yet been built. The first to be ordered, at Olkiluoto in Finland, commenced construction in 2005 and was expected to be operational in the first half of 2009; it is however well behind schedule, with completion now expected some time in 2012, and the extent of the cost overrun has not been publicly disclosed. However it seems probable that commitments to build several new nuclear power stations will soon be made – for the first time in over twenty years – both in the UK and the US. It will however be the best part of a decade before we have clarity over the real costs of new nuclear power stations.

Renewables

The remaining low-carbon route to meeting a substantial proportion of our electricity needs is through the use – on a large scale – of renewable sources of energy. As was noted in Chapter 1 the overall renewable energy resource is extremely large, with the flow of solar energy to the Earth totalling approximately 87 000 billion kW; which is about 7000 times larger than the total worldwide rate of energy consumption.[17] But though the total amount of solar power that we receive is extremely large it is spread over the whole surface of the Earth, and the solar power density on the Earth's surface typically averages just 100 W/m^2 at latitudes similar to that of the UK, increasing to an average of about 250 W/m^2 for regions that are much closer to the equator. The challenge then is to harness a small part of this very large but relatively diffuse solar energy resource at a cost that is comparable with the cost of electricity from fossil fuels.

One approach is to seek to convert the incident solar radiation directly to electricity, using photovoltaic (PV) panels, but natural processes give a wide range of additional options. For example some of the incoming solar energy gets captured in the leaves of plants where it is converted by photosynthesis and used to produce vegetable matter, which can then be used as fuel in biomass power systems. Also the greater amount of solar energy received by equatorial regions makes the air above them hotter than the air above polar regions; this gives rise to large-scale convection currents in the atmosphere and through this process some solar power gets converted into wind power. Then as winds flow over the surface of the seas some wind power gets converted into wave power. Solar energy also evaporates water from the seas and some of this is carried by winds and subsequently falls as rain over the land; as it then returns via rivers to the sea it can be used to provide hydroelectric power. The power in the tides provides another renewable energy resource, though in this case the power source is primarily the relative motion of the Earth and the Moon.[18]

With just two exceptions – hydroelectric power and PV systems – there was generally little interest prior to 1973 in using renewable sources of energy to help meet the growing demand for electricity. In developed countries most of the good sites for harnessing hydroelectric power had been utilised in the early decades of the twentieth century, and where hydroelectric potential remained there was much increased awareness of the negative consequences of building large dams.[19] And though the photovoltaic (PV) conversion of solar radiation to electricity has its origins in the nineteenth century its development as a practical source of electricity for use in remote locations only commenced in the 1950s. Much effort was then put into developing PV systems that could provide power for satellites, with the first PV-powered satellite put into orbit in 1958, and for applications such as this the very high cost of early PV systems was no barrier.[20]

The 1973 oil crisis made countries aware of the extent to which they were dependent on imported fossil fuels, and made them realise that they should no longer plan for economic development based on the assumption that fossil fuels would continue to be abundant and available at low cost. One result was an upsurge of interest in the potential of renewable energy technologies, and substantial resources were then put into their development. Other authors have reviewed the progress that has been made since across the full range of renewable energy options,[21] but the technology that has emerged as the one that is expected to make the largest contribution to meeting our electricity needs up to – and probably well beyond – 2020 is the technology for harnessing the power in the wind. This is illustrated by the fact that in the European Commission's 2007 Renewable Energy Roadmap,[22] which indicates how renewable energy technologies might provide 20% of overall European *energy* supply by 2020, wind power is expected to provide approximately 50% of the increase in electricity from renewable sources; biomass is expected to provide a further 35%, and all the other renewables (hydroelectric, PV, solar thermal, geothermal, tidal and wave power) between them contribute just 15%. And growing global concerns over the effect on food prices of using agricultural land for energy crops may well require that the proposed contribution from biomass be reduced.[23] Wind power's leading role is confirmed by the scenario proposed in 2008 by the UK Government to show how renewables could provide 15% of the total *energy* supply by 2020 (the UK share of the overall European target); this indicates that wind power is expected to provide almost 80% of the increase in electricity from renewable sources.[24]

Wind power's progress

In the immediate aftermath of the 1973 oil crisis wind power was just one of the many renewable energy options that came in for scrutiny, and several countries initiated national wind energy programmes that focussed on the rapid development of megawatt-scale wind turbines. This proved to be much more challenging than had been anticipated and these national programmes were on the whole expensive failures. In Denmark, however, a small number of wind power enthusiasts had a different approach, which in due course was to prove very successful. They focussed their efforts on relatively small three-bladed grid-connected wind turbines, such as those shown in Figures 5.12 and 5.13, which typically had a diameter of about 10 m and a maximum power output of about 30 kW. These performed well enough to be purchased in growing numbers, and by the early 1980s several Danish agricultural equipment producers had diversified into their manufacture.

Competition to supply the small but growing Danish market led to the development of progressively larger wind turbines, as it became clear that larger machines could produce electricity at lower cost. And this process was accelerated by the Californian wind boom of the early 1980s. The American-made wind turbines that were then available proved to be insufficiently robust, and Californian wind farm developers turned to Denmark and imported many thousands of Danish machines before the oil price slumped in 1986 and the boom ended. Turbine sales continued, but at a reduced level; and by 1990 reliable robust and reasonably cost-effective wind turbines with diameters up to about 35 m and power outputs up to about 400 kW, and typified by the machines shown in Figures 5.19 and 5.20, could be supplied by several manufacturers that were mostly – but not wholly – Danish. These included companies such as Vestas, Bonus and Enercon that are still industry leaders, though Bonus was bought by Siemens in 2004 and became Siemens Wind Power.

By the end of 1990 the installed global wind power capacity totalled just under 2000 MW, with 1500 MW provided by 15 000 wind turbines in California, 340 MW in Denmark and a total of 140 MW elsewhere. Though oil prices were low – and were to stay low until 2003 – the growing concerns over greenhouse gas emissions and their effect on the climate led countries such as Germany and Spain to enact legislation which specified that a premium price should be paid for electricity generated from renewable energy sources. And wind-powered electricity generation had progressed to the stage where it was ready and able to benefit from such legislation.

Over the years that followed the global installed wind power capacity increased year by year by approximately 25% with the result that by the end of 2007 the total installed capacity had risen to just under 94 000 MW; and preliminary data for 2008 indicates that a further 27 000 MW was installed, taking the global installed wind power capacity to approximately 121 000 MW.[25] The United States – which added just over 8000 MW in 2008 – has now reclaimed its position as the global leader in installed wind power capacity with a total of approximately 25 000 MW, and is followed by Germany with almost 24 000 MW; it should however be noted that the United States is 26 times larger than Germany. Spain with just under 17 000 MW has the next largest installed capacity, followed by China with 12 000 MW, and then by India with close to 10 000 MW. China's progress in recent years has been quite remarkable. Its installed capacity at the start of 2006 was just 1260 MW, but it more than doubled this by the end of the year; the capacity doubled again in 2007 to just under 6000 MW, and the installation of over 6000 MW in 2008 meant that once again – and for the third consecutive year – it had doubled its installed capacity in just 12 months. As was indicated in the previous chapter China has been rapidly building up its wind turbine manufacturing capability, and Chinese manufacturers now supply most of the turbines that are installed in the country. Denmark – with an installed wind power capacity of just over

3000 MW – has now been surpassed by Italy, France and the UK as well as by the five countries already mentioned, but Denmark is a small country (only one-eighth the size of Germany) and its wind turbines provide about 20% of its total electricity consumption, substantially more than in any other country.

Over the period 1990 to 2008 the commercial pressure to generate electricity at as low a cost as practicable led manufacturers to develop wind turbines that year by year grew steadily larger. Most new projects now use wind turbines with maximum power outputs in excess of one megawatt, typified by the 80 m diameter 2 MW rated Vestas machines shown in Figure 1.3 as well as by the 107 m diameter 3.6 MW rated Siemens machines shown in Figure 8.12. The use of three-bladed rotors with the blades upwind of a tubular steel tower has become almost universal; and considerable progress has been made in the design and manufacture of the very large fibre-composite blades that are now required, and which make increasing use of the higher performance – but more expensive – carbon fibres.[26] The biggest change since 1990, apart from the overall turbine size, is that most machines now use blades with full-span variable pitch, and the generator design is now of a type that allows the rotor to operate at variable speed. And the advent of reliable and competitively priced multi-megawatt wind turbines has made it possible for a number of countries – most notably Denmark and the UK – to start building commercial wind farms offshore, where for many the wind power resource is much larger than on land.

Cost-of-energy comparisons

As the size of wind turbines steadily increased through the 1990s and into the next decade their economics steadily improved, and the cost of wind-generated electricity reached a minimum in about 2003.[27] Increases in the cost of commodities such as steel, together with turbine supply shortages, subsequently led to substantial cost increases. However fossil fuel prices also increased substantially after 2003, and the rapidly growing demand for turbines continued.

The usual marker for fossil fuel prices is the price of oil, and though this averaged $62/barrel (at 2007 price levels) for the 12 years following the 1973 oil crisis it slumped to approximately $28/barrel in 1986, and remained at this relatively low level for the next 17 years.[2] However it became increasingly difficult to meet the continually growing demand for oil, and from 2003 onwards the price increased substantially. By mid 2008 it had climbed to a peak of just under $150/barrel, but the global financial crisis and recession that developed through 2008 then led to reduced demand and by the end of the year the price had fallen to below $50/barrel.

There is, inevitably, a great deal of uncertainty about how the price of oil will vary over the next decade and beyond. However there is the expectation that when the global economy comes out of the present recession the demand for energy will once again grow steadily, underpinned by rapidly growing demand in developing countries; there is also the expectation that it will become pro-gressively more difficult to increase oil production in line with this growing demand. The underlying trend for the future price of oil is therefore almost certainly upwards, but as a small mismatch between supply and demand can cause a large change in the oil price it is likely that there will be considerable volatility superimposed on this rising trend. As was noted in Chapter 1 the UK Government's central forecast is that the 2020 oil price will be $70/barrel, and the United States Energy Information Administration's forecast is similar[28] (though many commentators have suggested that by 2020 the oil price could be well in excess of $100/barrel). Whether buying oil or gas what the purchaser wants is their energy content, and gas prices are therefore linked to oil prices. Oil at a sustained price of $70/barrel would lead to a gas price of about 80 p/therm, and the cost of electricity from the gas-fired power stations[29] that are Britain's main source of electricity would then be about 7 p/kWh. How does this compare with the cost of wind-powered electricity generation?

Wind power costs have in common with nuclear power costs the fact that they are dominated by the repayment of their respective capital costs over a number of years. The level of repayments then depends on the rate of return required by the owner, and the period over which the cost has to be repaid, as well as being proportional to the capital cost. Public sector owners were traditionally satisfied with a 5% or 8% rate of return and would be content with repayment over the life of the power plant – which for a nuclear power plant might be 30 or 40 years. In the new world of liberalised electricity markets the private-sector owners of power plants require a substantially higher rate of return and much shorter repayment period.[30] With a wind project the cost of electricity also depends on the windiness of the location where it is sited. All these factors, the capital cost, the repayment period, the required rate of return and the site wind speed influence the cost of wind-powered electricity generation, and there is conse-quently no single figure that one can quote as the cost for comparison with the cost of electricity from a gas-fired power station.[31] However the price paid for wind-powered electricity generation in countries that have taken measures to encourage the use of renewables can provide a good generic indicator of wind power's cost; and these prices are shown in Figure 9.1 for Germany, the United States, China and the UK.[32] The figure also shows the 1990 cost of wind-generated electricity, adjusted to 2008 price levels, as well as the cost – discussed above – of gas-fired electricity generation using a modern, high

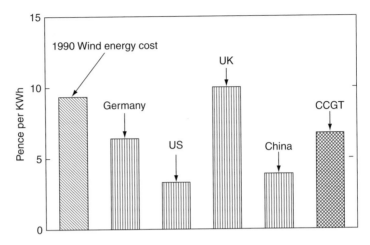

Figure 9.1 Wind energy costs in 2007/8.

efficiency, combined-cycle gas turbine (CCGT). As discussed previously there is no data for the cost of electricity from either a modern nuclear station or from a coal-fired power station with carbon capture and storage.

Back in 1990 the cost of wind-powered electricity generation was approximately 5.5 p/kWh, which equates to 9.4 p/kWh at 2008 price levels. As has been noted wind power costs came down steadily until about 2003,[27] but since then there have been significant increases. However the most recent costs for Germany, the United States and China, shown in Figure 9.1, indicate quite clearly that despite these increases the cost of wind energy is still well below 1990 levels. More importantly the figure also shows that wind energy costs compare favourably with the cost of electricity provided by gas-fired power stations burning gas at a price consistent with oil at $70/barrel. The economic downturn of 2008 has already caused sharp falls in commodity prices, and new entrant wind turbine manufacturers – as well as increased production from existing manufacturers – will greatly ease the turbine supply situation. One can therefore expect to see wind turbine prices, and consequently wind energy costs, stabilise or even reduce over the next two to three years. In the longer term (out to 2020 and beyond) there is considerable scope for further substantial cost reductions,[33] and it seems reasonable to anticipate that by 2020 the cost of wind-powered electricity generation will be well below the cost of electricity from the gas-fired power stations which today provide a large proportion of the electricity we consume, and which will continue to make an important contribution to meeting our electricity needs out to 2020 and beyond.

The marked difference between the wind energy cost of about 3.3 p/kWh shown for the US in Figure 9.1 and the cost of about 6.4 p/kWh shown for Germany is mostly due to the fact that wind farm locations in the United States are generally much windier than in Germany; this is illustrated by the fact that wind farm capacity factors in 2007 averaged 32% in the US, compared with just 21% in Germany.[34] Wind speeds in China are intermediate between those in Germany and those in the US, and the relatively low cost of Chinese-made wind turbines helps to ensure a low cost of energy. UK wind speeds are also intermediate between those in the US and those in Germany, as is confirmed by the fact that in 2007 wind farms in the UK had an average capacity factor of just over 27%.[35] The anomalously high cost of wind energy to the UK electricity consumer is a direct consequence of the fact that the financial mechanism used in the UK to encourage the use of renewables – the Renewables Obligation – is much less cost-effective than Germany's feed-in tariff or the Production Tax Credit in the United States. With a feed-in tariff the cost of wind-powered electricity generation in the UK could have been halved.[36]

Variability: keeping the lights on

Wind power is often criticised because the output from wind farms is variable, and on occasion will be small or even zero during periods of high demand.[37] Wind, it is then said, is unreliable and it would be a waste of money and resources to build thousands of megawatts of wind power capacity. What such statements fail to recognise, however, is the very important difference between energy and power. The problem we face is how we meet our future energy needs reliably, affordably, and with much reduced greenhouse gas emissions. Though the power provided by the winds does indeed vary substantially from one day to the next the winds are – year by year – a very reliable source of energy[38] and for many countries, including both the UK and the US, the wind resource is extremely large. Wind power's greenhouse gas emissions are near zero,[6] and wind turbine technology has, as we have seen, progressed to the point where the energy payback period is just a matter of months[39] and the cost of wind-powered electricity generation is comparable with the cost of electricity from gas-fired power stations. There is, moreover, the expectation that the cost of wind energy will reduce significantly over the period to 2020 and beyond, while the cost of gas-fired electricity generation – which for many years will continue to be an important source of electricity – is likely to increase substantially.

Does wind power's variability matter? Wind turbines typically generate electricity for about 75% to 85% of the hours in the year[40] and all the time they are operating they reduce the amount of electricity that has to be generated by fossil-fuelled power stations. This then reduces the quantity of fossil fuel consumed and gives corresponding carbon dioxide emission savings; and many detailed system integration studies over the years have shown that wind power's variability has very little adverse effect on either of these.[41] And these savings can be very substantial. For example in the scenario put forward by the UK Government in 2008 to show how the UK could provide 15% of its energy from renewables by 2020 it is proposed that wind power should by then produce about 82 billion kWh of electricity per year, equal to about 22% of the total UK electricity consumption.[42] Without the adverse effects of wind power's variability this would give a fuel saving equivalent to about 136 million barrels of oil per year;[43] wind power's variability reduces this by approximately 5% to about 129 million barrels per year. The corresponding reduction in emissions of the greenhouse gas carbon dioxide would total well in excess of 30 million tonnes per year.[44] With oil at \$70/barrel an energy saving of nearly 130 million barrels per year would be worth approximately £6 billion per year, and this is the annual sum that would be retained within the UK economy instead of being spent on energy imports. Beyond 2020 wind has the potential to make a contribution that is much larger than 22%, and the price of energy imports is likely to be substantially more than \$70/barrel, or its equivalent for gas.

Wind power's variability may have little adverse effect on its ability to provide substantial savings in both fossil fuel consumption and carbon dioxide emissions, but will it – as some critics suggest – make the lights go out? The answer is a clear but emphatic no, as has been shown by experience elsewhere and by many detailed system studies. Wind power already provides 20% of the electricity consumed in Denmark, and in Germany wind power's contribution is more than 7%. In the UK the Carbon Trust has looked in detail at the consequences of installing 40 000 MW of wind power capacity – enough to provide 31% of the electricity consumed – and states very clearly that this can be done without compromising security of supply; in its own words 'the lights will not go out'.[45] And the US Department of Energy has concluded that 'wind generation could reliably supply 20% of US electricity demand'.[46]

To understand why wind power's variability is not the big problem that some people suppose one needs to recognise, as was noted previously,[47] that the air over just a single region of England weighs many billions of tonnes; and when it is moving – like a supertanker, only more so – it cannot suddenly stop. Weather systems in fact take many hours to cross a country such as the UK, and the overall output from wind farms distributed around the country consequently changes relatively slowly. As the UK system operator, National Grid, has noted 'variations in wind generation up to 12 hours ahead are easy to forecast and plan for'.[48] When the overall wind power output slowly increases the output from fossil-fuelled power stations is reduced, and vice versa. And over the next decade or so, as the installed wind power capacity increases to provide 20% or more of the electricity consumed, and as some existing fossil-fuelled power stations come to the end of their lives and are shut down, it will be necessary to build some new fossil-fuelled power plants so as to ensure that there is no risk to the security of the electricity supply during the occasional periods when wind speeds are low over large areas of the country. These new fossil-fuelled power plants will typically be required to run for just a few hundred hours per year, compared with the many thousands of hours per year that the wind turbines will be generating, and will be low capital cost and flexible units such as open-cycle gas turbines.[49] The additional costs involved in maintaining the security of supply in this way, plus the costs required to provide some additional short-term balancing services, have been the subject of many comprehensive studies[50] and typically add substantially less than 1 p/kWh to the cost of wind-powered electricity generation.

Though wind energy's additional costs, as discussed above, are not included in the costs shown in Figure 9.1 the CCGT cost shown – for electricity produced by a modern gas-fired power station – does not include any allowance for the damaging consequences of its greenhouse gas emissions. The European Commission has estimated that its Emissions Trading Scheme will result in a price for carbon dioxide emissions of about €39/tonne in the period 2013 to 2020,[51] though this falls well short of any quantification of the damage that the greenhouse gas emissions are likely to cause. This would add approximately 1.4 p/kWh to the CCGT cost shown in the figure,[52] substantially more than wind energy's additional costs. The costs for wind-powered electricity generation shown in Figure 9.1 are of course for wind farms on land, and the cost of energy from offshore wind farms is at present substantially higher;[53] however the offshore wind industry is still at an early stage in the learning process, and significant cost reductions can be expected over the next decade.

Wind power's global growth

As indicated earlier the global installed wind power capacity increased from just under 2000 MW at the end of 1990 to 94 000 MW by the end of 2007, and provisional data for 2008 suggests that by the end of 2008 the total installed capacity was approximately 121 000 MW. This growth is shown in Figure 9.2,

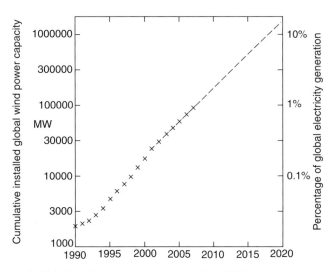

Figure 9.2 Global wind power capacity growth since 1990.

though note that the 60-fold increase in just 18 years has required the use of a non-linear vertical scale.[54] Sometime in early 2008 the global capacity passed the 100 000 MW (= 0.1 TW) threshold, sufficient to generate approximately 190 billion kWh (= 190 TWh) of electricity per year. Global electricity generation in 2008 was about 18 000 TWh,[55] so wind power then passed the point where it was supplying 1% of the electricity that is used globally. The right-hand scale of Figure 9.2 shows approximately the percentage of global electricity generation that corresponds to the installed capacity indicated on the left-hand scale; and it can be seen that wind power's contribution has climbed from 0.1% to 1% in just a decade.

Concern over the consequences of climate change is growing stronger year by year, and the nearly $150/barrel peak in the price of oil in mid 2008 has served to increase political pressure both in Europe and in the United States to reduce dependence on oil and gas imports. Wind power – as has been seen – is the most cost-effective of the high-resource renewable energy options, and it therefore seems likely that the installed wind power capacity will continue to show strong annual growth. Though the 25% year by year capacity increases that have been achieved in recent years cannot continue indefinitely there is, as yet, no sign that the annual growth is reducing. There are no fundamental material supply issues,[56] and the turbine shortages that have constrained growth in recent years have led to a considerable number of new entrant manufacturers; they have also stimulated existing wind turbine manufacturers and component suppliers to substantially increase their production capabilities. The dashed line in Figure 9.2 shows how the global wind power capacity would increase if the growth rate that has been achieved since 2001 is sustained, and this extrapolation suggests that the 1 000 000 MW = 1 TW threshold will be reached in 2018. If the global recession caused by the financial crisis in 2008 proves to be sustained then passing this threshold may well be delayed, but by 2020 or very soon after the global installed wind power capacity will pass 1 TW, and wind power will then be providing about 10% of global electricity generation.

In the period 1990 to 2008 wind turbines grew in size from about 35 m diameter to about 90 m diameter, and their rated power outputs increased from about 400 kW to about 3 MW. This trend to increased size was driven by the commercial pressure to reduce the cost of wind-powered electricity generation, for though the turbine cost – relative to the rotor swept area – is relatively insensitive to size[57] the bigger turbines have taller towers and benefit from the fact that the wind speed increases with height. Also having a smaller number of larger turbines in a wind farm of given capacity means that it extends over a smaller area, so reducing the cost of the access tracks and cable interconnections between the turbines; it also gives a reduction in the total cost of the foundations. However the logistic problems of trans-porting very long blades and large diameter ($\approx$ 4 m) tower base sections down the small and often winding roads that are common in rural areas are such as to suggest that for wind farms on land the wind turbine size will change little over the next decade. And the three-bladed rotor with its variable-pitch blades located upwind of a slender tubular tower is so well established – and has proved so successful – that any change to this rotor configuration over the next decade seems unlikely. The cost reductions that can be expected will be achieved in part through the increased production volume, aided by many incremental design improvements.[58]

With offshore wind farms there is much more scope for innovation. The costs of foundations and electrical infrastructure are much higher offshore than on land,[59] and are relatively insensitive to the turbine size. And as new and very large wind farms are developed at greater distance from the coast, and in deeper waters, these costs will increase further. There is consequently the expectation that the economics of offshore wind farms will be improved by using turbines that are larger than at present, possibly very much larger. And provided only that manufacturing facilities are by the coast the sea transportation of blades and towers imposes no real constraint on the size of these components. Most of the offshore wind farms that have been built in recent years have used turbines with diameters in the range 80 m to 107 m, and with power ratings in the range 2 MW to 3.6 MW, see Figures 1.3, 8.4 and 8.12. And up to 2008 the largest wind turbines used offshore have been the two 126 m diameter, 5 MW rated, Repower machines located 22 km off the north-east coast of Scotland, in water more than 40 m deep, which supply power to the nearby Beatrice oil platform. Manufacturers are however developing wind turbines that are substantially larger, and by 2020 or soon after it seems likely that there will be wind turbines with diameters up to about 200 m, and with power ratings up to about 15 MW, deployed offshore. With rotors this large it is by no means certain that the three-bladed configuration will continue to be the preferred option; other configurations, such as two-bladed rotors or possibly even vertical-axis rotors, may prove to be more economic. And though the UK and the US both have a large offshore wind resource where water depths are relatively shallow (less than about 30 m depth) they each have an even larger offshore wind resource in deeper waters. Wind turbine support structures that float, and are tethered to the seabed, are expected to be more economic than fixed structures when the water depth is greater than 100 m; and the development of such floating wind turbine systems has already commenced.

Wind power, as we have seen, already provides a little over 1% of global electricity generation and by about 2020 we can expect wind power to be providing about 10% of global electricity, with one million MW (= 1 TW) of capacity installed. What can we expect from wind power beyond 2020? In the United States, as has been discussed, a detailed assessment by the Department of Energy concluded that by 2030 an installed wind power capacity of 300 000 MW could provide 20% of US electricity.[60] This corresponds to a wind power density of 33 kW per square kilometre of land area, well below the levels already achieved by 2007 in Denmark, Germany and the Netherlands (see Table 8.2); and all three of these countries have population densities that are substantially larger than the US population density. Given the political will it is clear that the US could obtain much more than 20% of its electricity from the wind. In the UK the Government's suggested scenario for developing renewables through to 2020 indicated that wind power could by then provide 22% of UK electricity, though the Carbon Trust was more ambitious and proposed that by 2020 the UK could and should plan to use wind power – mostly offshore – to provide 31% of its electricity.[61] The UK offshore wind resource is in fact sufficiently large to make possible a contribution – beyond 2020 – that is well in excess of 31%. Given wind power's day-by-day variability is there a limit to the percentage contribution it can make? The answer is probably yes, but system integration studies suggest that the limit is well over 50% and the UK system operator, National Grid, has commented that 'if there is a limit to the amount of wind that can be accommodated, that limit is likely to be determined by economic/market considerations'.[62]

As was noted in the first chapter the global wind resource is extremely large, many times larger than the total global energy consumption, and we now have the technology to use this resource and generate electricity on a very large scale and at a competitive cost. Wind power will therefore have an important and growing role in helping to meet electricity needs affordably, and without damaging greenhouse gas emissions; and globally it seems reasonable to expect that wind power's contribution to our electricity needs will continue to grow strongly beyond the 10% level that will be reached by about 2020. In countries such as the UK and the US that have a particularly good wind resource wind power's contribution could by 2050 reach or even exceed 50%.[63] Wind power made a significant contribution to meeting energy needs in England and many other countries in past centuries; it seems set to make an even more significant contribution globally in the twenty-first century, helping to meet the much-increased energy needs of the much-increased global population in a clean, affordable and sustainable way.

Appendix A

The power output from wind turbines

As noted in Chapter 1 there is no fundamental difference between a wind turbine and a windmill. By convention we give the name *wind turbine* to modern designs, mostly used for generating electricity, and use the name *windmill* for older designs which were used not only for milling grain but for pumping water, sawing wood and a wide range of other applications. Modern wind turbines and traditional European windmills have in common the fact that the overall aerodynamic force acting on each blade (or sail) is primarily due to what aerodynamicists call *lift*; as described in Chapter 2 this overall force acts approximately at right angles to the relative wind, see Figure 2.6(b).

The overall aerodynamic force can be considered to be the sum of two component forces. The smaller of these is the *turning force*, which is in the plane of rotation of the blades and makes them – and the shaft to which they are attached – rotate; this provides the useful shaft power output of the turbine, which can then be used to drive a generator, or to turn millstones, etc. The second and much larger component of the overall force on each blade is the *downwind force*, and the blades and the tower and the foundations must all be made strong enough to withstand this. Newton's third law of motion tells us that action and reaction must be equal and opposite;[1] so corresponding to the downwind force exerted by the airflow on the blades there is an equal but opposite force exerted by the blades on the airflow, and this makes the air slow down as it approaches and then goes through the rotor. The air speed downwind from a wind turbine is therefore lower than the upstream wind speed and this wake of reduced air speed extends for several diameters downstream from the turbine.[2] The lower air speed in the wake means that it has less kinetic energy than the airflow upwind of the rotor; and if there were no aerodynamic losses the rate of reduction in the kinetic energy of the air flowing through the rotor would be equal to the wind turbine's shaft power output.

Figure A.1 shows schematically the airflow approaching the rotor of a wind turbine. Air within the dashed streamlines flows through the rotor, and is slowed down by the aerodynamic interaction with the blades; air outside the dashed streamlines passes round the rotor and is unaffected by its presence. Air speeds through the rotor are well below the speed of sound (which is over 300 m/s) and changes in the density of the air are therefore negligibly small.[3] As the airflow approaches the rotor, and is slowed down by the aerodynamic forces produced by the blades, the streamlines consequently diverge; the

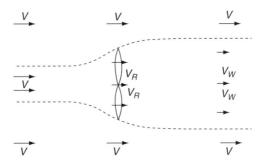

Figure A.1 Flow through an idealised wind turbine.

influence of the rotor extends downwind as well as upwind so the streamlines continue to diverge for a short distance, as the figure indicates.

The air mass flow rate m through the rotor is equal to the air density, multiplied by the air speed V_R as it goes through the rotor, multiplied by the area A of the circle swept by the rotating blades; i.e.

$$m = \rho V_R A \tag{A.1}$$

where by convention the Greek symbol ρ (rho) denotes the air density, which at a temperature of 15 °C and a pressure of 1013 millibars (a typical sea level value) is equal to 1.23 kg/m³ (see Chapter 3, note 38).

Upwind of the rotor the kinetic energy per unit mass is equal to $\frac{1}{2}V^2$, where V is the approaching wind speed. Downwind of the rotor, where the air speed in the wake is reduced to V_W, the kinetic energy per unit mass is reduced to $\frac{1}{2}(V_W)^2$. The rate of loss of kinetic energy for the air flowing through the rotor is equal to the mass flow rate m multiplied by the reduction in the kinetic energy per unit mass; *for an ideal wind turbine, with no losses, this would be equal to the power output P from the wind turbine*, giving

$$P = \rho V_R A \left\{ \frac{1}{2}V^2 - \frac{1}{2}(V_W)^2 \right\} \tag{A.2}$$

It may be shown[4] that the reduction in the air speed upwind of the rotor is equal to the reduction in the air speed downwind of the rotor, giving

$$V - V_R = V_R - V_W \tag{A.3}$$

This can be re-arranged to give

$$V_R = \frac{1}{2}(V + V_W) \tag{A.4}$$

Replacing V_R by $\frac{1}{2}(V + V_W)$ in Equation (A.2) then gives

$$P = \frac{1}{2}\rho A(V + V_W)\left\{ \frac{1}{2}V^2 - \frac{1}{2}(V_W)^2 \right\} \tag{A.5}$$

The Betz limit

The above expression for the power output from an ideal wind turbine (i.e. with no losses) is a maximum when the air speed in the wake is one third the upstream wind speed. Putting $V_W = \frac{1}{3} V$ into Equation (A.5) then gives

$$P_{max} = (16/27) \frac{1}{2} \rho A V^3 \qquad (A.6)$$

This maximum power output from an idealised wind turbine – with no aerodynamic or other losses – is known as the *Betz* limit, named after the German aerodynamicist Albert Betz who formulated this simple model for the flow through a wind turbine in 1920.[5]

In the absence of the wind turbine, the power in the wind flowing through the rotor swept area A is simply given by the mass flow rate ($\rho A V$) multiplied by the kinetic energy per unit mass, $\frac{1}{2} V^2$; i.e.

$$P_{wind} = \frac{1}{2} \rho A V^3 \qquad (A.7)$$

Importantly, this tells us that ***the power in the wind is proportional to the cube of the wind speed***. In other words a doubling of the wind speed would increase the power in the wind by a factor of eight. More typically a 20% increase in wind speed will give a 73% increase in the power in the wind; this could result from the difference between an exposed small hill and nearby more sheltered locations, hence the importance of careful local siting.

It is useful to have a means of comparing different wind turbines that is independent of their size, and also independent of the wind speed. A *power coefficient C_P* is therefore defined as the power output P from a wind turbine divided by the power in the undisturbed wind;

$$C_P = P \Big/ \left(\frac{1}{2} \rho A V^3 \right) \qquad (A.8)$$

From Equation (A.6) it can be seen that the Betz limit, i.e. the maximum power output from an ideal wind turbine with no losses, corresponds to $C_P = (16/27) = 0.593$.

How many blades? The effect of solidity

The simple model of the flow through a wind turbine rotor described above neglects any detailed consideration of how the blades interact with the airflow. Much more detailed models have been developed which allow for the number of blades, and their shape and aerodynamic characteristics, and these allow the performance of wind turbines, as well as the loads on the blades, to be calculated with reasonable accuracy.[6]

Getting the maximum power coefficient from a wind turbine requires the right degree of interaction between the blades and the air flow; it needs to be just enough to reduce the air speed at the rotor to about two thirds of the upstream wind speed. If the blades turn too fast there will be too much interaction, and the air flow approaching the rotor is slowed

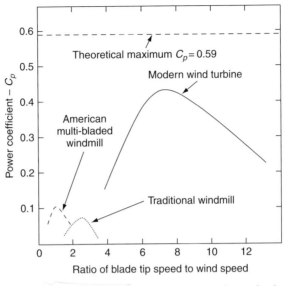

Figure A.2 Typical power coefficients and corresponding tip speed ratios.

too much; more air then flows round the rotor, rather than through it, and the power output is reduced. And if the blades turn too slowly there will be too little interaction with the air flow, which will also lead to reduced power output. The optimum rotational speed of the blades depends on the *solidity* of the rotor, which is a measure of the number of blades and their frontal area. The solidity is defined as the proportion of the rotor area A that is occupied by the blades, as seen by an observer standing well upwind and looking towards the turbine. For modern electricity-generating wind turbines[7] the solidity is usually in the range 4% to 5%, whereas for traditional windmills the solidity is usually in the range 15% to 30%; the multi-bladed windmills used for water pumping, see Figure 3.8, typically have a solidity of about 70%.

It is a popular misconception that the power output from a wind turbine could be increased by increasing the number of blades. The reality is that the energy in the wind flowing through the rotor swept area A can be intercepted efficiently with any number of blades, and one-bladed, two-bladed, three-bladed and four-bladed wind turbines have all been built and used in recent years.[8] The present consensus in favour of three-bladed wind turbines for electricity generation is primarily the result of structural, economic and aesthetic considerations. The solidity matters more than the number of blades and with high-solidity rotors the optimum interaction between the blades and the air flow is obtained, as one would expect, at relatively low rotational speeds; conversely for low-solidity rotors one needs substantially higher rotational speeds to get the optimum interaction. This is illustrated in Figure A.2, which shows how the power coefficient C_P varies with the tip speed ratio for a variety of wind turbine designs;[9] as its name implies the tip speed ratio is defined as the blade-tip speed divided by the upstream wind speed V. Modern wind turbines have greatly benefited from the progress made during the twentieth century in the development of aerofoil shapes for aircraft wings, which have very low drag relative to the lift that they produce;[10] the power coefficients of modern wind turbines are therefore about six times higher than those for traditional windmills.

Efficiency

Though the power coefficient gives an indication of the relative efficiency of different types of wind turbine (or windmill) it is wrong to refer to it (as some writers do) as 'the efficiency'. For example the Betz limit of $C_P = 0.593$ corresponds to the kinetic energy in the air flowing through the rotor being reduced from $\frac{1}{2}V^2$ to $\frac{1}{2}(V/3)^2$; in other words it corresponds to eight-ninths (89%) of the kinetic energy in the air flow being removed and delivered as a useful power output. Modern wind turbines can achieve power coefficients of $C_P \approx 0.45$, which means they are about 67% efficient in taking energy from the air flowing through their rotors. Traditional windmills with a power coefficient $C_P \approx 0.07$, see Appendix B, were about 10% efficient in taking energy from the air flowing through their rotors.

The lower C_P values for traditional windmills, and for American multi-bladed windmills, are due only in part to the fact that their blades (or sails) are less streamlined than the blades of modern wind turbines: the other main reason for their lower power coefficients is that their lower optimum rotational speed (which results from their higher solidity) leads to a higher torque on the blades.[11] Since action and reaction must be equal and opposite this means that the torque exerted by the blades on the air flow is also high, and this makes the air flow spiral as it flows through and then downwind of the rotor; at low tip speed ratios the energy dissipated in this swirling motion is substantial.

Equation (A.8) can equally well be written as;

$$P = C_P \left(\frac{1}{2} \rho A V^3 \right) \tag{A.9}$$

Superficially this appears to indicate that a wind turbine's power output is proportional to the cube of the wind speed, but the power coefficient C_P also varies with wind speed and modern wind turbines are designed so that their power is near constant above a wind speed of about 13 m/s (see Figure 6.2). The equation does however show that the power output is proportional to the rotor swept area, so that a doubling of the diameter gives a fourfold increase in the power output. Increased rotor diameters are usually associated with proportionately taller towers; the consequent increase in the rotor hub height then leads to an increase in the wind speed through the rotor, see Appendix C, which further increases the power output.

Rotor drag

The downwind force on the rotor, F_D, is equal to the rate of change of momentum of the air flowing through it; it is therefore equal to the mass flow rate through the rotor multiplied by the overall reduction in the velocity of the airflow, i.e.

$$F_D = \rho V_R A(V - V_W) \tag{A.10}$$

and a rotor drag coefficient, C_D, is defined as

$$C_D = F_D \Big/ \left(\frac{1}{2}\rho A V^2\right) \tag{A.11}$$

When the power coefficient C_P is a maximum we have seen that $V_R = \frac{2}{3}V$ and $V_W = \frac{1}{3}V$. Putting these values into Equation (A.10) gives $F_D = (4/9)\rho A V^2$, and the rotor drag coefficient is then $C_D = (8/9) = 0.89$. This is nearly as high as if the rotor were replaced by a solid disc (for which $C_D \approx 1.2$, see Hoerner[12]).

For modern wind turbines the maximum power coefficient is obtained at a wind speed of about 9 m/s, which is somewhat lower than the rated wind speed, i.e. the speed at which the turbine reaches its maximum power output. The latter is typically in the range 12 to 14 m/s; the power coefficient C_P is then equal to about 0.25, and the corresponding rotor drag coefficient C_D is about 0.4. For a turbine such as that shown in Figure 1.3, rated at 2 MW and with a rotor diameter of 80 m, the downwind force on the rotor when the wind speed is 14 m/s is then equal to about 240 kN (i.e. about 24 tons force). And though the density of air is low, the mass flow rate through the rotor is substantial; in this example it is nearly 80 000 kg/s (i.e. about 80 tonnes per second).

When the power coefficient C_P is a maximum, Equation (A.6) indicates that the power output $P = (16/27)\frac{1}{2}\rho A V^3$; and as noted above the corresponding rotor drag is given by $F_D = (4/9)\rho A V^2$. Combining these two expressions gives $F_D = 3P/2V$. Though this result is only strictly applicable when the power coefficient is a maximum (and $V_R = \frac{2}{3}V$) it can be used more generally to *estimate* the magnitude of the rotor drag. For the example given above, of a turbine generating 2 MW in a wind speed of 14 m/s, it indicates that the rotor drag is about 214 kN (or 21 tons force) which is within about 10% of the more precise calculation.

Appendix B

The performance of traditional windmills

Though many books have been written about traditional windmills, few give useful information on their power output. Bennett and Elton's 1899 *History of Corn Milling* is more informative than most, but given the difficulty in quantifying both the power output and the wind speed the information provided is typified by statements such as 'At Stockton, another of the old midland post-mills, it was possible to grind, with a fairly steady wind, very lately six bushels per hour with one pair of stones.'[1] Likewise for Little Dassett post mill, near Leamington, which was on high ground with good exposure to the south west, Bennett and Elton state that 'in a good fair wind, and with one pair of stones, twelve bushels of mixed grist[2] for pigs food can, on average, be ground in an hour. Wheat for bread takes longer, and the average is six to eight bushel per hour.'[3]

The power required to grind wheat

The bushel was a traditional measure of volume (equal to about 36 litres) and calculations made by Smeaton[4] allow us to quantify reasonably accurately the power needed for milling. In 1766 he had to determine whether there was sufficient water for a proposed water-powered mill to grind the equivalent of 32 bushels of wheat per week.[5] The head of water available was 9 feet 6 inches (2.9 m) and Smeaton stated that with this head he would need 3600 cubic feet of water (102 m³) to grind one bushel. This equates to a gross water power requirement of 805 W to grind one bushel of wheat per hour. Smeaton's own measurements indicate that an overshot waterwheel had an efficiency of about 65%;[6] losses in the bearings and gearing from the wheel through to the millstones would have been about 25%,[7] and the overall efficiency of the waterwheel, from the water power input through to the power available at the stones, would therefore have been just under 50%. Smeaton's figures therefore indicate that grinding one bushel of wheat per hour requires a power input to the stones of about 390 W. As a bushel of wheat weighs[8] about 28 kg this gives *a power requirement of approximately 14 W to grind 1 kg/hr of wheat between a pair of traditional millstones.* The energy used to grind the grain is consequently less than one half of one per cent of the grain's energy content.[9]

A cross-check on the magnitude of this estimate of the power required for milling is given both by Gimpel and by Lawton.[10] Gimpel states that in Roman times two slaves using a rotary hand-mill could grind 7 kg of corn per hour. Since a man or woman working physically hard all day can deliver about 60 W, see note 3 to Chapter 1, this indicates that about 17 W is required to grind one kilogram per hour. And Lawton quotes Thomas Pennant's 1772 book *A Tour of Scotland, and Voyage to the Hebrides* which reports on the continuing use of rotary hand-mills, or querns, on the Isle of Rhum; Pennant states that 'this method of grinding is very tedious: for it employs two pairs of hand four hours to grind only a single bushel of corn'. This indicates that about 120 W is required to grind a quarter of a bushel per hour, and therefore that about 17 W is required to grind one kilogram per hour. Gimpel's and Pennant's values are about 20% higher than the 14 W calculated from the information given by Smeaton, however both of the higher values relate to the use of rotary hand-mills which one would expect to be less efficient than the higher volume milling that windmills and watermills made possible.[11]

Bennett and Elton state that when they were writing, in the late nineteenth century, the post mill's 'average length of sailyard from tip to tip' was 'from 50 to 60 feet';[12] or in today's terminology the post mill's average rotor diameter was about 17 m. Their examples indicate that Stockton post mill gave an output of about 2.3 kW in a 'fairly steady wind', whilst Little Dassett gave about 2.7 kW in what they describe as a 'good fair wind'. What they were unable to do was to quantify the wind speed with any greater precision than the descriptions given above, and this is a problem that was not fully resolved until well into the twentieth century. A major part of the difficulty stems from the fact that the speed of the wind is not constant; friction with the ground causes turbulence and the consequent fluctuations in wind speed can be reduced – but not eliminated – by choosing to make measurements over smooth terrain. It has been found that most of the wind speed fluctuations due to turbulence have a period of less than 10 minutes;[13] current practice in determining the performance of modern wind turbines is therefore to measure the power and the wind speed averaged over a period of 10 minutes, and hundreds of such pairs of measurements are taken in order to characterise how the wind turbine's power output varies with wind speed. However it is only in recent decades that the necessary instrumentation and data-logging equipment has been available.

Smeaton's 1759 measurements using model windmills

When, in the mid eighteenth century, Smeaton wanted to quantify the performance of windmills his most practicable option was to make and test windmill models; and the test rig he used is shown in Figure B.1. Model windmill sails were mounted at the end of an arm which was made to rotate about the central vertical axis by a steady pull on the cord. The dumb-bell pendulum VX was used as a metronome to help ensure that the model sails were moved through the air at constant speed, which was typically about 2 m/s. The size of the apparatus is indicated by the scale at the bottom, which is 8 feet (2.4 m) long. The movement of the air past the sails made them turn (anticlockwise as seen in the figure), and as they did so they raised an adjustable weight via the pulley system shown; the torque produced by the sails and their power output could therefore be quantified. Smeaton tested a variety of sail configurations and though most had conventional rectangular sails (cloth covered and 0.46 m long by 0.14 m wide) he also tested sails whose breadth was increased towards the tip – as is shown in the figure – by the addition

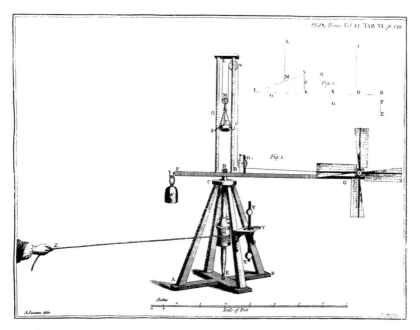

Figure B.1 Smeaton's windmill model testing apparatus. *Source*: Smeaton (1760).

of a triangular area attached to the leading edge of the rectangular sail.[14] In all cases the sail radius was 0.53 m, and with the rectangular sails that he used for most experiments the rotor solidity was 29%.

Early windmills, such as those shown in Figure 3.1, had sails which were effectively flat and untwisted, set at an angle of about 20° to the plane of rotation, see Figure 2.6. However experience had shown that there was benefit if this angle, which in traditional windmills is referred to as the 'weather', increased from just a few degrees at the tip to a substantially greater angle near the hub.[15] Smeaton experimented to find the most effective twist along the length of the sails, and concluded from his model tests that the best performance was given when the 'weather' increased from about $7\frac{1}{2}°$ at the tip to $22\frac{1}{2}°$ near the hub; his tests indicated that sails with this optimum twist would give about one third more power than sails that had no twist, and his results had a significant influence on windmills built subsequently. Smeaton's tests also led him to conclude, correctly, that for similar loading conditions the tip speed is proportional to the wind speed and that the power output is proportional to the cube of the wind speed.

In modern terminology the power output P from windmills is given – see Appendix A – by the expression $P = C_P(\frac{1}{2}\rho A V^3)$, where V is the wind speed, $A = \pi D^2/4$ is the area swept by the rotor whose diameter from sail tip to sail tip is D and ρ (Greek rho) is the air density (about 1.23 kg/m^3); C_P is the power coefficient which for large modern wind turbines will peak at $C_P \approx 0.45$. *Smeaton's results with model sails similar to those used in traditional windmills, i.e. rectangular and with a solidity of 29%, correspond to a maximum power coefficient $C_P \approx 0.23$ at a tip speed ratio of about 2.6.* His results are corrected for friction losses and this power coefficient therefore relates to the power produced by the sails and delivered to the windshaft. However Smeaton's test arrangement, as shown in Figure B.1, did not model the blockage caused by the body of the windmill. As the air approaches a windmill it is slowed down by the body of the mill, and for part of each revolution the sails must cross this region of reduced air speed; as there is less power in this slow-moving air this reduces the power output from the windmill. Smeaton's model test results can therefore be expected to be slightly optimistic.

Coulomb's 1781 measurements on working windmills

Though his windmill work has been almost forgotten the first performance measurements on full-size windmills were made barely twenty years after Smeaton by Charles Coulomb, the eminent French physicist and engineer.[16] He made his measurements near Lille, in Flanders, where he noted that there were a large number of windmills used to crush rapeseed and extract rapeseed oil. Coulomb comments that the construction of these oil-pressing windmills is the same as those used in Holland, where 'by trial and error they have achieved a high degree of perfection'. Their diameter, from sail tip to sail tip, is given as 76 'pieds', and the sails each have a width of just over 6 'pieds', of which 5 'pieds' is canvas covered and the remaining 1 'pied' is a light plank; the sails start at a distance of 6 'pieds' from the central windshaft. The pre-metric 'pied' equals 0.325 m so these dimensions correspond to a sail diameter of 24.7 m, a sail width of about 2 m and a solidity of 17%. The sails had a weather angle that was in the range 6° to 12° at their tips, increasing to 30° at their inboard ends.

In these oil-pressing mills the windshaft was pierced along its length by seven evenly spaced holes (perpendicular to the windshaft's axis of rotation) and through each of these there was a beam that was 1.14 m long; this was substantially longer than the windshaft diameter and where the two ends of each beam protruded from the windshaft they acted like cams to lift – and then let fall – up to seven stampers, twice every revolution. These stampers were iron-tipped wooden beams each about 6.8 m long and 0.3 m square. After some pre-processing the rapeseed was put in fabric bags, which were then placed between wooden spacer blocks in a sturdy wooden trough underneath the first two stampers. The spacer blocks included two wooden wedges; one – the pressing wedge – was inserted thick end up; the other – the releasing wedge – was inserted thick end down. The first stamper was used to hammer down on the pressing wedge, which would squeeze the oil out of the rape seed; a hole in the bottom of the trough allowed the oil to run out into a container. Then after about 50 blows the first stamper was disengaged and the second stamper engaged; this hammered down on the releasing wedge and after a few blows the bags of rapeseed – by now compressed to a cake – could be removed. The cake still contained some oil and was therefore put into a strong circular pot underneath one of the five remaining stampers, where it was pounded and broken up; it was then heated and given a second pressing. In light winds only one of these five pot stampers would be engaged, but in stronger winds the number used would be progressively increased. Oil-pressing windmills (and watermills) were widely used in the eighteenth century and their use continued throughout the nineteenth century; further details of the oil extraction process are given by Hills[17] and by Stokhuyzen.[18]

Coulomb recognised that the way these oil-pressing windmills worked allowed their power output to be accurately determined. He could observe how many stampers were in use, and could measure their weight and the height through which they were lifted, twice every revolution, by the cams on the windshaft. All he then needed to determine the windmill's power output was to measure the speed at which the sails were turning. However he then had the problem of measuring the wind speed, without the benefit of instrumentation which would not be developed until well into the nineteenth century. His solution was brilliantly simple. Two men were placed on small platforms 150 'pieds' apart in the upwind direction; the first released a very light feather, and the second measured the time it took to reach him. Repetition and averaging then gave an accurate measure of the average wind speed.

Measurements were made on an oil-pressing windmill in which the first two stampers (the pressing stamper and the releasing stamper) each weighed 500 'livres' and the remaining five pot stampers each weighed 1020 'livres'. The pre-metric 'livre' equals 0.49 kg, so this gives 245 kg for the first two stampers and 500 kg for the other five. The cam arrangement lifted each stamper 18 'pouces', and since 12 'pouces' = 1 'pied' this corresponds to a lift of 0.49 m. Coulomb states that the most favoured wind speed was 20 'pieds' per second (6.5 m/s), which with the sails fully spread was sufficient to operate the pressing stamper plus all five of the heavier pot stampers. (In stronger winds the sails would need to partially reefed.) With this load and at this wind speed the sails turned at 13 revolutions per minute (rpm), corresponding to a tip speed of 16.8 m/s; the tip speed ratio was therefore 2.6. Neglecting losses the power output was simply $(245 + 5 \times 500) \times 9.81 \times 0.49 \times 2 \times 13/60 = 5.69$ kW. Coulomb recognised that there were impact losses as each stamper was brought impulsively into motion by the cam hitting a peg near the top of each stamper, and calculated the magnitude of these; at 13 rpm they totalled 360 W. The total power available at the windmill's windshaft was therefore $5.69 + 0.36 = 6.05$ kW, which corresponds to a power coefficient $C_P = 0.075$.

Measurements were also made at two lower wind speeds. With the wind speed at 12 to 13 'pieds' per second (≈ 4.1 m/s) the windmill turned at 7 to 8 rpm, corresponding to a tip speed ratio of about 2.4. The power produced was enough to operate the pressing stamper plus two of the pot stampers, and so equalled 1.49 kW; adding to this the impact losses, which amounted to 31 W, gives a total for the power that was available at the windshaft of 1.52 kW, corresponding to a power coefficient $C_P = 0.077$. And when the wind speed was just 7 'pieds' per second (≈ 2.3 m/s) the windmill could operate just one pot stamper whilst turning at 3 rpm; the total power output was then 0.24 kW, corresponding to a power coefficient $C_P = 0.068$ at a tip speed ratio of 1.7. At this low wind speed when the load was removed the windmill turned at $5\frac{1}{2}$ rpm, corresponding to a tip speed ratio of 3.1.

Coulomb estimated the windshaft bearing losses by waiting for a calm day; he then set two of the sails horizontally and hung weights, first from one sail tip then from the other, to see what was needed to overcome friction in the bearings. He observed that a weight of 5 livres was required, but recognising that the losses on-load would be somewhat higher he used a weight of 6 livres in his calculations; in modern units his calculations gave a friction torque of 356 Nm. At 13 rpm – achieved under full load when the wind speed was 6.5 m/s – this gave friction losses of 485 W, equal to about 8% of the useful windshaft power output of 6.05 kW. Friction losses were more significant at lower wind speeds; for example when the wind speed was 4.1 m/s, turning the sails at 7 to 8 rpm and giving a

Table B.1 *Coulomb's windmill performance results*

	Wind speed m/s	Power kW	Power coefficient C_P	Tip speed ratio
Oil-pressing	2.3	No load	0	3.1
	2.3	0.24	0.068	1.7
	4.1	1.5	0.077	2.4
	6.5	6.1	0.075	2.6
Wheat grinding	5.9	5.8	0.098	2.5
	9.1	12.3	0.056	3.1

windshaft power output of 1.52 kW, friction losses would have been 280 W, representing about 15% of the power produced by the sails.

In addition to the many oil-pressing windmills around Lille there were a number of windmills used to grind grain. Coulomb noted that these had sails of similar design and length to those used on the oil mills, and made measurements on one. He noted that it would start to turn when the wind speed reached 10 to 12 'pieds' per second (≈ 3.6 m/s), and in a wind speed of 18 'pieds' per second (≈ 5.9 m/s) it would turn at about 11 or 12 rpm and grind between 800 and 900 'livres' of wheat per hour. With full sails and in a wind speed of 28 'pieds' per second (≈ 9.1 m/s) the windmill would turn at 22 rpm and grind up to 1800 'livres' per hour; however this wind speed was clearly about the maximum at which full sails could be used as Coulomb commented on how hot the flour was as it left the millstones, and observed that the miller had to 'refresh' the millstones from time to time by changing the type of grain that was ground.

The 800 to 900 'livres' per hour ground when the wind speed was about 5.9 m/s equates to 417 kg/hr. As noted previously 14 W is required to grind 1 kg/hr; 417 kg/hr therefore corresponds to a power to the millstones of 5.8 kW. Since Coulomb noted that the wheat-grinding windmills were the same size as the oil-pressing ones the corresponding power coefficient $C_P = 0.098$ at a tip speed ratio of 2.5. Similarly the 1800 'livres' per hour ground when the wind speed was about 9.1 m/s corresponds to a power output of 12.3 kW, which gives a power coefficient $C_P = 0.056$ at a tip speed ratio of 3.1.

Table B.1 lists Coulomb's results. For the oil-pressing windmill the two data sets at a tip speed ratio of about 2.5 indicate that the maximum power coefficient $C_P \approx 0.076$; and this is for the power available at the windshaft. The close agreement between the two data sets is probably fortuitous; the accuracy of measurement of the wind speed would – at best – have been plus or minus 5%, which would give plus or minus 15% uncertainty in the estimation of the power coefficient. For the wheat-grinding windmill the relatively high tip speed ratio of 3.1 when the wind speed is 9.1 m/s indicates that in this strong wind the millstones could not absorb as much power as the windmill could produce; the tip speed ratio consequently increases to a value close to its unloaded value, and the power coefficient would then be well below the maximum. The one other data set for the wheat-grinding windmill indicates that the maximum power coefficient $C_P \approx 0.098$ at a tip speed ratio of 2.5, and this is for the power delivered to the millstones. Losses in the bearings and gearing between the windshaft and the stones would probably have been about 15% (see note 7), which implies that $C_P \approx 0.11$ for the power available at the windshaft, though previous comments about the accuracy of the wind speed measurement – and its effect on the power coefficient – apply here also. However with the wheat-grinding windmill there is the additional uncertainty over the accuracy of the figure of 14 W per kilogram per hour of wheat ground that was used to estimate the power output. One can have greater confidence in the data from the oil-pressing windmill as its power output could be accurately calculated, and based on these measurements the following statement can be made.

Measurements made by Coulomb indicate that the performance of traditional wind-mills – as used in Flanders and Holland in the late eighteenth century – corresponds to a maximum power coefficient $C_P \approx 0.075$ (plus or minus about 0.015) at a tip speed ratio of about 2.5. These windmills had a diameter of about 25 m and a solidity of 17%, and in a wind speed of 6 m/s they would give a power output of about 5 kW.

Dutch measurements on windmills in the 1920s and 1930s

After Coulomb there was a lapse of over 140 years before measurement of the perform-ance of full-size traditional windmills was again seriously attempted. This followed the formation in The Netherlands in 1923 of De Hollandsche Molen (the Dutch Windmill Society), set up to campaign for the preservation of traditional windmills. One of its actions was to set up a Technical Committee to consider how the design of traditional windmills might be improved, and this led in due course to a series of measurements being made on traditional windmills, both before and after modification, mostly in the period 1926 to 1939. These are described at length in the book *Research Inspired by the Dutch Windmills*, written by the 'Prinsenmolen Committee' and published in 1958. They focussed their studies on Dutch drainage mills, and much of the work they report was done on one called the Prinsenmolen. The upper part of a drainage mill is virtually the same as the upper part of a smock mill or tower mill, see Figure 3.5, with the sails driving the windshaft which through two meshing gear wheels transfer the power output to the central vertical shaft. In a drainage mill this shaft continues down to ground level where a lantern pinion (or equivalent) on the vertical shaft meshes with a geared wheel fixed to a horizontal shaft, and at the other end of this horizontal shaft there is a large diameter scoop wheel. In essence the arrangement at the *bottom* of the drainage mill is the same as that shown for a watermill in Figure 2.7, except that the drive is *from* the vertical shaft via the gears and *to* the scoop wheel, which for drainage mills is housed in a close-fitting channel so that as it is turned by the power from the windmill it lifts the water from the level of the inlet channel to the higher level of the outlet channel. (The maximum lift is given by the radius of the wheel, but it is usually about half of this.)

The measurement programme commenced in 1926 but a succession of problems had to be overcome and it was only in 1936 and 1937 that measurements were successfully made on the Prinsenmolen. This large drainage mill stands on the bank of the river Rotte, near Rotterdam, and was originally built in 1648 to pump water from a polder into the Rotte. It has a sail diameter (from tip to tip) of 28.7 m; the four sails each have a width of about 2.8 m and the rotor solidity is 21%. The 'weather' angle on the sails varies from

Table B.2 *Prinsenmolen performance results*

	Wind speed m/s	Power kW	Power coefficient C_P	Tip speed ratio
1936	5.9	No Load	0	3.4
	7.6	17.9	0.102	2.4
	8.1	20.9	0.098	2.4
	9.3	24.6	0.075	2.6
	9.2	26.1	0.084	2.8
1937	6.4	10.0	0.094	2.6
	6.9	12.5	0.094	3.0

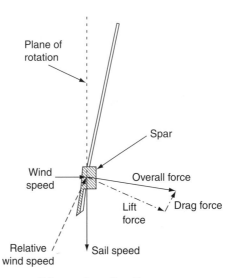

Figure B.2 Forces on a Prinsenmolen sail section.

near zero at their tips to about 20° near the hub, and Figure B.2 shows a typical sail cross-section; in front of the spar there is a leading board, while the sail area behind the spar is provided – as is usual – by canvas spread over a lattice wooden frame. Open water to the west and north of the Prinsenmolen favoured testing when the wind was blowing from these directions. Measurements made included the tangential force on the brakewheel teeth, from which the torque on the brakewheel and the windshaft power could be calculated. The volume of water pumped was determined using a measurement weir in the outlet water channel; knowing the height through which the water was lifted then allowed calculation of the water power output. The wind speed was measured at two heights, 10 m and 24 m, using cup anemometers.[19]

Six data sets are reported, four in 1936 and two in 1937, when all the instrumentation was operational and wind conditions were relatively uniform over each 30 minute test period.[20] In these tests the sail cloths were fully spread, and the results are summarised in Table B.2. The scatter in the C_P values corresponds to an uncertainty in measuring the average wind speed of only plus or minus 5%, and the Prinsenmolen test results give an average power coefficient of $C_P \approx 0.091$ at a tip speed ratio of about 2.6. This is in reasonable agreement with Coulomb's $C_P \approx 0.076$ at a tip speed ratio of about 2.5. In both cases the power output corresponds to the power available at the windshaft, and we can take the average of the Prinsenmolen and the Coulomb results, i.e. $C_P \approx 0.084$ at a tip speed ratio of about 2.6, as being reasonably representative of the performance of the larger traditional windmills that were in widespread use in the eighteenth and nineteenth centuries.

Pearce has calculated a power coefficient of $C_P = 0.10$ for the power produced by the sails of Wicken windmill, a 19 m diameter traditional corn mill built in 1813 near Ely, in Cambridgeshire.[21] Coulomb's measurements indicate that the friction losses in the windshaft bearings are about 10%, so the power coefficient for the power available from the windshaft would then be $C_P \approx 0.09$, in broad agreement with the value of $C_P \approx 0.084$ derived from the Coulomb and Prinsenmolen experiments. Wicken mill's optimum tip speed ratio is about 2.0; though this is lower than the optimum tip speed ratio of about 2.6 derived from the Coulomb and Prinsenmolen measurements this is consistent (see Appendix A) with the fact that Wicken mill has a substantially higher solidity,[22] i.e. 29% compared with 21% for the Prinsenmolen and 17% for the windmills that Coulomb tested.

Smeaton's anomalous results

The large discrepancy between the value of $C_P \approx 0.084$ obtained from measurements on traditional windmills and the value of $C_P \approx 0.23$ derived from Smeaton's careful model tests requires explanation. Smeaton's C_P value corresponds to the power *input* to the windshaft; since Coulomb's measurements indicate that the friction losses in the windshaft bearings are about 10%, the power coefficient for the power available *from* the windshaft is $C_P \approx 0.21$. Also, as noted earlier, Smeaton's test arrangement did not model the blockage produced by the tower of the mill. The air approaching any obstruction is slowed down as it gets close, and wind tunnel measurements of the airflow approaching a model of the Prinsenmolen's tower indicate that the speed of the air in front of the tower is reduced to about one half of the upstream wind speed.[23] The sails will produce little if any power as they pass through this region, which extends over an area equal to about 10% of the total rotor swept area. Allowing for this tower blockage effect would therefore reduce Smeaton's power coefficient by a further 10% to $C_P \approx 0.19$, but this is still more than double the $C_P \approx 0.084$ measured on full size windmills. The substantial difference that remains is almost certainly due to differences in the aerodynamic characteristics of the sails.

Figure B.2 shows a cross-section through one of the four sails of the Prinsenmolen at a distance of 9.5 m from the hub, i.e. about two thirds the distance from the centre to the sail tip. The *relative wind speed* arrow shows the direction of the airflow relative to the sail (see Figure 2.6(b) and the related description in Chapter 2) and the *overall force* produced by the airflow on this part of the sail is shown as the sum of a *lift force* (which by definition is perpendicular to the airflow) and a *drag force* (which by definition is parallel to the airflow). At this radial position the sail width – that is to say the distance from the leading edge (bottom) to the trailing edge (top) – is 2.75 m. Canvas sailcloth is spread (on the upwind side) over the lattice framework which extends from the spar to the trailing edge; and between the spar and the leading edge is a wooden board. The spar is approximately rectangular in shape and must be strong enough to withstand the loads on the sails even in gale force winds; it therefore gets progressively thicker towards the hub of the sails, and is thinner towards the tips.[24] In early windmills this spar was located near the middle of the sails, see Figure 3.1, in the reasonable expectation that the force either side would be about equal. However as sails got larger windmill builders learnt by experience what aerodynamicists only discovered very much later, which is that the overall force resulting from the airflow over the sail acts not through its middle, but through a point that is about one quarter the width of the sail behind the leading edge. Putting the spar near this quarter-width position eliminates the twisting forces that it would otherwise experience, and it is then much less likely to fail in strong winds. There is, however, a downside.

As aerodynamicists discovered when they systematically tested aircraft wing sections in wind tunnels in the early twentieth century (including sections approximating to flat plates, such as were used by some early aircraft) most of the *lift* results from reduced pressure on what for aircraft wings is their upper surface, but what for windmill sails is their downwind side (i.e. the right-hand side in Figure B.2); and this region of reduced pressure extends from the leading edge to about halfway along the section. Putting a large spar in the middle of this low pressure region can be expected to have a significant effect on the air flow, and on the section's lift and drag, and measurements made on behalf of the Prinsenmolen Committee confirm this. They arranged for models to be made of cross-sections of the Prinsenmolen's sails at several radial positions, one of which was two thirds the distance from the centre to the sail tip, corresponding to the section shown in Figure A2.2. These model sail sections were then wind tunnel tested. The results clearly show that the spar reduces the lift and increases the drag, and as the spar gets thicker towards the hub these adverse effects get larger.[25]

Smeaton gives no details of the thickness of the spars on his model sails, but if the proportions at the tip are similar to the Prinsenmolen's then the spar thickness would be about $3\frac{1}{2}$% of the sail width, i.e. about 5 mm. Though the Prinsenmolen's spars are tapered, and are several times thicker towards the hub, this would not be necessary for Smeaton's model sails. This is because for geometrically similar sail designs the maximum stress in the spar is proportional to the square of the operational wind speed, though independent of size.[26] The maximum wind speed in Smeaton's experiments was only 2.7 m/s, whereas the Prinsenmolen had to operate with sail cloths fully spread in gusts up to about 15 m/s. Smeaton's much lower wind speeds would allow him to use an un-tapered spar, and a spar thickness of 5 mm extending from hub to tip would have given quite adequate strength.

Wind tunnel tests of Prinsenmolen sail cross-sections reveal that with the thin spar that is used only near the tip of the Prinsenmolen's sails, but was probably similar to that used along the whole length of Smeaton's sails, the ratio of the lift force to the drag force is about 8 for angles of incidence[15] α in the range 4° to 8°, but reduces to 4 when α increases to 15° (comparable with measurements made on flat plates). However tests on the thicker section used in the Prinsenmolen at two thirds the distance from hub to tip, and shown in Figure B.2, indicate that for low angles of incidence, in the range 4° to 8°, the lift is reduced and the drag is increased with the result that the ratio of the lift to the drag is

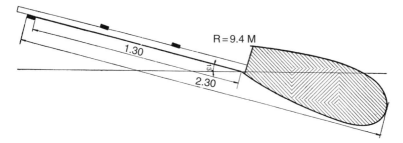

Figure B.3 Prinsenmolen 'streamlined' sail cross-section. *Source*: Prinsenmolen Committee (1958).

reduced from about 8 to about 3; as the spar gets thicker still towards the hub the ratio of lift to drag at these low angles of incidence gets even worse.

Figure B.2 shows that with a lift to drag ratio of about 3 the *Overall Force* has only a small component parallel to the sails' plane of rotation, and contributing to the windmill's power output. With a thinner spar giving a lift to drag ratio of 8 the *Overall Force* would have a much larger component in the plane of rotation and the power output would be substantially increased. The relatively high power coefficients measured by Smeaton with his model windmills can therefore be explained if, as seems probable, his sails had spars whose average thickness (compared with the sail width) was significantly smaller than on full size windmills, giving his model sails significantly better lift to drag ratios.[27]

Following the 1936 and 1937 measurement programmes on the unmodified Prinsenmolen windmill the Prinsenmolen Committee sought to reduce the drag caused by the tapered spars by covering them – and the leading edge boards – with aluminium sheet so as to give an enclosure whose rounded and partly streamlined shape resembled a truncated ellipse, as shown in Figure B.3. The width of this enclosure was near constant at about 0.9 m along the whole length of each sail, but its thickness reduced from about 0.45 m near the hub to about 0.2 m near the tip, following the tapering thickness of the enclosed spar; the traditional cloth covered lattice framework was retained to the rear of the spar. A further test programme was then carried out on the Prinsenmolen in 1939 and the results showed that this method of streamlining the spars almost doubled the windmill's power output in a given wind speed.[28] The maximum power coefficient was increased to $C_P \approx 0.16$ (for the power available from the windshaft), comparable with the $C_P \approx 0.19$ derived from Smeaton's model tests.

The power available from traditional windmills

As noted previously the measurements on traditional windmills made by Coulomb in Flanders in the late eighteenth century and on the Prinsenmolen in the Netherlands in the twentieth century gave very similar results and indicate that the peak performance of these windmills corresponds to a tip speed ratio of about 2.6, and to a power coefficient $C_P \approx 0.084$ based on the measured power available at the windshaft. In both cases the windmills had solidities of about 20% and their diameters were in the range 24 to 29 m. The

Prinsenmolen experiments indicated that losses from the windshaft through to the scoop wheel shaft were about 15%,[7] and losses from the windshaft to the millstones of a corn grinding windmill will be of similar magnitude. The power available at the *output* shaft of a traditional corn-grinding windmill – or a traditional drainage mill – consequently corresponds to a power coefficient equal to about 85% of 0.084, i.e. to $C_P \approx 0.07$.[29]

In summary we can conclude that the larger traditional windmills that were in widespread use in the eighteenth and nineteenth centuries typically had an output power coefficient $C_P \approx 0.07$ at a tip speed ratio of about 2.6.

For the smaller medieval windmills the maximum power coefficient would have been somewhat lower. Their smaller size (about half that of later windmills) meant that the power they produced was only about one quarter the power produced by later, larger, windmills. At this smaller size bearing and gearing losses would have been proportionately greater, a situation that would have been aggravated by the medieval windmill's more primitive bearing and gearing arrangements. And the medieval windmill's untwisted sails would also detract from its performance. The overall adverse effect on the performance cannot now be accurately determined, but is unlikely to have been greater than about one third. *We can therefore expect that medieval windmills would have had an output power coefficient $C_P \approx 0.05$; the optimum tip speed ratio would be little changed, at about 2.6.*

Average wind speeds over open countryside in southern and eastern England are typically between 5 and 6 m/s at heights in the range 10 to 20 m (above ground level),[30] which are typical of the hub heights for traditional windmills. Because the power in the wind is proportional to the cube of the wind speed the average power output P_{Av} corresponds to a wind speed that is higher than the average wind speed V_{Av}.[31] In fact, given the distribution[32] of wind speeds that is typical of many sites in Britain and elsewhere the *average* power output at a location where the wind speed *averages* 5 m/s actually corresponds to the windmill's power output when the wind speed is about 6 m/s, see Figure 3.7 and the related discussion in Chapter 3. For a medieval windmill with a diameter of about 13 m, as is discussed in Chapter 3, and with $C_P \approx 0.05$ this means that the average power output $P_{Av} \approx 0.8$ kW for a location where the average wind speed $V_{Av} \approx$ 5 m/s. Similarly the *average* power output at a location where the wind speed *averages* 6 m/s actually corresponds to the windmill's power output when the wind speed is just over 7 m/s; at such a location the medieval windmill would have an average power output $P_{Av} \approx 1.3$ kW. Windmills in the eighteenth and nineteenth centuries were both larger and more efficient ($C_P \approx 0.07$) and for a typical diameter of 26 m would give an average power output of about 5 kW at a location where the wind speed averaged 5 m/s, and an average power output[33] of about 8 kW where the wind speed averaged 6 m/s. The performance of typical early and late traditional windmills is summarised in Table B.3. The overall average power output of the several thousand medieval windmills would

Table B.3 *Power outputs from traditional windmills*

Sail diameter	$D \approx 13$ m; Medieval $C_P \approx 0.05$	$D \approx 26$ m; eighteenth/nineteenth centuries $C_P \approx 0.07$
Average power (for $V_{Av} \approx 5$ m/s)	0.8 kW	5 kW
Average power (for $V_{Av} \approx 6$ m/s)	1.3 kW	8 kW

have been intermediate between the values shown in the table, and would therefore have been about 1kW, which is equivalent to the muscle power output of about 15 to 20 men.[34]

An independent check on the accuracy of this estimated 1 kW average power output is provided by Langdon's analysis[35] of contemporary documents recording the quantity of grain taken as multure,[36] from which he calculated that the medieval windmill on average ground 1.9 quarters per day. Since 1 quarter = 8 bushels, and grinding 1 bushel requires 0.39 kWh, this corresponds to a daily energy output of 5.9 kWh, and therefore to an average power output of 0.74 kW over an 8 hour day. This is consistent with the estimation that a medieval windmill could give an average power output of 1 kW as there would not always have been sufficient grain to utilise the full available power output.[37] Though grain can be kept for more than a year, wholemeal flour does not keep so well (especially in hot weather), and the roughly constant monthly flour needs of the population served by a windmill meant that the volume of grain to be milled monthly was also approximately constant. Wind speeds are higher in the winter, with the result that about two-thirds of the energy available annually comes during the six winter months. A windmill that can meet local needs during the summer months would therefore have power to spare in the winter.

Comparison with drag-driven Persian windmills

It is interesting to compare the performance of traditional windmills, used so widely in Europe following their invention in or close to England in the twelfth century, with the performance of the older but very different vertical-axis windmills that were used in the Seistan region of eastern Iran (as is discussed in Chapter 2; see Figures 2.2 and 2.3). Wulff states that these could mill on average about 1 ton of grain in 24 hours, which corresponds to 42.4 kg/hr.[38] As noted earlier milling wheat in Western Europe requires about 14 W per kg/hr. In Seistan the millstones are not grooved[39] and the power require-ment may be somewhat different, but in the absence of any data specific to Seistan the figure of 14 W per kg/hr can be used to provide a rough estimate of the windmills' average power output. Milling 42.4 kg/hr would then indicate that the power produced was about 590 W, equivalent to the power output of about 10 men. Wulff also notes that the sails and the upper millstone both turned at 120 rpm (there is no gearing) and that the wind speed measured in the middle of the 'wind of 120 days' was 32 m/s. The radial distance from the centre of the rotor to the tip of the sails is typically 2.1 m (see Figure 2.3), which gives a tip speed of 26.8 m/s. The tip speed ratio is therefore about 0.84, and this value of slightly less than unity is what one would expect for a lightly loaded windmill driven round by drag. Wulff's illustrations show that the structure containing the windmills was about 6 m high and 6 m wide; given the power output, the wind speed and the relevant frontal area ($36 \, \text{m}^2$) allows the power coefficient to be calculated and its value is $C_P \approx 0.0008$. Though this is very much smaller than the power coefficient $C_P \approx 0.05$ of a medieval English windmill it must be recognised that when the 'wind of 120 days' is blowing at 32 m/s (which is Beaufort Force 11,[40] high enough in England to cause 'widespread damage') a higher efficiency could cause problems; in fact Harverson notes that in high winds the opening through which the wind blows on to the rotor would be partially blocked by a curtain of reeds.[41]

When Wulff visited Neh, in Seistan, in 1963 he saw a line of 50 windmills still working; when Harverson visited in 1977 they were all derelict. However Harverson did see similar windmills still working in Khorasan, about 400 km to the north, a region that also experiences the same seasonal 'wind of 120 days'. He noted[42] that the windmills there would grind an average of about 300 kg of wheat per day (and night), with a wind speed that was typically 25–30 m.p.h. (11–13 m/s) and with the runner stone turning at 25–30 rpm. The 12.5 kg/hr rate of grinding corresponds to a windmill power output of about 175 W – equivalent to the muscle power output of 3 men. The rotational speed gives a tip speed of about 6.1 m/s, and with the wind speed of about 12 m/s this gives a tip speed ratio of about 0.5, which is reasonable for a drag-driven windmill. The corresponding power coefficient is $C_P \approx 0.005$, smaller by a factor of ten than the power coefficient $C_P \approx 0.05$ of a typical medieval English windmill.

Appendix C
Wind characteristics

Wind speed variation with height

As is noted in Chapter 1 the winds we experience result from the fact that the Earth's equatorial regions receive more solar energy than the polar regions, and this differential heating gives rise to large-scale convection currents in the atmosphere.[1] Near the ground the speed of the wind is reduced by the Earth's surface roughness, and this so-called *planetary boundary layer* of slower-moving air extends up to a height of about 2000 m. The variation of wind speed with height within this boundary layer has been studied in some detail, and can be described mathematically in a variety of ways,[2] the simplest of which is an experimentally based power law, i.e. $(V_H/V_{10}) = (H/10)^\alpha$, where V_{10} is the wind speed at a height of 10 m (which is the usual reference height for meteorological measurements), V_H is the wind speed at height H (in metres) and the exponent α (alpha) depends on the surface roughness. Over the sea and surfaces of similarly low roughness $\alpha \approx 0.1$, over open countryside $\alpha \approx 0.14$, over countryside with scattered trees and hedges $\alpha \approx 0.24$; and over urban areas α may well exceed 0.3. Table C.1 shows how the magnitude of (V_H/V_{10}) varies with the height above the ground for the first three of these surface roughness categories; and it indicates, for example, that if a turbine sited in open countryside has a hub height of 60 m, then the wind speed at this height will be 1.29 times higher than the wind speed measured at the standard 10 m height. Since the power in the wind is proportional to the cube of the wind speed this means that the power in the wind at the 60 m hub height is 2.15 times greater than at the 10 m reference height.

Wind speed distribution

The annual energy output that can be expected from a wind turbine depends in part on how the turbine's power output varies with wind speed, see for example Figure 6.2; however one also needs to know how many hours per year the wind speed will be between 4 and 5 m/s, how many hours per year will it be between 5 and 6 m/s, etc. For this one needs to know the wind speed distribution, and for most sites in the UK (and many locations elsewhere) this can be approximated by the so-called Rayleigh distribution[3]. This tells us that the probability p that the wind speed is greater than (or equal to) a

Table C.1 *Wind speed variation with height*

Height H metres	V_H/V_{10}		
	$\alpha = 0.1$	$\alpha = 0.14$	$\alpha = 0.24$
10	1.0	1.0	1.0
20	1.07	1.10	1.18
40	1.15	1.21	1.39
60	1.20	1.29	1.54
80	1.23	1.34	1.65

specific value V_N is given by the expression $p = \exp[-(\pi/4)(V_N/V_A)^2]$ where V_A is the average wind speed. So if we want to know how many hours per year the wind speed will be above the 4 m/s speed at which (typically) a turbine starts to generate, for a location where the average wind speed at the turbine's hub height is 8 m/s, the Rayleigh distribution states that the probability is given by $p = \exp[-(\pi/4)(\,4/8)^2] = 0.822$; and as there are 8760 hours in a year this indicates that the wind speed will be above the starting wind speed for $0.822 \times 8760 = 7198$ hours per year. Most modern wind turbines are shut down when the wind speed exceeds 25 m/s (see the discussion in Chapter 6) and for the same average wind speed of 8 m/s we calculate that the probability is given by $p = \exp[-(\pi/4)(25/8)^2] = 0.00047$; the wind speed will therefore be above the shut-down speed for only $0.00047 \times 8760 = 4.1$ hours per year. These two calculations tell us that wind speeds will be within the wind turbine's normal operating range for 7194 hours per year, and since the availability[4] of modern wind turbines is typically 97% or higher we can therefore expect the turbine to be turning and generating electricity for about 6980 hours per year, i.e. for about 80% of the time.

For a location – typical for many traditional windmills – where the average wind speed at the hub height (that is the windshaft height) is 5 m/s the Rayleigh distribution indicates that winds exceeding 4 m/s will be experienced for a proportion of the year equal to $p = \exp[-(\pi/4)(4/5)^2] = 0.605$; this equates to 5300 hours per year. At the same location winds exceeding 5 m/s will be experienced for a proportion of the year equal to $p = \exp[-(\pi/4)(5/5)^2] = 0.456$, and this equates to 3995 hours per year; the difference $5300 - 3995 = 1305$ then gives the number of hours for which the wind speed is between 4 m/s and 5 m/s. Similar calculations for other wind speeds allow the number of hours in every 1 m/s wide band of wind speeds to be calculated, and the result is shown in the upper part of Figure 3.7.

More generally, when the annual average wind speed is equal to V_A, the upper part of Figure C.1 shows the number of hours that the wind speed will be within a succession of bands of wind speed, each of width $0.2V/V_A$; as might be expected wind speeds close to the average occur most frequently, and the wind speed only rarely exceeds three times its annual average value. The power in the wind approaching a turbine is given by $P = \frac{1}{2}\rho A V^3$ (see Appendix A) where $\rho \approx 1.23$ kg/m^3 is the air density and A is the rotor swept area. The energy in the wind for each narrow band of wind speeds is then given by multiplying the power in the wind at the central speed within each band by the

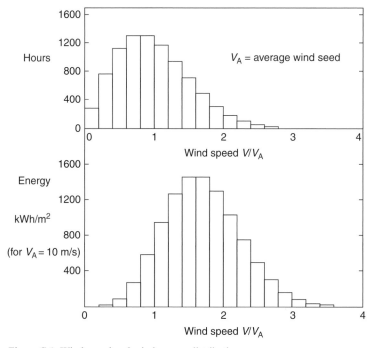

Figure C.1 Wind speed and wind energy distributions.

number of hours that the wind speed is within that band, and is shown in the lower part of Figure C.1. Note that the values in kilowatt-hours per square metre of rotor swept area given by the left-side scale correspond to a notional average wind speed of 10 m/s. Very few sites have wind speeds this high but the values shown can readily be scaled by the factor $(V_A/10)^3$ for other – more realistic – values of the average wind speed. As can be seen from the figure the most energetic wind speeds are about 60% higher than the average wind speed.

It was noted in Chapter 3 that because the power in the wind is proportional to the cube of the wind speed the average power is substantially higher than the power at the average wind speed (see Chapter 3 and the related discussion). Summing the energy in the wind for all the bands of wind speed shown in Figure C.1, and then dividing by the number of hours in the year, we find that the average power in the wind $P_{Av} = 1.91 \times \frac{1}{2}\rho A(V_A)^3$; in other words the average power is 91% higher than the power in the average wind speed.[5] It is worth noting that although wind speeds are below the annual average for 54% of the total time these light winds contain only 9% of the total energy; wind speeds greater than three times the annual average contain only $1\frac{10}{2}$% of the total energy, and the loss of energy that results from shutting down turbines when the wind speed exceeds 25 m/s is therefore very small.

Notes

Chapter 1 Wind power and our energy needs: an overview

1. Forbes (1958) p. 164 indicates that a Newcomen steam engine made in 1718 had a thermal efficiency of only 0.5%; however by 1792 the steam engines made by Boulton and Watt had an efficiency of 4.5%, almost ten times higher. By the 1830s continuing improvements had increased the steam engine's efficiency by a further factor of four, to about 17%. See also Smil (1994) pp. 161–4.
2. Whitt and Wilson (1982) pp. 42–52 give data on human power output over periods ranging from a few seconds to 200 minutes; the same reference, p. 177, gives data for power outputs over longer durations which are consistent with those measured in the eighteenth century both by Smeaton and Coulomb, see below.
3. John Smeaton (1724–92), the eminent eighteenth-century civil engineer, was one of the first to make systematic measurements of the power produced both by horses and by men, see Smith, N. (1981) p. 54. Smeaton found that a man pumping water for an 8 hour shift could average an output of 2400 ft.lb/min, which in today's units is just under 60 W. The French scientist Charles Coulomb (1736–1806) obtained a similar figure when he measured the output of a four man, pile-driving team, see Coulomb (1821) p. 309, and noted that this level of output could be sustained daily for many months. Weisbach and Du Bois (1886) p. 90 and Rankine (1866) p. 84 report similar figures for a man using a crank-handled windlass.

   Smeaton's measurements on horses led him to conclude, by 1765, that they could provide 22 000 ft.lb/min (497 W in today's units) over an 8 hour working day. When in 1783 James Watt wanted to specify the power output of his steam engines in terms of their horsepower equivalent he credited the horse with being capable of a sustained output of 33 000 ft.lb/min (746 W in today's units), but it seems he deliberately erred in favour of his steam engine customers as this power level corresponds to a strong horse over a fairly short period.
4. The UK Department for Business, Enterprise and Regulatory Reform indicates in its report on regional and local electricity consumption statistics, BERR (2008a), that in 2006 the 26.4 million domestic consumers in Great Britain used in total 117.8 billion kWh of electricity, an average of 4457 kWh per domestic consumer per year; this has been rounded here to 4500 kWh per year. An adult person working for eight hours daily and averaging 60 W could – with no holidays – provide 175 kWh per year, so the 4500 kWh average UK domestic electricity consumption is equivalent to the output that could be provided by about 25 domestic servants. In the United States electricity

consumption per household is about 11 000 kWh per year, see EIA (2001) and Gipe (1995) pp. 422–3; it is therefore equivalent to the output that could be provided by about 60 domestic servants.

5. Stern (2007) p. xv.
6. RCEP (2000) p. 18; Stern (2007) pp. 4–5.
7. And temperatures in the UK are rising more than twice as fast as the global average. Globally the ten warmest years averaged 0.42 °C above the 1961–90 average, whereas in the UK the ten warmest years averaged 0.91 °C above the 1971–2000 average; see www.metoffice.gov.uk/corporate/pressoffice/2007/pr20071213.html.
8. Stern (2007) pp. xiii, 65–70 and 104–5.
9. RCEP (2000) pp. 50–1.
10. DTI (2003) p. 11 and DTI (2007a) p. 8. The Climate Change Act became law in November 2008 with a legally binding commitment to reduce emissions of all greenhouse gases by 80% by 2050, see OPSI (2008).
11. DTI (2003), pp. 8–9.
12. Stern (2007) p. xv.
13. RCEP (2000) p. 9.
14. RCEP (2000) p. 19; Figure 2-II indicates that the global absorbed solar radiation averages 87 000 billion kW, and that people worldwide have a total primary energy consumption which averages 12 billion kW.
15. The vast majority of electricity consumers in the UK, as elsewhere in the developed world, are connected to the electricity distribution system (also known as the grid system) and buy their electricity for a price that is typically in the range 5 to 15 pence per kilowatt-hour (p/kWh), with high-volume industrial consumers paying a price near the low end of this range and low-volume domestic consumers paying a price near the high end. Present costs for electricity from PV systems are very much higher. The Cabinet Office Strategy Unit in its 2002 Energy Review noted that PV generation costs in the UK were about 70 p/ kWh, see Cabinet Office Strategy Unit (2002) p. 193; and the European Commission in its 2007 Renewable Energy Roadmap indicated that in the sunnier parts of Europe PV generation costs were about 65 eurocents/kWh, equal then to about 43 p/kWh, see EC (2007) p. 16. However in remote locations, where low-cost grid supplies are not available, people accept the need to pay a relatively high price for electricity for premium applications such as lighting, and PV systems are widely used: PV generation costs then compare very favourably with the approximately £100/kWh cost of electricity provided by batteries of the type used in portable radios or in torches.
16. There is no basic difference between a windmill and a wind turbine. Common usage is to refer to older machines for harnessing wind power – up to the early years of the twentieth century – as windmills, whether they were used for grinding grain, pumping water, sawing wood or any other use. More recent machines for harnessing wind power have mostly been designed to generate electricity, and are usually referred to as wind turbines. The turbines shown in Figure 1.3 were made by the Danish company Vestas, the world's largest manufacturer of wind turbines, who currently make and install about two thousand similar machines per year.
17. See Carter (2007). North Hoyle was Britain's first offshore wind farm but two wind turbines were installed 1 km off the Northumberland coast, near Blyth, in 2000; see BWEA (2003) p. 83.
18. See for example WMO (1981) pp. 5–9; also Hau (2006) pp. 451–9.
19. Smil (2003) p. 274; also Hau (2006) p. 452 and Eldridge (1980) p. 5.
20. The calculation assumes a Rayleigh distribution of wind speeds, see Appendix C.

21. Modern wind turbines are about 67% efficient in taking energy from the air flowing through their rotors, as is discussed in Appendix A. And though many people believe that more blades would give an increased power output the reality is that more blades would increase the cost but give no more power, as is discussed in Chapter 3 and in Appendix A. Hau (2006) p. 143 shows typical tapering blade shapes for several modern wind turbines; machines built in the past decade have usually had rotor solidities (the total blade area – as seen by an observer well upwind – divided by the area of the circle swept by the blades) in the range 4% to 5%.

22. A short energy payback period is a prime requirement for any energy conversion technology that is proposed for large-scale deployment. My first estimate in 1976 gave an energy payback period for megawatt-scale wind turbines of 13 months, see Musgrove (1976a). A more detailed assessment the following year by the Energy Technology Support Unit at Harwell for the Department of Energy indicated that the energy payback period was in the range 6 to 12 months, see Department of Energy (1977) p. 26. Most of the energy required for the construction of wind turbines goes into the steel that is used for the tower and within the nacelle and foundations, and this totals about 100 tonnes per megawatt of installed capacity. The energy required to produce steel is approximately 11 kWh/kg, see for example MacKay (2009) p. 325, so the total energy required for construction is about 1.1 million kWh. With a wind turbine capacity factor of 30% the energy output per megawatt of capacity is 2.6 million kWh per year, and this simplistic calculation indicates that the energy payback period is about 5 months. Recent and much more detailed assessments show that modern wind turbines typically take from 3 to 9 months to pay back – through their energy output – the energy consumed in their manufacture, installation, operation and eventual decommissioning; see Gipe (1995) pp. 420–2, Danish Energy Agency (1999) p. 20, Danish Energy Authority (2005) p. 21, Hau (2006) pp. 770–1 and the Danish Wind Industry Association at www.windpower. org/en/tour/env/enpaybk.htm.

23. The 2002 Energy Review which indicated that electricity from PV systems cost about 70 p/kWh (see note 15) also stated that the cost of electricity from wind turbines on good on-land sites was in the range 2.5 to 3 p/kWh, with the cost of offshore wind generation in the range 5 to 6 p/kWh, see Cabinet Office Strategy Unit (2002) p. 193. Likewise the European Commission, which in its 2007 Renewable Energy Roadmap gave a central cost of 43 p/kWh for PV generation, also indicated that the corresponding central cost for on-land wind was 4.3 p/kWh, see EC (2007) p. 16, with costs ranging from 2.5 to 5.7 p/kWh; for offshore wind the corresponding cost range was from 3.5 to 8 p/kWh. The House of Commons Environmental Audit Committee in its 2006 report on nuclear power, renewables and climate change quoted costs provided to it by several major power companies and for on-land wind these ranged from 3.1 to 5.4 p/kWh, with offshore wind costs in the range 5.5 to 8.4 p/kWh, see House of Commons (2006) p. 43. And the Sustainable Development Commission in its comprehensive 2005 report on Wind Power in the UK indicated that on-land wind cost around 3.2 p/kWh, with offshore wind costing around 5.5 p/kWh, see SDC (2005) pp. ii & 29–34.

   All energy project costs increased substantially in the period 2003 to 2008 – due in part to increased commodity prices – and offshore wind project costs more than doubled, see for example Carbon Trust (2008a) p. 39. Recent energy cost estimates have reflected these project cost increases, and the Carbon Trust's October 2008 report calculated that offshore wind would then cost close to 10 p/kWh, though it anticipates that by 2020 this will reduce to less than 6 p/kWh, ibid. pp. 6, 49 and 106–8. The global economic slowdown that developed through 2008 has already led to sharp falls in commodity prices, and wind energy project costs are consequently expected to decline from their recent relatively high levels.

24. The Sustainable Development Commission, SDC (2005) pp. 13–16, stated that the UK practical onshore wind energy resource is 50 TWh/yr, where one terawatt-hour (TWh) is equal to one billion kWh; it also stated that the UK practical offshore wind energy resource is 100 TWh/yr. And to put both these figures in perspective the total UK electricity consumption in 2007 was 351 TWh, see BERR (2008b) p. 25. The offshore wind energy resource estimate derived from a fact sheet produced several years earlier by the Department of Trade and Industry (DTI) which limited its assessment to areas that were relatively close to shore. A subsequent and more detailed assessment of the UK offshore wind resource was provided in the 2002 report *Future Offshore*, see DTI (2002) pp. 23–8; this indicated that within territorial waters (i.e. up to 22 km from the coast) and at water depths less than 30 m the resource is 697 TWh/yr. Including the UK share of the continental shelf but with the same 30 m water depth limit increased the estimated offshore wind resource to 1144 TWh/yr; and when the limiting water depth was extended to 50 m the overall offshore wind resource further increased to 3213 TWh/yr, nine times larger than the total UK electricity consumption. The use of waters deeper than 50 m would give a further large increase in the UK offshore wind resource, and wind turbines with floating support structures that are tethered to the seabed are already being developed.

   EWEA (2004) pp. 43–4 gives data which show that the UK wind resource, both onshore and offshore, is larger than that of any other country in Europe.

25. Oxera (2005) pp. 21–2, tables 3.4 and 3.5. This report was commissioned for the Energy Review, see DTI (2006) p. 197.

26. In its Renewable Energy Strategy consultation paper the Government suggested that wind power is likely to provide about 68% of the renewable electricity required to meet the UK's 2020 renewable energy target; see BERR (2008c) pp. 33–5. The Carbon Trust forecast that by 2020 wind power would provide about 80% of the renewable electricity generated in the UK; see Carbon Trust (2008a) p. 14.

27. The system integration studies reviewed by the UK Energy Research Centre in 2006 indicate that wind power's variability reduces fuel savings by between zero and 7% as compared with the theoretical maximum fuel savings, see UKERC (2006) p. 42. This is consistent with system integration studies that were undertaken in the late 1970s and early 1980s by the Central Electricity Generating Board (CEGB), which was then responsible for electricity generation and supply throughout England and Wales. Their studies had shown that if wind power was completely unpredictable the fuel savings would be 15% smaller than if it was completely predictable; making use of meteorological office forecasts gave fuel savings that were only about 5% smaller than when the wind power contribution was completely predictable, see Chapter 7 notes 20 & 21.

28. The Carbon Trust, in its October 2008 assessment of the contribution that wind power (and in particular offshore wind power) could make to UK electricity needs by 2020, concluded that the electricity system could by then accommodate 40 000 MW of wind power capacity (with three-quarters offshore), and this capacity would then supply 31% of UK electricity consumption; managing the consequences of wind power's variability would – it calculated – add about 0.9 p/kWh to the cost of wind-generated electricity. See Carbon Trust (2008a) pp. 26–8. Similarly the UK Energy Research Centre in its comprehensive report on 'The Costs and Impacts of Intermittency' indicated that for a wind energy contribution of about 20% to UK electricity supplies the consequences of wind's intermittency (or variability) would add about 0.5 to 0.8 p/kWh to the cost of wind-generated electricity; short-run

balancing costs would account for about 0.2 to 0.3 p/kWh of this total, while maintaining system reliability would add about 0.3 to 0.5 p/kWh. See UKERC (2006) p. vii and Gross *et al.* (2007).

Adding wind power capacity to an electricity power system does displace some conventional generation capacity, and this adds to the value of wind on the system. However wind power's capacity credit is substantially less than the total installed wind power capacity. For example in the scenario proposed by the Carbon Trust the 40 000 MW of wind power capacity installed by 2020 would have a capacity credit of 6000 MW; i.e. with 40 000 MW of wind power capacity installed the fossil-fuelled power capacity that is also required is 6 000 MW smaller than in a scenario without wind power, see Carbon Trust (2008a) p. 27.

29. Cabinet Office Strategy Unit (2002) pp. 193–9.
30. DTI (2006) pp. 182–9; see also the critique of both the wind and nuclear generation costs in Massy (2006) p. 50.
31. House of Commons (2006) pp. 43–6; see also pp. 29–30.
32. House of Commons (1990) pp. xiv, xv and xxiii gave costs in the range 6 to 8 p/kWh for Pressurised Water Reactors including Sizewell B, based on information provided by the electricity supply industry in the period leading up to its privatisation in 1990; see also House of Commons (2006) p. 29, which indicated a generation cost for Sizewell B in excess of 6 p/kWh. Some of the uncertainties with respect to new nuclear construction costs are discussed in ibid. pp. 29–30 and SDC (2006) pp. 9–10. Concerns relating to safety, radioactive waste disposal, decommissioning and nuclear proliferation are reviewed in House of Commons (2006) pp. 39–42 and SDC (2006) pp. 7–8 and 14–17.
33. See BP (2008) p. 16. A tabulation of oil prices since 1861 both in 'money of the day' and at 2007 price levels is given at www.bp.com/liveassets/bp_internet/globalbp/ globalbp_uk_english/reports_and_publications/statistical_energy_review_2008/ STAGING/local_assets/downloads/spreadsheets/ statistical_review_full_report_workbook_2008.xls.
34. DTI (2000) pp. 18–19. The $10/barrel and $20/barrel oil prices quoted from this report correspond to 1999 price levels; using the relevant BP price adjustment factor (see note 33) these become $13/barrel and $25/barrel at 2007 price levels.
35. BERR (2008c) p. 22, and EIA (2008) p. 2.
36. At the January 2009 exchange rate of £1 = $1.45 an oil price of $70/barrel equates to £48.3/barrel. As is noted in BP's Statistical Review of World Energy, BP (2008) p. 44, 1 tonne of oil equals 7.33 barrels and 1 tonne of oil or oil equivalent produces about 4400 kWh of electricity in a modern power station. The $70/barrel oil price therefore corresponds to an electricity price (fuel only) of 8.0 p/kWh.

The BP figure of 4400 kWh of electricity from one tonne of oil corresponds to an average power station efficiency of 38%. In the UK gas is the principal fuel for electricity generation, see BERR (2008b) p. 24, and modern gas-fired power stations have an efficiency that is close to 50%. Gas is priced in pence per therm, where 1 therm = 29.3 kWh, and the energy content of a barrel of oil is about 1590 kWh; so based solely on the energy content an oil price of $70/barrel corresponds (with £1 = $1.45) to a gas price of 89 p/therm. In practice oil, as a more versatile liquid fuel, commands a price premium; allowing for this a gas price of about 80 p/therm would correspond to a sustained $70/barrel oil price (given no change to the exchange rate). With this gas price the cost of electricity from a modern combined cycle gas turbine (CCGT) power station would be about 7 p/ kWh, see Ilex (2006) p. 48, plus whatever cost is added by Government to internalise – at least in part – the environmental consequences of the greenhouse gas emissions that result from the combustion of the fuel.

37. *Windpower Monthly* vol. 24 no. 4, April 2008, p. 82. The top five countries were Germany with 22 200 MW, the US with 17 000 MW, Spain with 15 100 MW, India with 7800 MW and China with 5900 MW; China more than doubled its installed wind capacity in 2007. See also Chapter 8 Note 4 and Figure 8.2.

38. As released by the Global Wind Energy Council (GWEC) in February 2009. Their figures indicate that in 2008 the US installed 8300 MW, followed by China with 6300 MW, India with 1800 MW, Germany with 1670 MW, Spain with 1610 MW, Italy with 1010 MW, France with 950 MW, the UK with 840 MW, Portugal with 710 MW and Canada with 520 MW; the global total capacity installed in 2008 was then 27 100 MW. The GWEC figures for the total installed capacity at the end of 2008 were as follows: US 25 200 MW, Germany 23 900 MW, Spain 16 800 MW, China 12 200 MW, India 9650 MW, Italy 3740 MW, France 3400 MW, UK 3240 MW, Denmark 3180 MW and Portugal 2860 MW; the global total installed capacity at the end of 2008 was then 120 800 MW. www.gwec.net/fileadmin/documents/PressReleases/PR_stats_annex_table_2nd_feb_final_final.pdf.

39. DUKES (2008) pp. 175–6.

40. BERR (2008c) pp. 3–11.

41. Carbon Trust (2008a) pp. 2–4, 26–7 and 72–6.

42. USDOE (2008) pp. 7–12, 17–18, 24 and 149; see also the US overview in Chapter 8.

Chapter 2 The first windmills

1. Landels (1978) pp. 16–26.
2. Landels (1978) pp. 26–7; see also Needham (1965) pp. 555–6 and the discussion by Shepherd (1994) pp. 4–6.
3. Schmidt (1899) pp. 203–7.
4. Flanders is a region and former principality bordering on the North Sea; now divided between France, Belgium and the Netherlands.
5. Needham (1965) pp. 556–8.
6. See Forbes (1956) pp. 615–6, Wulff (1966) p. 285, Needham (1965) pp. 557–8, Harverson (1991) pp. 11–13 and Shepherd (1994) p. 8.
7. Further details of these early Persian windmills are given by Hills (1994) pp. 11–14, Shepherd (1994) pp. 7–10, Wulff (1966) pp. 284–9, Harverson (1991) and Harverson (2000) pp. 22–8.
8. Bennett and Elton (1899) pp. 326–327. Hills (1994) pp. 16–22 gives details of a number of European drag-driven vertical-axis windmills, the oldest going back to the fifteenth century; most noteworthy were the three large vertical-axis windmills built in the late eighteenth century by Stephen Hooper at Sheerness, Margate and Battersea. Bennett and Elton's comments relate to the performance of windmills such as Hooper's, as compared with 'ordinary' windmills of similar size.
9. The data given by Harverson (1991) for the performance of the Persian vertical-axis windmill allows its power output and efficiency to be calculated, see the concluding paragraph of Appendix B. In a wind speed of about 12 m/s the power output was about 175 W, equivalent to the muscle-power output of about 3 men; the Persian windmill's efficiency was therefore smaller by a factor of about 10 than the efficiency of the traditional English windmill.
10. Vowles (1930) p. 8; see also Hills (1994) pp. 283–4.
11. King (1926) pp. 293–5; see also Golding (1955) Plate 1, and Needham (1965) pp. 558–9.

12. Needham (1965) pp. 559–61.
13. The Seistan – like the Sirocco, the Chinook and a few more – is a named wind; for meteorologists the Seistan is noted as a very strong northerly wind which lasts for about four months and can reach up to hurricane strength.
14. The earliest known illustration of a windmill is in the so-called Windmill Psalter, dating from about 1270 and believed to have been written in Canterbury; it is now in the Pierpont Morgan library in New York; see http://utu.morganlibrary.org/medren/single_image2.cfm?imagename=m102.002ra.jpg&page=ICA000004379.
15. There is an analogy here with a sailing dinghy when the wind is coming from the side; as the boat speed increases the relative wind direction (as seen from the boat) moves forward towards the bow, and a small flag on top of the mast is used to show this relative wind direction.
16. For aerofoils the lift, by definition, is the force perpendicular to the relative wind, and the drag is the force parallel to the relative wind. Good aerofoil sections, as used on aeroplanes and also on modern wind turbines, are designed so that the drag is much smaller than the lift, and the ratio of lift to drag is commonly well in excess of 20 to 1. At these high lift to drag ratios the overall aerodynamic force is just two or three degrees away from being perpendicular to the relative wind. Even simple flat plate sections can give ratios of lift to drag in excess of 5, see for example Riegels (1961) p. 237; at a lift to drag ratio of this magnitude the overall aerodynamic force is about 10 degrees away from being perpendicular to the relative wind.
17. Just as a sailing dinghy is pulled forward by aerodynamic lift when it is sailing with the wind coming from the side.
18. Bennett and Elton (1899) pp. 224–34.
19. Compiled for William the Conqueror in 1086 the Domesday Book provided a record throughout most of England of land ownership and other information relevant to the collection of taxes; it records manor by manor all items which contributed to a lord's income. It does not include the counties of Northumberland, Cumberland, Durham or Westmorland as they were outside the jurisdiction of the Crown, nor does it include the principal cities of London and Winchester. Sir Henry Darby and his colleagues have counted and mapped all the entries relating to mills, see Darby (1977) p. 361.
20. Bennett and Elton (1899) p. 229; see also White (1962) p. 87 who confirms that the abbey at Savigny was not founded until some years after 1105.
21. See for example *The Chronicle of Jocelin of Brakelond*, Butler (1949) pp. 59–60. Jocelin of Brakelond was a monk in Bury St. Edmunds whose chronicle was written soon after the year 1200. The story of Dean Herbert's windmill was retold and popularised by the Victorian writer Thomas Carlyle in his book *Past and Present*.
22. Holt (1988) p. 21.
23. Tithes were originally an obligation to pay a proportion (often one tenth) of the annual produce of agriculture (e.g. crops, wool, milk, young animals etc.) to support the local church and clergy. The 'profits of man's labour', most notably fishing and milling, were also tithed. Over time money payments were increasingly substituted for payments in kind, and the right to receive tithes could be bought and sold. Tithes were only finally abolished in England in 1936.
24. Hills (1994) p. 37; see also Brunnarius (1979) p. 1.
25. Holt (1988) p. 21; see also Bauters (1982).
26. White (1962) pp. 86–7.
27. Holt (1988) pp. 33–4.
28. In medieval times a manor was a parcel of lands granted to an individual, frequently including one or more villages together with neighbouring cottages and farmsteads and the land related to them. A manor carried with it not only land ownership and

rights in commons, woods and open fields, but also judicial and administrative authority.

29. The rate of multure, that is the fee for grinding, varied significantly in different parts of the country. It was based on volume and for the customary tenants who were obliged to use the manorial mill the rate was typically about one part in twenty; in some locations it could be as low as one part in thirty-two, and in the northern counties of Durham and Northumberland it was as high as one part in thirteen. For free tenants, who were not obliged to use the manorial mill, the rate of multure was lower; in Durham this lower rate was one part in twenty-four for free tenants, almost half the rate levied on the customary tenants. See Holt (1988) pp. 49 et seq. for a more detailed account.

30. Holt (1988) pp. 86 and 176–7.

31. The first Crusade had begun in 1096 and had led to the capture of Jerusalem in 1099, about 60 years before the first reliable documentary evidence of the windmill being used in England.

32. The distance from the eastern Mediterranean coast to the Seistan region of eastern Persia is about 2500 km, comparable with the 3500 km distance from England to the eastern Mediterranean. Harverson (2000) pp. 22 & 35–6 notes that Rex Wailes, one of the leading post-war experts on traditional windmills, also believed 'in the likelihood of the oral transmission of the general idea of the windmill from East to West, where it would be interpreted in vertical form'.

33. See Landels (1978) pp. 18–26 for further details of Vitruvius and Roman watermills. The watermill described by Vitruvius had what today we would call an undershot waterwheel; overshot waterwheels are more efficient but require a more costly installation.

34. Wailes (1954) p. 150.

35. The inventor of the post mill may also have been inspired by familiarity with the Norse mill (also known as a Greek mill) as well as with the undershot (or Vitruvian) watermill. The Norse mill is a primitive vertical-axis waterwheel used by peasant communities across Europe (and further east) since Roman times, see Reynolds (1970) pp. 57–63, Walton (1974) pp. 115–26 and Forbes (1956) pp. 589–95. It has no gearing, and the upper end of the vertical shaft directly drives one of the pair of millstones. At the lower end of the vertical shaft there are a number of radial, flat wooden planks, set at an angle to the shaft so that when a stream of water is allowed to fall onto them the shaft is driven round. Though a Norse mill is less efficient than an undershot (Vitruvian) watermill it is much less costly, and as Holt (1988) p. 120 points out it flourished everywhere in Europe that peasants were legally able to operate their own mills.

36. Russian windmills have continued to use flat wooden sails until recent times, see Wailes (1957) pp. 93 and 98. Hills (1994) pp. 83–4 notes the continuing use of removable wooden boards in Sweden and Finland as well as Russia, and suggests that this is because they are easier to handle in freezing conditions.

Chapter 3 Seven centuries of service

1. Langdon (2004) p. 252.

2. As noted in Chapter 2 German soldiers on the Third Crusade built a windmill in Syria in 1190. It is therefore almost certain that windmills were being used in Germany at least three decades earlier than the first recorded German windmill, which was built on the Cologne town walls in 1222. Other first recorded dates should be treated with

similar caution; the first use of the windmill in a given area could well be decades earlier than the first recorded date.

3. Hills (1994) pp. 40 and 53; also Forbes (1956) p. 620.
4. The Low Countries is the region of north-west Europe which today comprises most of Belgium, the Netherlands and Luxembourg.
5. Dyer (1989) p. 158 shows that for most people bread accounted for almost three-quarters of the calorie intake, and on p. 153 indicates that the average annual cereal consumption was about 12 bushels of wheat and barley per person; Langdon (2004) p. 153 uses the same figure. Braudel (2002) p. 121 indicates a similar figure of four hectolitres (equivalent to 11 bushels) per person per year even as late as the eighteenth century. As Braudel (2002) p. 133 points out 'the triumph of bread arose because grain … was the least expensive foodstuff in relation to its calorific content. In about 1780 it cost eleven times less than meat, nine times less than freshwater fish, six times less than eggs, three times less than butter and oil.'
6. The bushel is a traditional measure of volume, equal to 4 pecks, or 8 gallons, or 36.4 litres (though prior to 1826 there were small regional variations, see Appendix B note 5). A bushel of wheat weighs about 28.6 kg, a bushel of barley about 25.5 kg and a bushel of oats about 19.0 kg, see Brunnarius (1979) p. 32. Other units of volume that are widely used in publications relating to traditional windmills are the sack, equal to 4 bushels, and the quarter which is equal to 8 bushels.
7. Gimpel (1979) p. 21 states that using a hand mill one can grind about $3\frac{1}{2}$ kg of grain per hour, a figure confirmed by Lawton (see Appendix B note 10) who indicates that in the late eighteenth century in the Hebrides – where hand mills were still in use – it took 8 man-hours to grind a bushel.
8. Langdon (2004) p. 120.
9. See for example Golding (1955) p. 68, Halliday (1990) p. 44 or Meteorological Office (1976). See also www.bwea.com/noabl/index.html, which gives a map showing average wind speeds across the UK. The Department for Business, Enterprise and Regulatory Reform (formerly the Department of Trade and Industry) website www.berr.gov.uk/energy/sources/renewables/explained/wind/windspeed-database/page27326.html gives access to a wind speed database from which the wind speeds at various heights above ground level can be determined for any UK post code.
10. A man working hard all day can average an output of just under 60 W and the average daily output from a horse is about 500 W, see Chapter 1 note 3. However if a horse mill is used to grind grain there will be power losses in the gearing and bearings. Comparison with the similar losses in windmills (discussed in Appendix B) suggests that horse mill losses are likely to be in the range 15% to 20%.
11. Langdon (2004) p. 179.
12. Langdon (2004) p. 141 estimates that mills on average were worked for 240 days per year after excluding Sundays and major feast days and making allowance for repairs.
13. Langdon (2004) pp. 145–6 reviews recent estimates which range from 4 million to 7 million. He also points out that not all the grain consumed would have been ground; some, such as oats and barley, would instead have been boiled and consumed as porridge. Ten thousand mills would then have given sufficient milling capacity for a somewhat larger population.
14. The Beaufort scale is named after Francis Beaufort, 1774–1857, who was appointed Hydrographer to the Navy in 1829. His original scale of wind force was devised in 1805, though it was based on a 12-point scale that had been used for over a century and which had been used by Daniel Defoe in his book about the great storm of 1703. Beaufort's scale was officially adopted by the Navy in 1838, though it has been

modified and revised on a number of occasions since. The current specification is given in Table 3.1; see also www.metoffice.gov.uk/weather/marine/guide/beaufortscale.html.

15. Hills (1994) p. 8.
16. Langdon (2004) p. 109.
17. Watts (2006) p. 28 indicates that evidence for the use of brakes goes back to the fourteenth century in Flanders, though they were probably used – in Flanders and elsewhere – well before then. See also Langdon (2004) p. 123.
18. See Appendix B; also Drees (1976) and Shepherd (1994) pp. 26–7. Data collated by Drees indicates that the change in the position of the main spar took place between about 1550 and 1650, a period in which windmill diameters approximately doubled (from about 13 m to about 25 m).
19. Smeaton (1760) p. 145; his remarks on the weathering of English windmills with 'Dutch' sails are on p. 163.
20. Coulomb (1821) p. 302. Charles Coulomb (1736–1806) is best known for his experimental investigation of the force acting between electrically charged objects and the electrical unit of charge, the Coulomb, is named after him. Details of the windmills on which Coulomb made performance measurements are given in Appendix B, see also note 25 below.
21. See Appendix B, note 15.
22. The sails on the Prinsenmolen windmill, whose performance is reviewed in Appendix B, have a moment of inertia about their axis of rotation that is approximately equal to 250 000 m^2kg, see Prinsenmolen Committee (1958) p. 81. The energy stored in this rotor divided by its power output, which gives a measure of how long it would take to slow down if the wind suddenly stopped, is in the range 10 s to 20 s for wind speeds in the range 10 m/s to 5 m/s. Its optimum tip speed ratio is 2.6 and its no-load tip speed ratio is 3.4, see Table B.2 in Appendix B. A sudden lull in the wind from, for example, 9 m/s to 6 m/s would briefly raise the tip speed ratio from 2.6 to 3.9 before the rotor had time to slow down; at this tip speed ratio the rotor would, briefly, act like a propeller.
23. See Langdon (2004) p. 111; also Yorke (2006) pp. 81–4.
24. Langdon (2004) p. 162 indicates that in 1300 a single millstone made from millstone grit cost about 6 shillings (£0.30). The more favoured French or German stones each cost about 30 shillings (£1.50).
25. Coulomb reported his windmill performance measurements in 1781 to the French Academie Royale des Sciences, see Coulomb (1785); the windmill performance measurements made by Dutch engineers in the 1930s are reported in the Prinsenmolen Committee's book *Research inspired by Dutch Windmills*. Both sets of measurements are reviewed and discussed in Appendix B, and both sets of measurements indicated that the optimum tip speed ratio was about $2\frac{1}{2}$. The Dutch and Flemish windmills on which the measurements were made had solidities of about 20%, where the solidity – see Appendix A – is defined as the proportion of the rotor swept area that is occupied by the blades. If the windmill's solidity is higher than about 20% the optimum tip speed ratio will be somewhat lower than $2\frac{1}{2}$ and the sails will turn more slowly; Pearce (see Appendix B) notes that for Wicken windmill, which has a solidity of about 30%, the optimum tip speed ratio is about 2.
26. Smeaton (1938) p. 3.
27. Bennett and Elton (1899) p. 266.
28. See for example Beedell (1975) pp. 34 and 48. Some later post mills, including Stanton mill, in Norfolk, even had three pairs of stones, see Watts (2006) p. 9.
29. Smith, D. (1981) p. 65.

30. In Flanders and Holland, but not in England, iron was also widely used for the sail spars.

31. Smeaton (1837) vol.1, pp. 108–9 and vol. 2, plates X and XI, describes this windmill which was known as Chimney windmill. The cap and sails were removed in the 1920s but the tower remains in use as offices. Smeaton's 1782 drawing shows neither the sack hoist nor the centrifugal governor. These would have been added by John Farey, see Skempton (1981) pp. 242–5, in 1812 when preparing his article on windmills, with accompanying illustrations, for Rees' Cyclopaedia.

32. The arrangement described for supporting and locating the cap is typical for eighteenth-century smock mills and tower mills. For details of other arrangements see Hills (1994) pp. 68–72; see also Yorke (2006) p. 32.

33. See Hills (1994) pp. 53–5 and Coutant (2001) pp. 72–89.

34. Some tower mills, such as the one at Bembridge on the Isle of Wight, had the pinion which meshes with the gear ring on the outside of the lower curb turned via an endless chain which hung down from the cap and reached almost to the ground. When the cap had to be turned the miller would pull on this chain. In other tower and smock mills the cap-turning gearing could be operated from inside the mill.

35. See for example Blom (1999). Many twentieth-century Greek windmills, including the several thousand that were used in Crete on the Lasithi plain for pumping water, had triangular canvas sails similar to the jib-sails used by sailing ships. Sometimes referred to as Cretan windmills, they typically had 6 to 12 radial wooden spars supported by a bowsprit, with a jib-sail attached to each spar. As Blom indicates medieval Greek windmills had sails similar to those used in north-west Europe, with canvas spread over a wooden lattice support structure; jib-sail windmills seem to have been a nineteenth-century development. See also Calvert (1971) and Haldoupis (1990).

36. Beedell (1975) pp. 61–4 describes in more detail how the centrifugal governor was used to adjust the gap between the stones. See also Hills (1994) pp. 95–9, Yorke (2006) pp. 34–6 and Watts (2006) p. 36.

37. See Hills (1994) pp. 104–8; also Reynolds (1970) pp. 92–4, Beedell (1975) pp. 27–31 and Yorke (2006) pp. 27–9.

38. The density of air at standard temperature and pressure, i.e. at 0 °C and a pressure of 1013 millibars (= 101.3 kPa = 760 mmHg) is 1.293 kg/m^3. Its value is directly proportional to the pressure and inversely proportional to the temperature (measured in degrees kelvin, where 0 °C = 273.2 K); so at standard pressure and 15 °C the density of air is 1.225 kg/m^3. In the International Standard Atmosphere (which describes average conditions at mid-latitudes) the sea level temperature is 15 °C and the corresponding density is 1.225 kg/m^3; the density reduces to 1.17 kg/m^3 at an altitude of 500 m, and to 1.11 kg/m^3 at an altitude of 1000 m. At winter temperatures the slightly higher air density will give windmills a few per cent more power – at a given wind speed – than in the summer, but the effect is minor; and for traditional windmills in southern and eastern England the effect on the air density of the site altitude above sea level is even smaller.

39. For example if the size and design of the sails were such that a windmill's shaft power output was 1 kW in a wind speed of 1 m/s, then in a wind speed of 2 m/s it could give an output of 8 kW, and in a wind speed of 3 m/s it could give an output of 27 kW. Over a two-hour period in which the wind blew for the first hour at 1 m/s and for the second hour at 3 m/s the average wind speed would be 2 m/s; the average power output would however be (1 + 27)/2 = 14 kW, nearly double the 8 kW that corresponds to the average wind speed.

40. The figure assumes a Rayleigh distribution of wind speeds, which is reasonably representative for most UK sites, see Appendix C.

41. See for example Hills (1994) p. 8.
42. See Stokhuyzen (1962) p. 12; his statement is made in a Dutch context, but wind speeds in the Netherlands are similar to those in southern and eastern England, and the wind speed information given by him (p. 96) corresponds to an average wind speed of about 5.3 m/s. The windmill's calculated average power output of about 6 kW exceeds (by 20%) what is needed to mill the grain for 2000 people, but this is to be expected as there would not always be sufficient grain to utilise the full output from the windmill. Though grain can be stored for more than a year the ground wholemeal flour would not keep so well, especially in hot weather; Braudel (2002) p. 142, writes – in relation to France in the late eighteenth century – that 'flour hardly kept any time. It had to be ground almost daily.' The roughly constant monthly flour needs of the local population meant that the volume of grain to be milled monthly was also approximately constant. Wind speeds are higher in winter and about two-thirds of the energy available annually comes during the six winter months. A windmill that can meet local milling needs during the summer months would therefore have power to spare in the winter.
43. See Stokhuyzen (1962) p. 13; also Hills (1994) pp. 65 and 116.
44. See Stokhuyzen (1962) pp. 28–33.
45. Hills (1994) p. 122.
46. Smeaton (1938).
47. See notes 20 and 25.
48. Hills (1994) pp. 115–64.
49. The disease that caused what we now call the Black Death was most probably bubonic plague and its variant, pneumonic plague. It is spread by fleas from rats and is endemic in parts of eastern Asia. Between the twelfth and the seventeenth centuries there were more than a dozen outbreaks in England but the two most notable were the epidemics in 1348–9 and in 1665, the former being referred to as the Black Death and the latter as the Great Plague. The Black Death arrived by ship in Weymouth, on the south coast, in the summer of 1348 and soon spread over the whole country. In the 18 months that the plague was active in England nearly half the population died, see Cunliffe *et al.* (2004) pp. 104–5. See also Braudel (2002) pp. 82 et seq.
50. Forbes (1958) p. 163 states that by 1800 there were 496 Watt engines at work in Britain. 'One or two were rated at 40 hp, but the average capacity was only 15 to 16 hp', i.e. about 12 kW.
51. See Gregory (2005) p. 136; also Hills (1994) pp. 154 and 160 who indicates that in the late eighteenth century there were about 700 drainage mills in the Fens, plus about 140 in the Norfolk Broads area.
52. Hills (1994) p. 75.
53. See Hayes (2003) pp. 10–15, Armytage (1976) pp. 90–91 and Hills (1994) p. 79.
54. See Mosse (1967). As is discussed in Appendix B grinding 1 kg/hr of wheat requires approximately 14 W, so grinding 160 bushels per hour (equal to about 4600 kg/hr, see note 6) would require about 63 kW; this is consistent with the nominal power output from the steam engines, i.e. 100 hp = 74.6 kW.
55. See Daudet (1978); the relevant short story is *Old Cornille's Secret*.
56. UK wheat and flour imports averaged 0.13 million tons per year through the 1820s and 1830s. In the 1850s – following the repeal of the Corn Laws – imports averaged 1.0 million tons per year, rising further to 2.5 million tons per year in the 1870s and to 3.5 million tons per year in the 1880s, see Mathias (1969). Jones (2001) pp. 54–5 gives data which shows that in the period from 1875 to 1890 51% of wheat and flour

imports were from the United States, 14% were from Russia and 10% were from India. By the late 1880s nearly half the wheat imported from the US came as flour.

57. Wood (1995) pp. 58–9.
58. In fact from medieval times through to the late eighteenth century most people ate *maslin* bread, a coarse brown bread made from a mixture of unsieved wheat and rye flours. The germ contains fats which when exposed to the air through milling soon go rancid. Unsieved flour consequently does not keep well, see note 42. If the germ and bran are removed the white flour that remains keeps much longer.
59. The successor to both these guilds is The Worshipful Company of Bakers, whose website gives an interesting summary of their history, see www.bakers.co.uk/about-history.php4.
60. See Wailes (1954) pp. 140–4, Watts (1998) pp. 24–7 and Reynolds (1970) pp. 53–5.
61. Watts (1998) p. 8.
62. Described in some detail by Jones (2001); see also Watts (1998) pp. 7–10.
63. Jones (2001) p. 129 states that early roller-milling installations required 'an average of 2.5 h.p. per cwt. of wheat ground per hour'. A hundredweight (cwt.) equals 50.9 kg (112 pounds), so this corresponds to a power requirement of 37 W to grind 1 kg/hr of wheat; this power requirement does of course include all the auxiliary grading and purifying machinery. As noted in Appendix B low milling between stones requires about 14 W per kg of wheat per hour; this excludes the power needed for bolting though this was relatively small.
64. See Jones (2001) p. 119 et seq., also Watts (2004) p. 189. One prominent pioneer of nutrition who favoured wholemeal bread was Dr Thomas Allinson, who in 1892 bought a stone-grinding mill in London and subsequently started a bakery; wholemeal bread bearing his name is still widely available.
65. SPAB Mills Section, 37 Spital Square, London E1 6DY. See www.spab.org.uk/html/spab-mills/.
66. See Hills (1994) p. 77; also Golding (1955) p. 17. Wind speeds in Germany are generally lower than in England or the Netherlands, so German windmills of similar size would have had a smaller milling capacity than English or Dutch windmills. Golding, whose book on *The Generation of Electricity from Wind Power* was first published in 1955, comments that at that time there were still 400 corn-grinding windmills in use in the province of Brandenburg.
67. Stokhuyzen (1962) pp. 100–2.
68. See Hills (1994) pp. 225–6, Stokhuyzen (1962) p. 61 and Prinsenmolen Committee (1958) p. 29 and pp. 35–44; see also Appendix B, note 28.
69. Stokhuyzen (1962) p. 105.
70. Hills (1994) pp. 236–56.
71. Baker (1985) pp. 5–44.
72. As noted in Appendix A the solidity is defined as the proportion of the area swept by the windmill blades that is actually occupied by the blades, as seen by an observer standing upwind. Modern wind turbines have a solidity of about 5%; for traditional windmills the solidity was usually in the range 15% to 30%; for multi-bladed water-pumping windmills the solidity was typically about 70%. A high solidity leads to optimum performance at low rotational speeds, see Figure A.2, and to a high starting torque.
73. Webb (1931) pp. 334–5.
74. Gipe (2004) pp. 247–57 and Johnson (1985) pp. 279–87 give further practical details of windpumps. The Aermotor website shows clearly how a cylinder pump works. www.aermotorwindmill.com/Links/Education/Index.asp. Aermotor were one of the leading manufacturers of windpumps in the late nineteenth century and their windmills are still manufactured.

75. See Hills (1994) pp. 240–8, and Baker (1985) pp. 7–13; see also Gasch and Twele (2002) pp. 324–5.
76. Gipe (2004) pp. 261–4 gives data on the volumes of water required, both for domestic purposes and for livestock.
77. Perry's whirling-arm test rig was in essence very similar to Smeaton's, see Appendix B and Figure B.1, though Perry's was powered by a 60 kW (80 h.p.) steam engine and was large enough to test rotors that were 1.5 m (5 ft.) in diameter. Ball (1908) pp. 305–17 describes Perry's work in some detail and gives a scale drawing of his test rig; see also Hills (1994) pp. 252–5.
78. This corresponds to a power coefficient $C_P \approx 0.1$; see Appendix A, Figure A.2 and note 9.
79. See Chapter 1, note 3.
80. See Gipe (2004) p. 404.
81. See Johnson (1985) p. 3; also Eldridge (1980) p. 19, Kovarik *et al.* (1979) p. 11 and Righter (1995) p. 28.
82. Gipe (2004) p. 251.
83. Walton (1974) p. 155.

Chapter 4   Generating electricity: the experimental years, 1887 to 1973

1. Bowers (1982) p. 106.
2. Ibid. pp. 113–24; also Hills (1994) p. 265.
3. Bowers (1982) pp. 107–8 and 126–8.
4. Sir William Armstrong founded the Armstrong engineering company that was responsible for many innovations and became world renowned. His country home Cragside, in Northumberland, was in 1880 the first to be illuminated using incandescent lamps, with the power provided by a hydro-electric power station in the grounds. It is now owned by the National Trust and is open to the public.
5. Bowers (1982) pp. 135–40.
6. Thompson (1910) vol. 2 pp. 770–1.
7. Blyth (1894) p. 174.
8. Blyth (1888) pp. 363–4; also Blyth (1894) pp. 173–81.
9. Figure 4.1 is reproduced from Ball's 1908 book, p. 291. Blyth's 1894 paper (which was read to the Royal Scottish Society of Arts on 25 January 1892) shows a similar photograph, taken from a viewpoint only slightly different, in which the semi-cylindrical boxes have the same width but a greater height, i.e. about 3 m.
10. Blyth's vertical axis rotor resembles an enlarged cup anemometer, as first developed by Robinson in 1846 and still widely used for measuring wind speeds, see Golding (1955) p. 105; see also Appendix B, note 19. Blyth acknowledges the Robinson anemometer as the inspiration for his rotor configuration.
11. Bennett and Elton (1899) pp. 326–7 comment on the lower efficiency of drag-driven vertical-axis windmills (which they refer to as horizontal windmills). A simple calculation of the power output from a windmill like Blyth's, with the assumption that 2 'boxes' are always being pushed downwind, indicates that the power coefficient C_P would have been in the range 0.01 to 0.02, as compared with $C_P \approx 0.07$ for a traditional windmill.
12. Ball (1908) p. 293 includes a photo of the Montrose asylum windmill. Papers relating to James Blyth, including a copy of the Mavor & Coulson handbill, are in archive GB 249 OM/17 at Strathclyde University. A listing of the contents of this archive is given at www.strath.ac.uk/archives/collections/ua/omint/a-d/om_17int/.

13. As recorded in the biographical note on James Blyth (1838–1906) written in 1986 by his grandson D. B. Allan, who as a four year old in 1914 witnessed the demolition of the Marykirk windmill; Strathclyde University archive (see note 12) item OM17/20. See also Price (2005).

14. Lanoy (1944) p. 48 refers to the attempt made by the Duc de Feltre in 1887 to provide the Cap de la Hève lighthouses near Le Havre with electricity from a windmill, but does not give any indication that it was successful. Rogier (1999) states that de Goyon installed a 12 m diameter 'Halladay Standard' multi-bladed windmill at Cap de la Hève in the summer of 1887, and experimented through to the onset of winter with his system for providing electricity for the lighthouses there, but indicates that he had problems with the ability of the windmill to withstand storms as well as electrical problems; there is no indication that these problems were resolved. The Scientific American Supplement for 3 August 1889, vol. 28 pp. 11326–7, contains an article 'Windmill for producing electric light at Cape de la Hève' which gives considerable detail of the equipment used by de Goyon but makes it clear that when the article was written it was not yet operational. The James Blyth archive at Strathclyde University (see note 12) contains copies of correspondence in 1989 between D. B. Allan (Blyth's grandson) and the French 'Directeur des Ports et de la Navigation Maritimes' which confirms that authorisation was given in 1887 for the windmill trial (and includes copies of the relevant 1887 letters), but states that no documentation had been found in their archives giving any results.

15. The Brush Electric Company subsequently merged with the Thompson-Houston Electric Company in 1889 and with the Edison General Electric Company in 1891 to form the General Electric Company, which is still one of the world's major engineering companies.

16. *Scientific American*, **43**, no. 25, 20 December 1890, pp. 383 and 389. The front page of this issue includes the illustration of the windmill shown in Figure 4.2 plus several other illustrations, one of which shows the tower internal arrangements. These illustrations can be seen at http://commons.wikimedia.org/wiki/Image:Brush-windmill.jpg.

17. Auxiliary vanes such as that used on the Brush rotor were also widely used on multi-bladed water-pumping windmills, as an alternative to the furling arrangement described in Chapter 3 (where the force trying to turn the rotor out of the wind resulted from having the windshaft offset from the vertical axis around which the rotor turned to face the wind); see Chapter 3 note 75.

18. Assuming a power coefficient $C_P \approx 0.1$ (see Appendix A, Figure A.2 and note 9) the peak output of 12 kW corresponds to a wind speed of about 9.5 m/s, which is Force 5 on the Beaufort scale (see Table 3.1). The rotor speed of 10 rpm gives a tip speed of 8.9 m/s; the corresponding tip speed ratio of 0.9 is typical for high solidity multi-bladed rotors.

19. Though Ball (1908) p. 323 notes that early incandescent lamps (which had carbon filaments operating at a lower temperature than modern tungsten filaments) required about 3.5 W per candlepower of light output. A 20 candlepower lamp would therefore require about 70 W and the 100 lamps in everyday use in Brush's mansion would, if all were on at the same time, consume about 7 kW.

20. Righter (1995) p. 54 indicates that the Brush mansion was connected to the public electricity supply system when this became available in 1900; the windmill continued in use until 1908, providing power to Brush's basement laboratory.

21. One London-based company that made electricity-generating windmills in the 1890s was Rollason's Wind Motor Company, which made a drag-driven vertical-axis windmill invented – unusually for the time – by a woman, Sarah Jane Rollason,

see US patent 529,197 of 1894; also *The Engineer*, 20 April 1894, p. 337, which can be seen at www.theengineer.co.uk/Articles/294349/April+1894+Victorian+wind+turbine.htm.

22. Ball (1908) p. 321.
23. Brush presumably avoided similar problems by using a very large capacity battery, so that the fluctuating charging currents were relatively low, but this was an expensive solution. He was also helped by the fact that his large multi-bladed windmill would have acted like a flywheel and its inertia would have helped to smooth the rotor speed variations.
24. Gas engines were internal combustion engines fuelled by gas. If a piped public gas supply was available (as it had been in major cities since the mid nineteenth century) this could be used. In country areas there would be no piped public supplies, but many owners of large country homes would by the 1860s have installed their own equipment to produce gas from coal, so that they too could have the benefit of gas lighting. Then from the 1880s onwards if they wished to install electric lighting they had the option of either using the gas they produced in a gas engine to drive a dynamo, or replacing the gas-producing plant with a steam engine to drive a dynamo. Bowers (1982) pp. 126–8 gives an interesting comparison between the cost in the 1880s of lighting a large country house with gas and lighting it with electricity.
25. Southern England (i.e. south east England plus south west England) has an area of 43 000 km^2 and the Earth's boundary layer extends up to a height of about 2000 m, see for example Hassan and Sykes (1990) p. 15; the volume of air over southern England up to this height is therefore 86 000 km^3. The density of air is 1.23 kg/m^3 (see Chapter 3 note 38) and this volume therefore weighs 106 billion tonnes (with 1 billion = 1000 million). Even up to a height above ground level of only 100 m the air over southern England weighs 5 billion tonnes.
26. See for example Hassan and Sykes (1990) pp. 12–13.
27. Though many modern grid-connected wind turbines are designed to turn at almost constant speed, see the later discussion in relation to the Balaclava and Smith–Putnam machines shown in Figures 4.6 and 4.8.
28. The relatively high inertia of large traditional windmills (see Chapter 3, note 22) helped to smooth rotor speed variations.
29. Ball (1908) p. 327.
30. As discussed in Chapter 3 the fantail was invented by Edmund Lee in 1745 and used by Smeaton, see Figure 3.5; self-reefing shuttered sails were invented by Andrew Meikle in 1772 and further developed by William Cubitt in 1807.
31. Ball (1908) pp. 327–8 gives further details, including a diagram showing the transmission arrangement. See also la Cour (1901), la Cour (1905) p. 69; and Powell (1910) pp. 5–7 and 66–76.
32. See Rasmussen and Øster (1990) p. 8; also the Danish Wind Industry Association website at www.windpower.org/en/pictures/lacour.htm which notes that the school room windows had to be replaced several times, as the hydrogen would occasionally be contaminated with oxygen – with explosive consequences.
33. See Golding (1955) p. 16, la Cour (1901) p. 389 and la Cour (1905) pp. 33–4.
34. Arnfred (1964) p. 377; also the website of the Poul la Cour museum at www.poullacour.dk/engelsk/vision.htm.
35. Bowers (1982) pp. 141–50 and p. 166 discusses the issues that led to AC being preferred for public electricity supply systems.
36. A five-bladed, 7 m diameter, 5 kW rated Agricco was one of five different makes of electricity generating windmill that were tested in 1925 at Oxford University's Institute for Research in Agricultural Engineering at Harpenden, about 40 km north

west of London; see Cameron Brown (1933), who notes that in each case the battery cost was comparable with the cost of the windmill. The la Cour, Lykkegaard and Agricco windmills all used gearing at the tower top to drive a central vertical shaft that delivered the shaft power output to ground level, where the electricity generator was located. They also all used fantails for yaw orientation.

37. Flettner (1926) pp. 96–109; see also Putnam (1948) pp. 98–9.
38. Though Gleick (2003) pp. 81 and 221 points out that Isaac Newton described the effect and explained its cause in 1672, after watching tennis players from his room at Trinity College, Cambridge.
39. Savonius (1931); also Golding (1955) pp. 197–8, Le Gouriérès (1982) pp. 124–7, Johnson (1985) pp. 17–19 and Shepherd (1994) pp. 38–9.
40. See Chapter 3, note 72.
41. Sektorov (1934); also Putnam (1948) pp. 105–6, Golding (1955) pp. 180–1 and 220 and Shepherd (1994) p. 41.
42. As indicated in Chapter 1 (note 16) there is no fundamental difference between a windmill and a wind turbine. The convention is that modern designs used to generate electricity are called wind turbines; traditional designs used for corn grinding, water pumping, sawing wood, pressing oil seeds etc. are called windmills, as are the late nineteenth century and early twentieth century designs used for electricity generation. There is a 'grey area' in between but most windmills for electricity generation built after about 1930 are called wind turbines; they generally have blade cross-sections influenced by experience with aircraft wings and are usually more efficient.
43. See Gipe (2004) pp. 126–8 and Hau (2006) pp. 324–38; for more electrical detail see Freris and Infield (2008) pp. 112–21 and 139–44.
44. In the UK and most of Europe the grid frequency is 50 Hz (i.e. 50 cycles per second) and a basic 2-pole induction generator will, at zero power, turn at 3000 rpm; at full power the rotational speed is about 1% higher. Modern wind turbines commonly use a 4-pole induction generator, which at zero power turns at 1500 rpm; this reduces the speed ratio between the generator and the wind turbine's slowly turning rotor, and so reduces the gearbox cost. In the US the grid frequency is 60 Hz; a 4-pole induction generator designed for the American market will therefore turn at 1800 rpm under zero power conditions, and about 1% faster at maximum power.
45. Varying the pitch means that the blades can be turned about their lengthwise axis, so as to increase or decrease the interaction with the airflow.
46. The incandescent mantle was developed by von Welsbach, an Austrian, in 1891 and has been in widespread use for oil and gas lamps ever since. It is sold as a small silk or cotton bag, impregnated with thorium and cerium nitrates, which is attached over the flame of an oil or gas lamp. When lit for the first time the fabric burns away to leave a rigid, net-like, thorium and cerium oxide structure which in the heat of the flame radiates an intense white light, comparable with the light from an incandescent electric light bulb.
47. Righter (1995) pp. 77–104 describes developments in the United States in the 1920s and 1930s; see also Gipe (2004) pp. 168–70 and 119–22, Johnson (1985) p. 4 and Kovarik *et al.* (1979) pp. 11–13.
48. Jacobs (1964) p. 337.
49. The situation is analogous to a spinning skater who goes faster when his arms are by his side than when they are outstretched.
50. Gipe (2004) pp. 136–7 give further details of the Jacobs centrifugal governor, which Jacobs described as a 'flyball' governor.
51. Putnam (1948) p. 1.

52. In operation the downwind force on each blade tends to make it incline – or cone – backwards, but centrifugal forces on the inclined blade oppose and limit this movement. The bending moment at the blade root due to the downwind force on the blade is consequently reduced by the opposing bending moment produced by the centrifugal forces acting on the inclined blade; see Putnam (1948) pp. 111–2.
53. The Smith–Putnam wind turbine used hydraulic actuators located near the hub of each blade to change the blade pitch, see Putnam (1948) pp. 116–7.
54. See Hau (2006) pp. 320–4 and Walker and Jenkins (1997) pp. 46–8; for more electrical detail see Freris and Infield (2008) pp. 98–108 and 145–7.
55. See Putnam (1948) pp. 125–34.
56. Putnam (1948) pp. 12–13 and 132–3 explains the cause of the blade failure. As is often the case several factors contributed to the failure, but a principal cause was the stress concentration that resulted from an abrupt change of section in the box spar, aggravated by the notch-effect of undercut field welds at the same location.
57. Rasmussen and Øster (1990) p. 9; see also Golding and Stodhart (1949) p. 6.
58. Claudi-Westh (1975); also Claudi-Westh (1976) and Rasmussen and Øster (1990) pp. 9–10. The F. L. Smidth windmills were designed to operate at variable speed (45–90 rpm for the 17.5 m diameter design and 29–58 rpm for the 24 m diameter version) and had speed-increasing gearboxes (with ratios, respectively, of 15 to 1 and 25 to 1) so that for both models the DC generator speed was in the range 700–1400 rpm.
59. Golding (1955) p. vii; see also pp. 2–3.
60. J. Juul (1956) p. 65.
61. Juul (1956) pp. 65–6; also Juul (1964) p. 229. Juul's Bogø windmill had a rotor diameter of 13 m; its DC generating F. L. Smidth predecessor had a rotor diameter of 17.5 m.
62. Juul (1964) pp. 229 et seq.
63. Fluctuations in the downwind force on each blade are due in part to the turbulence in the wind but there is, in addition, a cyclical variation due to the fact that the wind speed increases with the height above ground level. The Gedser mill had a tower height of 25 m so that the blade tip at its uppermost position was 37 m above the ground; half a revolution later it would be just 13 m above the ground. The difference in wind speed between these two heights is about 15% (see Appendix C) and the force on the blade tip therefore varies cyclically by about 30%. Another cause of cyclical variations in the downwind force results from the fact that as the air approaches the tower it is slowed down, and each blade has to pass through this zone of reduced wind speed every revolution.
64. By using hanging cables Juul avoided the use of slip rings, which for high powers are costly and a source of unreliability. The cables could tolerate 10 complete turns of the nacelle; any more and a relay would be activated to shut the turbine down. Juul (1964) p. 237 noted from his experience that yawing would cause about 10 complete turns per year. In practice the cables were unwound, if necessary, during the regular monthly inspection. Juul's cable arrangement was trouble free and hanging – or pendant – cables became standard on most of the commercial wind turbines built after 1973.
65. Why should the power output be limited at high wind speeds? Because the power in the wind increases as the cube of the wind speed one could design a wind turbine that gave a power output (at the windshaft) of 100 kW at a wind speed of 7 m/s and which then gave a 27-times higher power output, i.e. 2700 kW, in a 3-times higher wind speed of 21 m/s. Wind speeds close to 7 m/s occur frequently (see for example

Figure 3.7) but wind speeds as high as 21 m/s are rare; little extra energy would therefore be gained by having a gearbox and generator that could deliver 2700 kW, despite their very much higher cost. In fact having a gearbox and generator with such a high rating would actually *reduce* the overall energy output; this is because most of the time the wind turbine would be producing somewhere close to 100 kW and at such a low part-load (only about 4% of the 2700 kW rating) the efficiency of the gearbox and generator would be very low. See for example Johnson (1985) pp. 139 and 141, also Hau (2006) pp. 299, 355–6 and 490.

66. See Appendix B, Figure B.2; as the wind speed increases the angle between the relative wind and the plane of rotation increases, and if the blade is fixed to the hub this means that the angle between the relative wind and the blade also increases. This latter angle is called the angle of incidence by aerodynamicists and given the symbol α (alpha) – see Appendix B, note 15. On the Gedser mill, as on modern wind turbines, the blade cross-section has an aerofoil shape similar to that of an aeroplane wing. When α exceeds about 15° the air can no longer flow smoothly around both sides of the aerofoil, but separates from the downwind side (for an aeroplane wing it is the flow over the upper side that separates) resulting in a loss of lift and an increase in drag. See Hau (2006) pp. 108–10, Sharpe (1990) pp. 68–9 and Burton *et al.* (2001) pp. 168–71; also the website www.windpower.org/en/tour/wtrb/powerreg.htm.

67. Juul (1964) p. 235.

68. When the decision was made in 1967 not to refurbish the Gedser mill it was dismantled and put into storage. Following the 1973 oil crisis, when interest in wind energy revived, it was refurbished in 1977 and put back into service. Over the following two years a comprehensive series of measurements was made and the results are summarised in DEFU (1981). This indicates (pp. 86–7) that the maximum power coefficient $C_P \approx 0.34$ in a wind speed of 9.6 m/s, and at the 200 kW rated power output $C_P \approx 0.18$.

69. See Yergin (1991) pp. 393 and 792–3, Yeomans (2005) pp. 21–2 and also BP (2008) p. 16. The latter indicates that in 1947/8 the price of oil (in 2007 US dollars) was $17/barrel; by the early 1960s it was just over $12/barrel and by 1970 the price was down to $10/barrel.

70. Rasmussen and Øster (1990) p. 11.

71. See Bonnefille (1976), Argand (1975), Divone (1994) pp. 75–6 and Le Gouriérès (1982) pp. 204–11 and plates 9–11 (after p. 128).

72. Self-aligning means that the rotor is unconstrained in yaw; with the blades downwind of the tower the rotor will, like a weather vane, automatically respond to changes in the wind direction. With such a passive yaw arrangement the cost of an active yaw system – as used on the Gedser mill – can be avoided, though for modern wind turbines this cost is only a very small percentage of the total. A disadvantage of passive yaw is that the blades must be downwind of the tower; they therefore have to pass through the tower's reduced velocity wake every revolution, which makes the loading on the blades (and the rest of the structure) more demanding and tends to increase their cost. Blades passing through the wake of the tower also produce a pulsating, low-frequency, noise which can cause nuisance for a considerable distance. Almost all modern wind turbines use active yaw systems, with the blades upwind of the tower.

73. See Armbrust (1964), Hütter (1976), Divone (1994) p. 79 and Hau (2006) pp. 40–44. Measurements on Hütter's 100 kW wind turbine indicated that in light winds, corresponding to a tip speed ratio of about 14, a power coefficient $C_P \approx 0.44$ was achieved. The maximum power output was obtained when the wind speed was

about 9 m/s, corresponding to a power coefficient $C_P \approx 0.22$ and a tip speed ratio of about 8.

74. See Garrad (1990) pp. 138–42, also Hau (2006) pp. 203–4 and 265–9. Allowing the blade experiencing the greater downwind force to tilt backwards reduces the bending moment at the blade's hub end, and so reduces the maximum blade stress.

75. Potentially damaging vibrations caused, under some operating conditions, by the aerodynamic forces acting on blades that are flexible in both bending and twisting; see Garrad (1990) p. 132 and Hau (2006) pp. 390–2, also Eggleston and Stoddard (1987) pp. 231–50.

76. See Stodhart (1976) Divone (1994) pp. 76–7, Warne and Calnan (1977) pp. 964–5 and 976 and Golding (1955) pp. 221 and 323.

77. Centrifugal forces tend to stiffen the blades.

78. Figure 4.14 shows the Enfield–Andreau machine just after its erection and prior to the installation of a tail cone which enclosed the hub ends of the blades and extended forward to cover the cowl that housed the low speed shaft bearings. Golding (1955) plates 26 & 27 shows the Enfield–Andreau machine with the tail cone fitted; see also Hau (2006) p. 38. Andreau had previously built an 8 m diameter wind turbine to demonstrate his novel design, see Golding (1961) p. 55 and Figure 44; also Golding (1955) pp. 217 and 221.

79. The British Electricity Authority was established in 1948 and was responsible for the generation of electricity and its bulk transmission for most of the UK. It was abolished in 1957 and most of its functions were transferred to the newly formed Central Electricity Generating Board, see Bowers (1982) p. 163.

80. See Delafond (1964) p. 289; also Hütter (1976) p. 1–70.

81. Stodhart (1976) p. 1–24.

82. The evolution of the design of this wind turbine – which included extensive model testing in the high density wind tunnel at the National Physical Laboratory – is described in some detail in Elliott (1975); the wind turbine was built by R. Smith Ltd.

83. The blade-tip air brake design on the Gedser mill is clearly superior to the design of the air brakes on the Isle of Man machine, which relied on rupturable shear pins for their operation; see Warne and Calnan (1977) p. 976. In addition Juul had developed for the Gedser mill a sophisticated control system which allowed it to run unattended, automatically stopping if the wind speed dropped too low for power generation (i.e. below about 5 m/s) and re-starting when the wind speed subsequently increased, see Juul (1964) pp. 237–9.

Chapter 5 The evolution of the modern wind turbine, 1973 to 1990

1. See for example Yergin (1991) p. 558. 'The outcome of the Six-Day War [in 1967] seemed to confirm how secure the supply of oil was … After all, the oil was there, it was endlessly abundant, and it was cheap. It flowed like water. The surplus had lasted for almost twenty years, and the general view was that it would continue indefinitely, a permanent condition. That was certainly the way things looked to most people within the oil industry.'

2. See the comprehensive history of the oil industry by Yergin (1991), in particular chapters 29 and 30.

3. Yergin (1991) p. 845.

4. Yergin (1991) pp. 567–8, 591 and 625; free world oil demand excludes the Soviet Union and China.

5. The prices quoted are in 'money of the day'; allowing for inflation the prices in 2007 US dollars would be as follows: \$9.7 per barrel in 1970, rising to \$13.6 by mid 1973, then to \$23.9 in October 1973 and to \$54.9 in December 1973; see BP (2008) p. 16, also Chapter 1 note 33.

6. The Day of Atonement, probably the most important holiday of the Jewish year when most Jews refrain from work, fast and attend synagogue services.

7. See Yergin (1991) chapters 33 and 34. The 1981 'money of the day' price of \$34 per barrel corresponds to \$77.8 per barrel in 2007 US dollars, see note 5 above.

8. The accident at the Three Mile Island nuclear power station near Harrisburg, Pennsylvania, in March 1979 led to meltdown in the reactor core and radiation leakage, see http://americanhistory.si.edu/tmi/. Many orders for nuclear power stations were subsequently cancelled, and dozens of partially built nuclear power stations were left unfinished, see Righter (1995) pp. 150–2.

9. Other renewable energy options that were initially thought to be more promising included solar photovoltaics, solar thermal and geothermal, all of which received substantially more funding than wind energy through the 1970s; see Taylor (1983) p. 4.

10. Divone (1994) pp. 82–5, Johnson (1985) pp. 7–9 and BWEA (1982) pp. 170–1.

11. Divone (1994) pp. 116–7, Johnson (1985) pp. 7–11 and BWEA (1982) pp. 170–1.

12. Divone (1994) pp. 118–20, Johnson (19850 pp. 10–11 and BWEA (1982) pp. 172–4.

13. Boeing 747 'Jumbo Jets' built in the period 1969 to 1989 had a wingspan of 59.6 m (later models are slightly larger).

14. The small wind turbines, with diameters in the range 10 to 15 m and power ratings in the range 20 to 55 kW, that were then (1980–2) being installed in Denmark were averaging annually about 420 kWh/m^2. By the end of the decade the turbines being installed in Denmark typically had diameters in the range 25 to 30 m and power ratings in the range 150 to 300 kW, and were averaging about 780 kWh/m^2; see Figure 6.3, also Danish Energy Agency (1999) pp. 30–1.

15. Gipe (1995) p. 103.

16. The tubular tower of the Mod-2 was dynamically 'soft', with a natural frequency intermediate between the blade rotation frequency (1P) and the blade passing frequency (2P), unlike the earlier Mod-0, Mod-0A and Mod-1 machines which all had 'stiff' towers with natural frequencies above the 2P blade passing frequency; see Johnson (1985) p. 12, BWEA (1982) p. 172 and Hau (2006) pp. 408–11. 'Soft' towers are usually less expensive and much lighter than 'stiff' towers.

17. See Divone (1984) p. 21, Divone (1994) pp. 128–30, BWEA (1982) pp. 172–4, Lowe and Wiesner (1983) and Musgrove (1984) pp. 73–4.

18. See Chapter 4, note 74.

19. The corresponding power coefficient was then $C_P \approx 0.34$ at a tip speed ratio of 6.9. As is noted in Appendix A modern wind turbines are designed to give their maximum power coefficient at a wind speed that is somewhat lower than the rated wind speed.

20. See Divone (1994) pp. 130–1, Lowe and Wiesner (1983) pp. 533–6 and Carlin *et al.* (2002) pp. 23–4. Another 'third-generation' wind turbine, designated Mod-5A, was designed by General Electric but was never built, see Divone (1984) p. 22.

21. See EWEA (2004) pp. 21–2; also Hau (2006) pp. 342–4, Freris and Infield (2008) pp. 143–4 and Carlin *et al.* (2002) pp. 8–18. The wound rotor induction generator (also called a doubly fed induction generator or DFIG) allows speed variation over a significant speed range but only a small proportion of the overall power output has to pass through a power electronic converter. It is less costly (and more efficient) than

full range variable speed systems in which all the output power has to pass through a power electronic converter.

22. The capacity factor is the total energy output over a period, usually a year, divided by the energy output that would be obtained if the rated power output was sustained throughout the whole period.

23. Gipe (1995) p. 106.

24. Gipe (1995) p. 71 and Divone pp. 112 and 122.

25. Gipe (1995) p. 106.

26. Musgrove (1984) pp. 75–6 and pp. 83–4, Gipe (1995) pp. 104–7 and Righter (1995) pp. 175–9. The slender tubular tower was designed to be 'very soft', that is to say its natural frequency was even lower than the 1P rotation frequency (see note 16).

27. Divone (1994) pp. 86–92.

28. Gipe (1995) pp. 72–5; Righter (1995) p. 161.

29. The innovative designs included ducted wind turbines, blade tip vanes, the Madaras tracked system using Flettner rotors, oscillating wings and other devices (including some that are often re-invented); see Vas and South (1980), BWEA (1982) pp. 283–301, Eldridge (1980) pp. 32–49 and Hau (2006) pp. 74–8.

30. Golding (1955) p. 19 and Putnam (1948) pp. 98–104 both refer to Darrieus's work on conventional propeller-type wind turbines. His patent for vertical-axis wind turbines was filed in France in October 1925, see Darrieus (1926). Details of this French patent, as well as the corresponding British patent issued in April 1927 and the US patent issued in December 1931 are given at http://v3.espacenet.com/textdoc?DB=EPODOC&IDX=NL19181C&F=0. It is not clear whether Darrieus ever built one of the novel vertical-axis wind turbines (VAWTs) that he patented, but Argand (1975) p. 180 shows a 7 kW straight-bladed Darrieus VAWT that was tested as part of the French wind power programme in the early post-war years; the same wind turbine is shown by Le Gouriérès (1982) plate 13 (facing p. 129), see also p. 138.

31. See Divone (1981) pp. 72–3 and 87–90, Musgrove (1984) pp. 77–8 and 84–6, Divone (1994) pp. 95 and 98–103 and Gipe (1995) pp. 170–5.

32. The general picture, with the relative wind meeting the blade at a small angle so as to give an overall force almost perpendicular to the relative wind and with a small component in the same direction as the blade's motion, is essentially the same as was discussed previously when considering how traditional windmills worked; see Figure 2.6 and the associated text in Chapter 2, and also Figure B.2 in Appendix B.

33. See Chapter 4, note 66.

34. See Chapter 2, note 16.

35. Compliance in the drive-train can be provided via the slip of an induction generator or by using an hydraulic coupling. Musgrove (1990a) p. 300 shows that with 2% slip and a medium-sized VAWT the power fluctuations are reduced to about 14% of the average power output, which is comparable with the fluctuations caused by turbulence. Larger wind turbines have a greater flywheel effect and therefore allow even better power smoothing provided the generator is of a type that permits some rotor speed variation, such as the doubly fed induction generators referred to in note 21.

36. Divone (1994) p. 99.

37. See Divone (1984) p. 28 and Musgrove (1990a) p. 297.

38. See Divone (1994) pp. 99–103 for further details.

39. Braasch (1979).

40. Canada, where the Darrieus design had been re-discovered in the late 1960s, continued with the development of large vertical axis machines and in 1986

constructed the 4 MW rated Éole, which was 96 m high and had an equatorial diameter of 64 m, see Divone (1994) pp. 126–7, Carlin *et al.* (2002) p. 29 and Hau (2006) pp. 49 and 53. It became fully operational in early 1988 and over the next five years generated 13 million kWh, corresponding to a capacity factor of 7%. It was shut down permanently after the failure of the bottom bearing in 1993.

41. See Gipe (1995) pp. 104 and 109, Musgrove (1984) pp. 82–3 and Divone (1994) pp. 124–5.

42. See Musgrove (1984) pp. 45–6, Divone (1994) pp. 125–6, Gipe (1995) pp. 104 and 107–8, Hau (2006) pp. 45–6, 51, 158, 230 and 265–7 and Carlin *et al.* (2002) p. 21.

43. See Musgrove (1984) pp. 46–7 and 124, Divone (1994) p. 126, Hau (2006) pp. 49, 52 and 707 and Gipe (2004) pp. 102–4.

44. Hau (2006) p. 543 indicates that noise emissions increase by about the 5th power of the blade-tip speed, and a 25% reduction in tip speed reduces aerodynamic noise emissions by a very significant 6 dB(A). The 98 to 126 m/s tip speed of Monopteros compares with 80 m/s for Smith–Putnam's wind turbine, 38 m/s for Juul's Gedser mill and 84 m/s for the Boeing Mod-2; wind turbines in production in the early 2000s typically have tip speeds in range 60 to 80 m/s, see EWEA (2004) p. 20.

45. The Italian company Riva-Calzoni in co-operation with MBB also produced a range of single-bladed wind turbines through the 1980s, and nearly one hundred of their 33 m diameter, 250 kW rated machines were installed at various locations in Italy. Production was discontinued in the early 1990s.

46. See Chapter 4, note 68.

47. See DEFU (1981); also Pedersen and Nielsen (1980), Musgrove (1984) pp. 30–2 and 118, Divone (1994) pp. 121–2 and Gipe (1995) pp. 104 and 111. The original method of blade construction proved unsatisfactory and after a few years both the Nibe machines were retrofitted with blades made from wood.

48. See Gipe (1995) pp. 104 and 113; also Hau (2006) p. 56.

49. See Eldridge (1980) pp. 66–74, Carlin *et al.* (2002) pp. 29–30, Jensen and Bjerregaard (1980), Le Gouriérès (1982) pp. 213–5, Gipe (1995) pp. 104 and 111–12 and Gipe (2004) pp. 353–5. The November 1978 first issue of *Windirections*, the newsletter of the newly formed British Wind Energy Association, carried an interview with some of the people who had been involved in the Tvind wind turbine's construction. After the formation of the European Wind Energy Association in 1982 *Wind Directions* (note the spelling change) became its newsletter. The Tvind article, plus the rest of the contents of the first issue of Windirections, can be seen at www.ewea.org/index.php?id=59&no_cache= 1&sword_list[]=wind&sword_list[]=directions&sword_list[]=1978.

50. See Chapter 4, note 66 and the discussion of *stall control* in Chapter 4.

51. An interesting film history with the title 'Vestas – when reality exceeds imagination' was produced in 2006 by a Danish TV station and can be seen (with English sub-titles) on the Vestas website www.vestas.com/en/media/videos/vestas-films.aspx. Henrik Stiesdal was employed by Vestas and was subsequently responsible for the design of their first pitch-regulated wind turbine, which went into production in 1985. After Vestas' financial difficulties in 1986 and their major re-structuring Stiesdal joined Bonus, and led the development of their wind turbines as they evolved to multi-megawatt size over the next two decades. Bonus was bought by Siemens in December 2004.

52. The two aerodynamic braking options discussed relate to stall-controlled wind turbines. Pitch-controlled wind turbines (which only came into widespread use in Denmark following their introduction by Vestas in 1985) can achieve aerodynamic braking simply by pitching their blades through a larger than usual angle.

53. Spoilers would be fitted on the lifting side of blades, i.e. the downwind side, about midway between the leading edge and the trailing edge. Increased centrifugal force during incipient overspeed would make them deploy and they would then both increase the drag and reduce the lift. See Hau (2006) pp. 110–12.

54. The subsidy was reduced to 15% in 1986 and removed completely in 1989, having served its purpose in helping to stimulate a substantial domestic market for wind turbines.

55. See Petersen (1980) pp. 6–34.

56. Økær Vind Energi went bankrupt in 1981, but prior to this Grove-Nielsen had licensed production of his 'AeroStar' blades to Coronet, a Danish company that made fibreglass blades (and which later changed its name to Alternegy). Erik Grove-Nielsen's website www.windsofchange.dk/WOC-bladestory.php gives a detailed and well-illustrated account of the evolution of wind turbines in Denmark from the mid 1970s.

57. See Petersen (1980) pp. 36–9.

58. See Danish Energy Agency (1999) p. 30.

59. See Musgrove (1984) pp. 32–9; wind turbine prices are those that prevailed in April 1982.

60. In a class 2 roughness location, which corresponds to farmland with scattered buildings and hedges, the annual average wind speed would be about 4.7 m/s and the annual average energy output from a 15 m diameter 55 kW rated wind turbine would be approximately 65 thousand kWh.

61. See Gipe (1995) pp. 57–64.

62. One point of difference with later turbines is that as the turbine size goes up the gearbox becomes disproportionately larger. This is because the rotor swept area – and therefore the power output – goes up as the square of the rotor diameter, but the blade-tip speed remains much the same, regardless of size (and is usually in the range 60 to 80 m/s, see note 44). A consequence of the constant blade-tip speed is that the rotor speed goes down as the rotor diameter goes up. The torque input to the gearbox is therefore proportional to the *cube* of the rotor diameter. This much-increased torque, as well as the need for a higher gear ratio, requires an increase in the size of the gearbox relative to other components within the nacelle.

63. See Righter (1995) pp. 190–222, Gipe (1995) pp. 30–6 and Musgrove (1984) pp. 78–81; see also Asmus (2001) pp. 57–128.

64. Though if the wind turbine performance was too poor there was the risk that the Internal Revenue Service would seek to recover the tax benefits, see for example Righter (1995) pp. 226–7 and 328.

65. Spera (1994) pp. 611–24 gives details of the various wind turbine designs (including small outline drawings) and also indicates the approximate numbers of each that had been installed by the mid 1980s. Information on the total number of turbines and the capacity installed year by year is given by Lynette (1988) p. 16, Righter (1995) p. 215, Lynette and Gipe (1994) p. 204, Gipe (1995) p. 35 and Lindley and Musgrove (1991) p. 7.

66. Gipe (1995) p. 179 (Figure 6.22) indicates that the tower head mass (i.e. the rotor plus the nacelle) of the U.S. Windpower 56–100 wind turbine (100 kW rated, 17 m diameter) divided by its rotor swept area was just over 10 kg/m^2; comparable Danish wind turbines were over 20 kg/m^2.

67. For investors to be able to claim the all-important tax credits the wind turbines they were investing in had to be operational by midnight on 31 December. Turbine deliveries and installation were therefore concentrated in the final months of each year, with frantic activity towards the end of December.

68. As discussed earlier in this chapter Danish wind turbines with a diameter of about 15 m cost 310 000 kroner in 1982, see note 59. Turbines this size in Denmark would typically be rated at 55 kW but when adapted for sale to the US (where the grid frequency is 60 Hz, as compared with Europe's 50 Hz) the rating was usually increased to 65 kW. In late 1983 and late 1984 the exchange rate was typically about $1 = 10 kroner, so the Danish price equates to about $31 000. However this price does not include the cost of access tracks, the reinforced concrete foundation or the grid connection; in Denmark these would typically add about 25% to the turbine price, see Danish Energy Agency (1999) p. 25, to give an installed price per turbine of about 390 000 kroner – equivalent to about $39 000. Gipe (1995) p. 231, table 7.3 indicates that in California these Danish wind turbines were sold on to investors for prices ranging from $96 000 to $180 000, allowing those who developed wind farms before the expiry of the tax credits (in 1985/1986) to make very substantial profits.

69. Danish export numbers come from Vindmølleindustrien (1997) pp. 2–3.

70. Micon was formed in 1982 by a former Nordtank employee; it made more extensive use of sub-contractors, which helped it to rapidly expand its production in response to the surge of demand from California.

71. In October 1984 one dollar would buy 11.1 Danish kroner but just three years later, in October 1987, the dollar would buy only 6.9 kroner, making imports from Denmark 60% more expensive.

72. See Righter (1995) pp. 220–2 and 233, Asmus (2001) pp. 97–8 and 109–10 and Gipe (1995) pp. 29–30 and 33–4.

73. The high oil prices that followed the 1973 crisis stimulated improved energy efficiency as well as more intensive exploration and production from areas such as Alaska and the North Sea. Between 1979 and 1983 oil consumption reduced by 6 million barrels per day while non-OPEC production increased by 4 million barrels per day; see Yergin (1991) p. 718. To sustain the price of oil OPEC countries agreed to reduce their output, but the main burden was borne by Saudi Arabia and by the summer of 1985 its production had dropped below the output from the North Sea. When Saudi Arabia decided that it needed to regain market share – and the political influence that went with it – by increasing its production the oil price plummeted from $32/barrel in late 1985 to less than $10/barrel just a few months later. Some stability was restored when a reference price of $18/barrel (equivalent to $34/barrel in 2007 dollars, see note 5) was agreed later in 1986; see Yergin (1991) pp. 746–64. Oil prices were to remain low (except for a brief spike following Iraq's invasion of Kuwait in August 1990) for the next 17 years, averaging $28/barrel in 2007 dollars, i.e. less than half the $62/barrel average for the period 1974 to 1985 (but more than double the average price of $13/barrel – in 2007 dollars – that had prevailed in the post-war years 1945 to 1973); see Chapter 1 note 33.

74. 59 MW was installed in 1988, with Danish turbines accounting for 20 MW. 134 MW was installed in 1989, including 70 MW with Danish machines; and in 1990 a further 175 MW was installed, including about 80 MW with Danish turbines. See notes 65 and 69.

75. See for example Gipe (1995) pp. 271–89, SDC (2005) pp. 57–8 and 82–6, and EWEA (2004) pp. 190–200. Gipe (1995) pp. 251–71 also discusses at some length the general issue of visual acceptability.

76. Asmus (2001) p. 239 states that in the US 97 million bird deaths per year are the result of collisions with buildings with plate glass facades. In the UK the British Trust for Ornithology estimates that about 33 million birds deaths per year are the result of collisions with windows see www.bto.org/gbw/science/windowstrikes. htm; another 55 million birds are killed by cats every year.

77. Gipe (1995) pp. 342–52; see also Asmus pp. 239–42.
78. The quotation is from the DTI (2001) p. 3.
79. See Carlin *et al.* (2002) p. 26; also Asmus (2001) pp. 157–61 and 176–83 and Righter (1995) pp. 270–2.
80. Kenetech suggested that variable speed operation would increase the energy output by about 10%, but Hau (2006) p. 514 indicates that a 3 to 5% increase is more realistic. Variable speed operation also allows power output fluctuations due to turbulence to be smoothed (in effect using the rotor as a flywheel) and so reduces loads in the drive train. Against these advantages one has to set the cost and complexity of the power electronics used.
81. See Lindley and Musgrove (1991) p. 7, also Gipe (1995) p. 13. The generation of 2.4 TWh from an installed capacity of 1500 MW corresponds to a capacity factor of 18%.

Chapter 6 Progress and economics in Europe, 1973 to 1990

1. See Danish Energy Agency (1999) pp. 29–31, and Amannsberger and Hau (1991) pp. 5–6. Though the capacity at the end of 1990 was 343 MW the average capacity through the year was 303 MW; hence the calculated 23% capacity factor. Denmark's total electricity consumption in 1990 was just under 34 TWh, so the 0.61 TWh (or 610 million kWh) generated by wind turbines was 1.8% of the total.
2. Lindley and Musgrove (1991) p. 14 give details of about 20 wind farms built in Denmark between 1985 and mid 1990. Gipe (1995) pp. 65 and 289–91 gives further details of the Taendpipe/Velling Maersk wind farm.
3. See the discussion of stall control in the context of the Gedser mill in Chapter 4, including note 66 to that chapter.
4. The short-period power fluctuations caused by turbulence (to which pitch-controlled turbines are more sensitive) are effectively random and when large numbers of wind turbines are connected to the grid they nearly cancel one another out. (More precisely the magnitude of the short-period power fluctuations relative to the average power output from n wind turbines is inversely proportional to the square root of n.)
5. Allowing the rotor speed to vary by just a few per cent greatly reduces the magnitude of the power fluctuations that the gearbox and other components must withstand. Vestas subsequently developed an arrangement, which they call Optislip, which provides a limited variable speed capability. (It uses a wound rotor induction generator with a variable external resistance; see Chapter 5, note 21.) This was followed in the later 1990s by a more advanced arrangement, which Vestas call OptiSpeed, which allows speed variation over a 2 to 1 range.
6. See for example Risø (1989) pp. 31–3.
7. Though it should be noted that the performance of stall-controlled wind turbines is more prone to degradation due to the accumulation of dirt and dead insects on the leading edge of their blades. For example Lissaman (1994) pp. 299–302 notes that dirty blades could reduce by nearly 30% the energy output of the 16 m diameter, 65 kW rated, Micon turbines that were used in California in large numbers in the mid 1980s. The problem was dealt with by regularly washing the blades of affected turbines with a specially adapted water bowser. Later, larger, turbines on taller towers were found to be less affected as with the blades further above the ground there were fewer insects and less dust. In Denmark (and in the UK) where rainfall is more frequent there has been no need for artificial washing.
8. Henrik Stiesdal, who was responsible for the design of the first Vestas variable-pitch machine, notes (personal communication, 2008) that the driver for his decision in

1984 to go for pitch control was evidence from tests on the Nibe A machine (see Chapter 5) that had shown the occurrence of blade vibrations due to unsteady stall. Though this had not been a problem with the turbines up to about 55 kW rating that had then been built there was concern that blade vibrations caused by unsteady stall could become a serious problem with larger turbines, and at that time the design tools needed to quantify the problem did not exist. The safe design choice was to use variable-pitch blades and so avoid stall. Experience would later show that stall-vibration problems on large stall-controlled wind turbines could be avoided by fitting stall strips to the blades; see Hau (2006) pp. 113–5. By then Vestas had established leadership in the manufacture of competitively priced variable-pitch machines.

9. The calculation assumes that the distribution of wind speeds throughout the year corresponds to the Rayleigh distribution, see Appendix C.
10. See Chapter 4, note 65.
11. On very windy sites, such as upland locations in the UK, wind turbines need to survive even higher wind speeds, ranging up to about 70 m/s.
12. For a high-solidity rotor such as the Savonius rotor discussed in Chapter 4 (see note 39), or for the ducted rotors that are sometimes advocated (see Chapter 5, note 29), the need to withstand wind speeds of 60 m/s or more does add substantially to the cost of the structure and its foundation.
13. Danish Energy Agency (1999) p. 24, Figure 10.
14. Musgrove (1984) pp. 34–5.
15. Lindley and Musgrove (1991) p. 45 gives March 1990 prices for 11 Danish wind turbines.
16. Danmarks Statistik, Copenhagen, provides historic data on the Danish consumer price index; month by month changes since 1980 are given at www.dst.dk/Statistik/seneste/Indkomst/Priser/Forbrugerprisindeks.aspx.
17. See Danish Energy Agency (1999) p. 25 Figure 12 and p. 31 table A4. Figure 9 on p. 22 indicates that there was no material change through the 1980s in the average windiness of the sites where turbines were being deployed. The average site had – in the terminology of Risø National Laboratory – a roughness class of about $1\frac{1}{2}$; this corresponds to an average wind speed of 6.6 m/s at 50 m height above the ground, and to an average wind speed at 10 m height of about 5.0 m/s.
18. See Appendix C. First-generation wind turbines typically had tower heights of about 18 m but the larger-diameter fourth-generation machines had tower heights of about 30 m. At comparable locations, in open countryside, the hub height wind speed of the larger machines will therefore be $(30/18)^{0.14} = 1.07$ times higher, and the power available per unit of rotor swept area will therefore be about 24% higher.
19. See Danish Energy Agency (1999) p. 19 table 5 and p. 25 Figure 14; see also Hau (2006) p. 739.
20. Danish utilities usually do their cost-of-energy calculations assuming a rate of return of 5% and a cost recovery period of 20 years, see for example Danish Energy Agency (1999) p. 20 Figure 8; see also EWEA (1999) pp. 66–7. In the UK through most of the post-war years the Government specified that public sector infrastructure investments, such as new power stations, should target to achieve a 5% rate of return, net of inflation; however this UK target rate of return was increased to 8% from April 1989. Private sector investors usually require a somewhat higher rate of return with cost recovery over a period that is substantially less than 20 years.
21. The formula for the annual charge a that is required to repay one unit of capital cost over n years and to provide a rate of return p is given by the standard formula that is also widely used for mortgage repayment calculations, i.e. $a = p + p/[(1+p)^n - 1]$,

see for example Hau (2006) pp. 755–6. With $n = 20$ and $p = 0.05$ this gives $a = 0.080$; with $n = 20$ and $p = 0.08$ this gives $a = 0.102$.

22. See Danish Energy Agency (1999) pp. 19–20; as might be expected larger wind turbines have, relative to their output, lower operation and maintenance costs.

23. Allowing for inflation in Denmark from 1990 to 2007 these 1990 wind energy costs become (at 2007 price levels) 0.60 kroner/kWh for a 5% return and 0.71 kroner/kWh for an 8% return. The average UK pound to krone exchange rate in 2007 was £1 = 11.0 kroner; these 1990 Danish wind energy costs were therefore equivalent (at 2007 price levels) to 5.5 pence/kWh for a 5% return and 6.5 pence/kWh for an 8% return.

24. EWEA (1991) in *Time for Action* p. 22 gives energy costs in ECU/kWh; conversion to pence/kWh uses the conversion factor 1 ECU = £0.7 given on p. 52 of the report.

25. House of Commons (1990) vol. 1 pp. 14–19.

26. Denmark has a land area of 42 400 km^2 with a mid 2008 population of about 5.5 million; by comparison the UK has a land area of 242 000 km^2 with a population of 60.9 million, Germany has a land area of 349 000 km^2 with a population of 82.4 million, France has a land area of 546 000 km^2 with a population of 60.9 million and Spain has a land area of 500 000 km^2 with a population of 40.5 million. The Netherlands has a land area of 33 900 km^2, which is somewhat smaller than Denmark, but its population of 16.7 million is more than three times larger. See the CIA World Factbook www.cia.gov/library/publications/the-world-factbook/geos/uk.html. The United States has a land area of 9.16 million km^2, with a population of 304 million.

27. See Musgrove (1984) pp. 60–3.

28. See Chapter 5 note 65.

29. See EWEA (1991) p. 57 and EWEA (1999) p. 203.

30. See EWEA (1991) p. 57, EWEA (1999) p. 200 and Lindley and Musgrove (1991) pp. 15–17.

31. Nor anywhere else, apart from the United States – which as discussed in Chapter 5 had about 1500 MW operating at the end of 1990 – and India, which had about 35 MW, see EWEA (1999) p. 202.

Chapter 7 UK progress, 1973 to 1990

1. Department of Energy (1976), pp. 4 and 95; also pp. 23, 27, 59 and 62–4; this suggested that wave energy could make a contribution of 30 TWh/year by the year 2000, rising to 100 TWh/year by 2025. For solar heat a contribution of 12 TWh by the year 2000 was considered feasible, divided equally between domestic water heating and domestic space heating. The energy output that could be provided by a barrage across the Severn Estuary was estimated to be about 20 TWh/year. Note that the report actually quotes potential energy contributions in terms of mtce (million tons of coal equivalent); the conversion factor appropriate at the time was 1 mtce = 2 TWh, see Department of Energy (1977), pp. 36 and 49.

2. House of Commons (1977) p. 425; see also pp. 310–13.

3. See DUKES 2007, Table 5.1.2. The growth in electricity consumption was forecast to be about 2% per year, so the 215 TWh/year in 1975 was expected to grow to about 380 TWh/year by the end of the century; the actual electricity consumption in the year 2000 was 330 TWh.

4. See Department of Energy (1977), pp. 1–2, 25, 36 and 49–50. This report on the prospects for the generation of electricity from wind in the UK assumed that the proposed ten thousand 1 MW wind turbines would have a diameter of 46 m. Three thousand one hundred would be sited on hill tops and with a capacity factor of 36%

would generate 9.7 TWh/year; the remaining six thousand nine hundred would be sited in flat, coastal, locations where the lower wind speeds would result in a capacity factor of only 11%, and the output from these turbines would therefore total just 6.7 TWh/year. See also the ERA evidence to the Select Committee on Science and Technology, House of Commons (1977) pp. 154–71.

5. The Central Electricity Generating Board (CEGB) was responsible for both electricity generation and supply throughout England and Wales prior to the 1990 privatisation of the electricity supply industry. Their pessimism about the potential energy contribution from wind turbines is noted in House of Commons (1977) p. xxi, see also pp. 202–3.

6. House of Commons (1977) p. 423; see also pp. 425–8.

7. The relative rankings of the five principal renewable energy options (wave, solar, geothermal, tidal and wind) are given in House of Commons (1977) p. 313; see also Department of Energy (1976) pp. 23 and 27. Over the following six years, 1977 to 1983, the Department of Energy's wave energy expenditure totalled £16.2 million, while its expenditure on wind energy totalled £5.0 million. 1983/4 was the first year in which wind energy funding (£2.5 million) exceeded that for wave energy (£1.0 million). Data on wave energy and wind energy funding in the period 1973 to 1976 is given in House of Commons (1977) pp. 426 and 428; expenditure in the period 1976 to 1979 is given in Department of Energy (1979) vol. 1 p. 55, and expenditure in the period 1979 to 1986 is given in Department of Energy (1987) p. 3.

8. See for example Musgrove (1976b) and my 2 August 1977 evidence on 'wind energy systems and their potential in the UK' to the Windscale enquiry (on the proposed extension to the nuclear re-processing plant) which was reprinted in *Wind Engineering*, Musgrove (1977a). See also Musgrove (1978a), Musgrove (1978b), Musgrove (1978c) and Brown (1978).

9. The variation of wind speed with height is discussed in Appendix C, and the average wind speed of 9 m/s at the 70 m hub height corresponds to a power law exponent $\alpha = 0.11$. The calculated 860 W/m^2 average power in the wind assumes a Rayleigh distribution of wind speeds; 92 kW/m is then the power in the wind up to the turbines' tip height of 107.5 m.

10. See Stoddard (2002), also Inglis (1978) pp. 59–67.

11. North Sea gas production grew steadily and by 1995 was sufficient to meet the total UK consumption; and for just under a decade Britain was a net exporter of gas. Gas production peaked in 2000 and by 2004 Britain had again become dependent on gas imports. See DUKES (2007) pp. 98–101.

12. Around the coast of Scotland and Northern Ireland the 20 m depth contour is mostly within just a few kilometres from the shore.

13. See Department of Energy (1979) vol. 1 pp. 58 and 68 which noted that wind turbines sited offshore could contribute about 50 TWh/year. Wind power's cause was helped by the support of Sir Martin Ryle, the Astronomer Royal and Professor of Radio Astronomy at Cambridge University who had been awarded the Nobel Prize for Physics in 1974 in recognition of his pioneering radio astronomy research. Ryle had argued that the on-land wind potential was much larger than had been estimated by the ERA and the Department of Energy; see Ryle (1977) and subsequent correspondence.

14. In an address to staff at Fawley power station, Southampton, on 4 July 1978 which was subsequently published by the CEGB, see CEGB (1978). Reference to Glyn England's Fawley speech is also made in the article *Wind over the Waves* on p. 3 of the November 1978 first issue of *Windirections*, the newsletter of the newly formed British Wind Energy Association; see Chapter 5, note 49. Further assessment by the

CEGB led to their estimate in 1981 that the gross wind power resource in UK waters less than 30 m deep was about 240 TWh/year, comparable with the total UK electricity demand. Even with generous allowance for the reduction in siting areas due to shipping, fishing and sea bed unsuitability the net offshore wind power resource was estimated to be 100 TWh/year, and the assessment also concluded that the offshore wind power resource was several times larger than the wave power resource; see Rockingham *et al.* (1981).

15. Musgrove (1978b) p. 213 gives further details of the derivation of the £410/kW capital cost estimate. £410/kW at 1976 price levels is equivalent, at 2007 price levels, to about £2100/kW. Milborrow (2007) indicates that offshore wind farm costs in 2006 were in the range €1600/kWh to €1900/kWh, and similar figures were given in 2005 by the Sustainable Development Commission, SDC (2005) p. 126; with £1 ≈ €1.45 in 2006 this gave an offshore wind farm cost in the range £1100/kW to £1300/kW. Shortages both of wind turbines and the vessels required to install them subsequently saw offshore wind farm costs double by mid 2008, see BERR (2008c) p. 88 and Carbon Trust (2008a) pp. 39 and 108, though the latter anticipates that there will be substantial cost reductions over the next decade.

16. Department of Energy (1979) vol. 1 p. 57 notes that the estimated capital costs for the wave energy devices then being considered were in the range £4000/kW to £9000/kW, and their forecast capacity factor was about 20%. The cost of supplying electricity to the grid from these wave power devices was expected to be in the range 20 to 50 p/kWh. See also Department of Energy (1979) vol. 2 pp. 112–3.

17. The average fuel cost in 1978–9 for the CEGB's most modern power stations (i.e. those built between 1965 and 1977) was 1.08 p/kWh for coal-fired power stations, and 1.13 p/kWh for oil-fired power stations; see CEGB (1980).

18. In hydro-electric pumped storage systems water is pumped from a low-level reservoir to a high-level reservoir (usually many hundreds of metres higher) during periods of power surplus; then in periods of power shortage the water is allowed to flow back down to the lower reservoir through turbines which generate electricity.

19. By 1979 the Department of Energy was using the conversion factor 1 mtce = 2.36 TWh (cf. note 1), see Department of Energy (1979) vol. 1 p. 68. The conversion from tonnes of coal burnt to tonnes of carbon dioxide emissions assumes that the carbon content of the coal is about 70%, which is a reasonably representative figure.

20. Taylor and Rockingham (1980) review and compare the results of system integration studies undertaken by the CEGB, Imperial College, the University of Reading and others; see also BWEA (1982) pp. 201–38.

21. BWEA (1982) pp. 210–12.

22. BWEA (1982) p. 220.

23. In the late 1970s the CEGB system had a total installed capacity of approximately 60 000 MW, provided by about 160 power stations (whose average capacity factor was just 40%); see Musgrove (1978b) p. 212.

24. Department of Energy (1979) vol. 2 pp. 116–7; see also Bedford (1980).

25. See Lindley and Stevenson (1981), Twidell *et al.* (1995) and BWEA (2003) p. 25. Construction of the LS-1 cost £12.2 million, equivalent to about £27 million at 2007 price levels. In the five-year period from 1988 through to 1992 it operated for a total of 7300 hours and generated a total of 10.1 million kWh. The forecast 12.5 m/s annual average wind speed at the turbine's hub height of 45 m was later revised down to 10.7 m/s.

26. Simpson and Lindley, (1980). See also BWEA (1982) pp. 239–70, which reviews early offshore wind power studies in the UK, the US and Sweden; as this notes (p. 259) the capital cost estimate from the American study was £566/kW, little more

than one-third of the UK (Taylor Woodrow) estimate and somewhat lower than my estimate (of £410/kW in 1976) which when adjusted for inflation through to January 1980 became £650/kW. The 1970s was a decade of high inflation in the UK so adjustments for inflation are very necessary.

27. House of Commons (1990) p. ix.
28. Department of Energy (1988) pp. 5 and 18.
29. Though the Department of Energy and the CEGB did support further desk-based studies of both the offshore wind energy resource and of suitable wind turbine and foundation designs, see Burton and Roberts (1985).
30. I was at that time a lecturer in the Engineering Department at the University of Reading and the 3 m diameter prototype vertical-axis wind turbine (VAWT) was built as a third-year student project in 1975–6. In the figure I am on the right and my first wind energy Ph.D. student, Ian Mays, is on the left. Ian subsequently became Managing Director of Renewable Energy Systems Ltd., a leading international developer, constructor, owner and operator of wind farms which is wholly owned by Sir Robert McAlpine & Sons Ltd.
31. The performance of the straight-bladed vertical-axis wind turbine is determined by the component of the wind speed perpendicular to the blade, not the wind speed itself, so as the blade inclination to the vertical increases the aerodynamic lift force on the blade decreases, see Musgrove (1977b).
32. Scaling up a wind turbine rotor design does not affect the optimum tip speed ratio, and at the same site (or one with a similar wind regime) the tip speed will consequently be unchanged. Appendix B, note 26 explains why the blade-bending stress that results from aerodynamic forces is unchanged by scaling, provided the tip speed is unchanged. Similar reasoning shows that the blade stress that results from centrifugal forces is also unchanged by scaling, but the blade stress resulting from the blade's own weight increases linearly with the blade size. See also Gasch and Twele (2002) pp. 223–7, who in addition show that the ratios between a rotor's exciting frequencies and its natural frequencies are unchanged by scaling.
33. Mays *et al.* (1988); also Powles *et al.* (1988). BWEA (2003) p. 38 shows the VAWT 25 with the blades fully inclined.
34. With a vertical-axis wind turbine the maximum angle between the relative wind and the blade occurs when the blade is crossing the wind, see Figure 5.6. If the rotor speed is constant this maximum angle steadily increases as the wind speed increases until the flow over the blade stalls. This is similar to what happens with a fixed-pitch horizontal-axis wind turbine; see the description of the Gedser mill in Chapter 4 and the related note 66. However with a vertical-axis wind turbine the angle between the relative wind and the blade varies cyclically as the blade goes round the vertical axis (and is zero when the blade is going directly upwind or directly downwind); consequently stall – when it occurs – happens cyclically twice per revolution. With an H-configuration vertical-axis wind turbine there was the expectation that the flow would stall simultaneously – and abruptly – along the whole length of the blade, potentially causing severe dynamic loading. In the event it was found that cyclic stall was a much more benign phenomenon than had been anticipated.
35. Morgan and Mays (1990); see also Mays *et al.* (1990). The designation VAWT 850 in the titles of these two references refers to the 850 m^2 rotor swept area of the 35 m diameter test machine, which for simplicity is referred to here as the VAWT 35. The report by Mays *et al.* notes that the VAWT 35 was self-starting, a characteristic that was also noted on the 20 m diameter machine installed on the Isle of Scilly, see Anderson *et al.* (1990). This ability to self-start was unexpected as smaller vertical-axis wind turbines with similar configurations and solidity do not self-start. The

difference in behaviour is most probably a consequence of fact that the larger blades of the larger turbines operate at substantially higher Reynolds numbers (see Appendix B, note 27), which would increase the stall angle and give a corresponding increase in the maximum aerodynamic lift.

36. See Chapter 5, note 44.

37. There was an insufficient quantity of load-carrying spanwise glass fibres at the blade mid-span position where failure occurred.

38. Bedford and Page (1996). The somewhat higher forecast cost of vertical-axis wind turbines is primarily the result of the higher optimum solidity and the consequent lower rotational speed. The latter means that torque levels are higher than for horizontal-axis wind turbines with similar power outputs, though the H-configuration vertical-axis wind turbine readily allows the use of a large-diameter power transmission system, such as a very large multi-pole generator or a large diameter 'open' gear ring driving multiple conventional generators. The higher solidity also means that the blade cross-section is much larger than for comparable horizontal-axis wind turbines, so that bending stresses caused by aerodynamic and centrifugal forces can be accommodated without the need to use the more exotic and expensive fibre-composite materials. The cross-arm weight is substantial but for very large vertical-axis wind turbines there are a variety of structural options for keeping the weight and cost down. Diameters well in excess of 200 m seem feasible, corresponding to rated power outputs well in excess of 20 MW.

39. These are very briefly summarised in BWEA (2003) pp. 20–3; see also BWEA (1982) pp. 124–34.

40. McGuigan (1978) pp. 79–82, Nickols (1979) and *Windirections* no. 3, December 1979, p. 3.

41. Stevenson and Somerville (1983), and Somerville (2000); see also the website of the Fair Isle Electricity Company. www.fairisle.org.uk/FIECo/index.htm.

42. Young and McLeish (1983), and BWEA (2003) p. 24. See also *Windirections* vol. 2 no. 3, January 1983, pp. 1–2 and *Windirections* vol. 4 no. 3, December 1984, p. 10.

43. Bullen *et al.* (1988).

44. Brown (1984), Jamieson and Hunter (1985) and Shearer and Brown (1986).

45. *Windirections* vol. 5 no. 4, April 1986, pp. 7–9; also vol. 6 no. 2, October 1986, p. 4.

46. *Windirections* vol. 7 no. 1, July 1987, p. 15; vol. 8 no. 4, Spring 1989, pp. 9–10; and vol. 9 no. 2, Autumn 1989, pp. 1, 5 and 6.

47. *Windirections* vol. 3 no. 2, October 1983, pp. 12–14; also Armstrong *et al.* (1984).

48. *Windirections* vol. 6 no. 3, January 1987, pp. 1, 13 and 20–21; also Elliott *et al.* (1990). The Wind Energy Group that built the MS-1 and LS-1 machines comprised Taylor-Woodrow, British Aerospace and GEC Energy Systems; the first two of these made the decision to develop commercial, medium-sized, wind turbines and formed the company, Wind Energy Group Ltd., which took forward the development of the MS-2 and MS-3 machines; see also BWEA (2003) p. 37.

49. *Windirections* vol. 8 no. 2, Autumn 1988, pp. 1, 4 and 12.

50. See for example SDC(2005) pp. 13–16; also EWEA (2004) pp. 40–4.

51. See Chapter 6, note 26.

52. See Musgrove (1998) pp. 21–2 and Clare *et al.* (1984) pp. 437–41; also *Windirections* vol. 3 no. 1, July 1983, p. 4.

53. As noted in Chapter 5 the price paid in Denmark in the early 1980s for the output from wind turbines was about 0.4 kroner/kWh, equivalent then to about 2.7 p/kWh; see Musgrove (1984) pp. 37 and 111.

54. Musgrove (1990b) p. 2; also Clare *et al.* (1984) pp. 441–3 and Leicester (1989) p. 3.

55. Operation and Maintenance (O&M) costs include servicing, consumables, repairs, insurance, site lease payments etc. and in the early 1980s totalled about 1p/kWh. (As noted in Chapter 5 a 55 kW Danish wind turbine in the early 1980s cost about £21 000 and on a good site would deliver about 90 thousand kWh/yr; O&M costs, see Chapter 6, would total about 4.6% of the capital cost.) And Gipe (1995) p. 234 indicates that O&M costs were about $0.02/kWh in the early 1980s, equivalent to about 1.3 p/kWh.
56. Musgrove (1990b) p. 2, and Clare *et al.* (1984) p. 442; also *Windirections* vol. 7, no. 4, Spring 1988, p. 27.
57. Musgrove (1990b) p. 2.
58. Marshall (1988); also *Windirections* vol. 7, no. 4, Spring 1988, pp. 22–4.
59. The CEGB's high-voltage transmission system would be allocated to another company, the National Grid Company, which was initially co-owned by the twelve regional supply companies. Parallel arrangements were made for privatisation of the electricity industry in Scotland. See Musgrove (1990b); also DTI (1995) pp. 7–16.
60. Electricity contracted and supplied under NFFO would be purchased at a premium price by the Non Fossil Purchasing Agency (NFPA) who would sell it to the electricity supply companies at the average price for electricity generated from fossil fuels. NFPA would recover the deficit through a levy – the fossil fuel levy – on all electricity generated from fossil fuel sources. In the six years following 1 April 1990 this levy added between 10% and 11% to electricity prices, and provided the nuclear power stations with a subsidy worth about £1 billion per year. See SDC (2006) p. 36. And Hansard, 29 January 1996, column 511 indicates that the levy payments for nuclear power averaged £1.19 billion per year in the period 1990 to 1995.
61. See Chapter 6, notes 20 and 21; also Musgrove (1990b) p. 3.
62. The European Commission indicated that they had imposed the 1998 deadline because the NFFO package presented to them for approval was predominantly a measure to provide support for nuclear power stations (and all but one were already operational). They made it clear that they would not object to support for renewables being provided over a longer period. See House of Commons (1990) p. xxxiv, para. 113.
63. *Windpower Monthly* vol. 6 no. 12 December 1990, pp. 6–7.

Chapter 8 Development and deployment, 1990 to 2008

1. See for example Weart (2003) pp. 154 et seq.; also RCEP (2000) pp. 1–32.
2. EWEA (1991). This report on *Wind Energy in Europe* came in six parts; a main report with the sub-title *Time for Action*, a summary report with the sub-title *A Plan of Action* and four more detailed wind energy status reports.
3. See EWEA (1991) *A Plan of Action* pp. 4 and 6; the mid 1991 total of 510 MW included 360 MW in Denmark, with 55 MW in both the Netherlands and Germany and with the remaining 40 MW divided between Spain, Greece, Italy, Belgium, Portugal and the UK.
4. At the end of 1990 the US had the highest installed capacity with 1500 MW, followed by Denmark with 343 MW, the Netherlands with 40 MW, Germany with 35 MW, India with about 30 MW, the UK with 8 MW and Spain with 7 MW; the European and global totals – as given in Table 8.1 – were 440 MW and 1980 MW respectively. By the end of 2000 Germany had the highest installed capacity with 6110 MW, followed by the US with 2560 MW, Spain with 2400 MW, Denmark with 2300 MW, India with 1220 MW, the Netherlands with 450 MW, the UK with 410 MW, Italy with 390 MW, China with 340 MW and Sweden with 230 MW; no other country had over 200 MW. The European and global totals were 13 000 MW and 17 700 MW. Table 8.2 lists the

installed capacities in the ten leading countries at the end of 2007; no other country had more than 2000 MW but another 15 had installed more than 200 MW. The European and global totals were then 57 100 MW and 93 900 MW. Data on the installed capacity at the end of the year 2007 for all countries with active wind programmes, plus global and European totals, were given in *Windpower Monthly* vol. 24 no. 4, April 2008, p. 82. Similar data for earlier years – from 1995 onwards – were given annually in earlier issues of *Windpower Monthly*, commencing in April 1997. Data for the years up to 1995 were given by EWEA (1999) pp. 200–3 and by EWEA (2004) p. 117. For the UK in 2007 the more precise installed capacity data given in DUKES (2008) p. 196 has been used (though with the addition of the 10 MW Beatrice offshore wind project – see note 93 – which is not included in the DUKES total as it is not grid-connected). The land areas stated in Table 8.2 are as given by the CIA World Factbook, see Chapter 6 note 26, and population densities have been calculated using population data from the same source.

5. The stated average oil price of $28/barrel (for the period 1986 through to 2002) was at 2007 price levels, see Chapter 1 note 33.
6. The American Energy Information Administration's *International Energy Outlook 2008* expects the oil price through to 2030 to be in the range $60/barrel to $115/barrel, at 2006 price levels; see EIA (2008) Highlights p. 2.
7. Danish Energy Agency (1999) p. 7 and *Windpower Monthly* vol. 10 no. 3, March 1994, p. 15. Also Krohn (1998) pp. 1–2.
8. Gipe (1995) pp. 60–2 and Danish Energy Agency (1999) p. 12.
9. See for example Stiesdal and Kruse (2000).
10. See *Windpower Monthly* vol. 11 no. 3, March 1995, pp. 39–42; also *Windpower Monthly* vol. 16 no. 3, March 2000, pp. 38–9.
11. As with Juul's Gedser mill – see Chapter 4 note 64 – the cables hang freely so that the nacelle can be turned to face into the wind without the use of slip rings, which are costly and a source of unreliability. The microprocessor controller that is used to control the wind turbine's operation keeps track over the months of how many complete turns the nacelle makes, and when necessary turns the nacelle back so as to unwind the cables.
12. The foundation is usually a flat disc of reinforced concrete, which for turbines the size of those used at Carno is typically about 11 m diameter and 1 m thick. As this is buried beneath one to two metres of soil it has a central cylindrical upstand – also made from reinforced concrete – whose diameter is slightly larger than the base of the tower. The tower is joined to the foundation using long vertical bolts that are embedded in the upstand, as can be seen in Figure 8.14. On very weak soils piled foundations will be used. See Hau (2006) pp. 431, 433 and 445–8; also EWEA (2004) pp. 66–7.
13. EWEA (1999) pp. 110–11.
14. *Windpower Monthly* vol. 17 no. 3, March 2001, p. 32; also Krohn (1998) p. 10 and Danish Energy Authority (2002) p. 6.
15. *Windpower Monthly* vol. 7 no. 9, September 1991, pp. 22–4 and *Wind Directions* vol. 16 no. 3, April 1997, pp. 12–14. See also EWEA (1999) pp. 41–2 and Hau (2006) pp. 620–1.
16. Danish Energy Authority (2002) p. 7; also Danish Energy Authority (2005) p. 20.
17. Danish Energy Authority (2002) pp. 7–10; also Danish Energy Agency (1999) pp. 9–10 and Danish Energy Authority (2005) pp. 3–4 & 7.
18. Following a succession of changes affecting the ownership of Danish utilities the Horns Rev wind farm is now co-owned by the Danish company Dong Energy (40%) and the Swedish company Vattenfall (60%). Comprehensive details of the project,

including its environmental assessment, are given on its website www.hornsrev.dk/engelsk/default_ie.htm. See also Hau (2006) pp. 623–6 and 634–6.

19. See Chapter 6 note 5, and Chapter 5 note 21.

20. An earlier variant of this wind turbine with a 2 MW power rating and a 76 m diameter rotor had previously been used for the 40 MW Middelgrunden wind farm, which was built in 2000 just 2 km offshore from Copenhagen and was the first offshore project to use multi-megawatt wind turbines. Following the changes affecting the ownership of Danish utilities referred to in note 18 the Nysted/Rødsand wind farm is now co-owned by Dong Energy (80%) and E.on Sweden (20%); comprehensive details of this project – including its environmental assessment – are given on its website www.dongenergy.com/Nysted/EN/index.htm.

21. In high wind speeds the flow over the blades stalls, as is usual with stall-controlled wind turbines, but in an arrangement introduced by Bonus – and known as active stall – the blade pitch is slowly adjusted so as to keep the average post-stall power output nearly constant; the power characteristic is then similar to that shown in Figure 6.2(b); see Stiesdal (1999).

22. See *Windpower Monthly* vol. 21 no. 4, April 2005, p. 52; *Windpower Monthly* vol. 22 no. 3, March 2006, p. 78 and *Windpower Monthly* vol. 23 no. 9, September 2007, p. 82.

23. *Windpower Monthly* vol. 23 no. 1, January 2007, pp. 55–6; see also Danish Energy Authority (2006).

24. Danish Energy Authority (2005) pp. 22–3. In Denmark the grid system operator provides the offshore transformer and the cable connection to the shore. If the wind farm owner had to pay these costs they would add about 0.045 kroner/kWh to the offshore wind energy generation cost, giving a total of 0.495 kroner/kWh; at the average exchange rate in 2002 of £1 = 11.8 kroner this was equivalent to about 4.2 p/kWh.

25. The financial viability of both Horns Rev and Nysted with a feed-in tariff of 0.45 kroner/kWh was helped by the fact that utilities are accustomed to taking a long-term view of their power generation investments, and their required rate of return has historically been relatively low, see Chapter 6 note 20. As Danish Energy Authority (2005) p. 23 indicates the premium price of 0.45 kroner/kWh applies only for the first 42 000 full-load hours. A capacity factor of 40% corresponds to 3500 full-load hours per year, and the premium price will therefore only be paid for about 12 years; after this the price paid for the output is capped at 0.36 kroner/kWh, equivalent (at the 2002 exchange rate) to 3.1 p/kWh. The Horns Rev and Nysted wind farms cost about £1200/kW to build; see SDC (2005) p. 30 and *Windpower Monthly* vol. 21 no. 1, January 2005, p. 33; see also the Horns Rev website, whose address is given in note 18.

26. Vestas (Denmark) with 20% market share in 2007 was followed by GE Wind (US) with 15%, Gamesa (Spain) with 14%, Enercon (Germany) with 13%, Suzlon (India) with 9%, Siemens (Denmark) with 6%, Acciona (Spain) with 4%, Goldwind (China) with 3.7%, Nordex (Germany) with 3.0% and Sinovel (China) with 3.0%; see *Windpower Monthly* vol. 24 no. 5, May 2008, p. 96.

27. See EWEA (2004) pp. 21–2. With pitch-controlled wind turbines the maximum downwind force on the blades occurs close to the rated wind speed, which is usually in the range 12 to 14 m/s; with stall-controlled turbines the downwind force on the blades (and therefore on the whole structure) is of similar magnitude at wind speeds close to rated, but continues to increase in higher wind speeds. With large blades the cost consequences of this increased downwind force become significant.

28. See Chapter 6 note 5, and Chapter 5 note 21.

29. Grid code requirements and compliance are discussed in some detail in *Windpower Monthly* vol. 21 no. 9, September 2005, pp. 47–62; see also Hau (2006) pp. 673–9.
30. *Windpower Monthly* vol. 24 no. 3, March 2008, p. 66.
31. The feed-in tariff applies for just the first 50 000 full-load hours, see note 25, and the exchange rate used is the average for 2005, £1 = 10.90 kroner. The Siemens 2.3 MW turbines for the Horns Rev extension are very similar to those shown in Figures 8.4 and 8.5, though the rotor diameter has been increased to 93 m and the power output is controlled by varying the blade pitch and using power electronics to provide wide range variable speed (as in the Siemens SWT-3.6–107). See Danish Energy Authority (2005) p. 23; also *Windpower Monthly* vol. 24 no. 1, January 2008, p. 39.
32. Danish Energy Authority press release dated 25 April 2008; also *Windpower Monthly* vol. 24 no. 5, May 2008, p. 60. The feed-in tariff applies for just the first 50 000 full-load hours, and the exchange rate of £1 = 9.74 kroner is the average for early 2008.
33. Danish Energy Authority (2002) p. 11; note that the quoted costs relate to a class 1 roughness location and for less windy sites the costs would be higher. Note also that the 0.48 kroner/kWh cost indicated for 1990 is at 1999 price levels; in *real* money terms (i.e. net of inflation) the cost-of-energy reduction over the decade was about 40%. Though inflation in Denmark through the 1990s was relatively low the cumulative loss of value of the kroner between 1990 and 1999 was 19%; the 'money of the day' cost of energy in 1990 for a class 1 roughness location was about 0.40 kroner/kWh.
34. Danish Energy Authority (2005) p. 23 and *Windpower Monthly* vol. 23 no. 11, November 2007, p. 46. In 2006 the average exchange rate was £1 = 10.94 kroner = €1.47.
35. Wollmerath (1992); also Gipe (1995) pp. 37–40.
36. These external costs are discussed at some length in EWEA (2004) pp. xiii, 110 and 147–71; see also Gipe (1995) pp. 428–34, and Freris and Infield (2008) pp. 205–9.
37. EWEA (1999) pp. 192–3 and 200, Gipe (1995) pp. 38–40, Breukers (2006) pp. 188–91, *Wind Directions* vol. 15 no. 1, October 1995, pp. 21–3 and *Windpower Monthly* vol. 14 no. 10, October 1998, p. 40. Though the main driver for growth was the feed-in tariff the availability of low interest rate loans for periods of up to 20 years was also an important factor; *Windpower Monthly* vol. 23 no. 3, March 2007, p. 58 gives further details of these loans.
38. EWEA (1999) p. 166 and *Windpower Monthly* vol. 17 no. 3, March 2001, pp. 29–30.
39. Breukers (2006) pp. 198–9 and 210; also *Windpower Monthly* vol. 20 no. 11, November 2004, pp. 59–62.
40. See Carlin *et al.* (2002) pp. 24–5; also *Wind Directions* vol. 17 no. 4, July 1998, pp. 21–2, EWEA (1999) p. 112 and Hau (2006) pp. 260–1, 276, 345–7 and 603–4.
41. Though Tacke started to develop wind turbines in 1984 in a joint venture with Renk, a company in the MAN group. MAN discontinued all their wind activities in 1990 but Tacke chose to continue, and formed Tacke Windtechnik; see EWEA (1999) p. 113 and *Wind Directions* vol. 16 no. 3, April 1997, pp. 22–3.
42. A turbine whose output was more than 150% of the output from a reference turbine at a standard site where the average wind speed is 5.5 m/s (as measured at a height of 30 m) would receive the higher price of DM 0.178/kWh for the minimum period of 5 years. A turbine in a less windy location whose output was lower than this 150% level would receive the higher price for a period extended by 2 months for each 0.75% that the output fell short of 150%; so for example if the output from a turbine was just 120% of the reference turbine's output the higher price would be paid for an additional 80 months, i.e. for a total period of 11 years and 8 months. See *Windpower*

Monthly vol. 16 no. 1, January 2000, pp. 17–18 and *Windpower Monthly* vol. 16 no. 3, March 2000, pp. 20–21; also www.bmu.de/files/pdfs/allgemein/application/pdf/eeg_en.pdf.

43. In Germany, as in most European countries, the euro replaced previous national currencies with effect from January 2002. For wind projects built in 2004 the EEG revision reduced the higher price payable (for the initial period of at least 5 years) from 8.80 to 8.70 cents/kWh; at the average 2004 exchange rate of £1 = €1.475 this corresponds to a reduction from 6.0 to 5.9 p/kWh. The lower rate payable for the remainder of the 20-year feed-in tariff period was reduced from 5.9 to 5.5 cents/kWh, which corresponds to a reduction from 4.0 to 3.7 p/kWh. For projects completed in 2005 the prices would be 2% lower, and for subsequent years through to 2013 prices would reduce by a further 2% per year. See *Windpower Monthly* vol. 20 no. 9, September 2004, p. 37; also BMU (2007) p. 10. As a consequence of increased wind turbine prices the EEG was further amended in June 2008, and with effect from January 2009 the price payable for the initial period of at least 5 years was increased by 17% to 9.2 cents/kWh, see *Windpower Monthly* vol. 24 no. 7, July 2008, p. 49; at the average 2008 exchange rate of £1 = €1.26 this equates to 7.3 p/kWh.

44. *Windpower Monthly* vol. 24 no. 3, March 2008, pp. 54–5. Preliminary data indicates that a further 1700 MW was installed in Germany in 2008, see Chapter 1 note 38.

45. See Chapter 5 note 80.

46. Hau (2006) pp. 537–9 and 543; see also Legerton (1996).

47. BMU (2007) pp. 10, 15 and 23; see also *Windpower Monthly* vol. 23 no. 8, August 2007, p. 6.

48. Denmark's wind power density of $74 \, \text{kW/km}^2$ benefits from the fact that it has 400 MW of offshore wind power capacity. Denmark's on land capacity of 2720 MW corresponds to a wind power density of $64.4 \, \text{kW/km}^2$, only marginally larger than Germany's $63.6 \, \text{kW/km}^2$.

49. The CIA World Factbook referenced in Chapter 6 note 26 also states each country's total electricity consumption; the per-capita consumption in Germany is 6620 kWh/year and in Denmark it is just 7% lower, at 6182 kWh/year. The population density is $236/\text{km}^2$ in Germany and $129/\text{km}^2$ in Denmark, as noted in Table 8.2.

50. BMU (2007) p. 22.

51. *Windpower Monthly* vol. 23 no. 11, November 2007, pp. 51–6 lists the 19 projects, which range in size up to 400 MW. Germany's Exclusive Economic Zone (EEZ), which by international agreement can extend out to 200 nautical miles ($\approx 370 \, \text{km}$), is strongly constrained by the competing rights of Denmark and the Netherlands; see the map, ibid. p. 53. The total available sea area is therefore restricted to about $40\,000 \, \text{km}^2$, and is mostly in the North Sea.

52. BMU (2007) pp. 10–11 and 22; see also *Windpower Monthly* vol. 24 no. 3, March 2008, p. 108 and *Windpower Monthly* vol. 24 no. 7, July 2008, p. 49. Offshore wind farms completed in 2016 will be paid 13 cents/kWh for their first 12 years; for projects completed subsequently this price will reduce by 5% per year.

53. In Germany the law requiring TSOs to provide and pay for the cable connections from offshore wind farms to the grid system on land (with the costs passed on to all electricity consumers) came into effect in December 2006; see *Windpower Monthly* vol. 22 no. 12, December 2006, pp. 35–6 and *Windpower Monthly* vol. 23 no. 11, November 2007, p. 52. The corresponding arrangements in Denmark are summarised in Danish Energy Agency (2005) pp. 16–17.

54. EWEA (1999) pp. 180 and 202 and *Windpower Monthly* vol. 11 no. 2, February 1995, p. 15. In December 1994 £1 = 206 pesetas.

55. Ragwitz and Huber (2005). Also *Windpower Monthly* vol. 21 no. 3, March 2005, pp. 35–7; *Windpower Monthly* vol. 22 no. 3, March 2006, pp. 55–6; *Windpower Monthly* vol. 23 no. 3, March 2007, pp. 61–2; *Windpower Monthly* vol. 24 no. 3, March 2008, pp. 52–4.
56. See Chapter 5 note 21.
57. Spain has a population of 40.5 million and Germany has a population of 82.4 million, see the CIA World Factbook referenced in Chapter 6 note 26; land areas are given in Table 8.2, as are the corresponding population densities of $81/km^2$ and $236/km^2$. The CIA World Factbook also indicates that electricity consumption in Spain is 243 billion kWh/year, equivalent to a per-capita consumption of 6000 kWh/year; this is about 10% lower than in Germany, see note 49.
58. *Windpower Monthly* vol. 24 no. 1, January 2008, p. 40.
59. See Chapter 7 note 62.
60. Edwards (1993); see also BWEA (2003) p. 50 and Gipe (1995) pp. 43, 278–281, 409–411 and 436–7.
61. Lindley *et al.* (1993).
62. Lindley *et al.* (1993). Also Gipe (1995) pp. 261–3 reviews in some detail the Cemmaes Inquiry and the conflicting perceptions of the visual change that a wind farm entails.
63. National Wind Power was then 50% owned by National Power, one of the two generating companies that between them had inherited the CEGB's portfolio of conventional power stations; the other 50% was owned in equal shares by Taylor Woodrow and British Aerospace, the two parent companies of the Wind Energy Group.
64. Ovenden (9.2 MW) and Kirkby Moor (4.8 MW) are both in England, in Yorkshire and Cumbria respectively; Chelker (1.2 MW) is also in Yorkshire and used four WEG MS-3 turbines, the same as at Cemmaes. Hannah (1997) lists all the wind farms built in the UK up to the end of 1997 and gives details of when they commenced operation, what turbines they used, etc.
65. At the height of the storm the sustained ten-minute average wind speed reached 40 m/s, with gusts up to about 60 m/s. Though the blades came off two turbines there was no risk to people. Anyone unwise enough to go onto the Mynydd y Cemais ridge that night would have been quite literally blown away. As Hoerner (1965) p. 3–14 notes the terminal velocity of the human body (when the drag on the body equals its weight) is typically in the range 45 to 55 m/s.
66. See *Windpower Monthly* vol. 14 no. 4, April 1998, pp. 31 and 34. WEG made their own wood-epoxy blades using technology they had developed, but which was inspired by the early wood-epoxy blades made by Gougeon Brothers in the US in the late 1970s (see Chapter 5). NEG Micon continued with the production of wood-epoxy blades, following their purchase of WEG, and used them for some of their own turbines. After the merger in 2004 between Vestas and NEG Micon the wood-epoxy blade making factory on the Isle of Wight was substantially expanded, and now employs several hundred people making blades for Vestas wind turbines.

Though the main business of Renewable Energy Systems Ltd (RES) has been as a developer, constructor, owner and operator of wind farms it was also responsible through the 1990s for the development of a 52 m diameter, 1 MW rated, three-bladed wind turbine with full-span variable-pitch blades. A pre-production demonstration machine was built and commenced generation at Slievenahanagan in Northern Ireland in 1999; see Noakes *et al.* (2000). By then, however, the volume manufacture of commercial wind turbines was well established in

Denmark, Germany, Spain and elsewhere, and RES did not proceed to put their machine into production.

67. The start date for each 15-year power purchase contract was the date when the project was completed and generation commenced. However each NFFO and SRO Order had a defined end date, and projects that were more than usually delayed in planning could end up with a contract that would run for substantially less than 15 years.

68. There were in fact two wind bands in each of the NFFO3, NFFO4 and NFFO5 Orders but for simplicity the table only gives data for the main band, which was for larger projects. The indicated contract prices are the average in the band for that year: in NFFO3 the range was from 4.0 to 4.8 p/kWh; in NFFO4 it was from 3.1 to 3.8 p/kWh; and in NFFO5 it was from 2.4 to 3.1 p/kWh. See BWEA (2003) pp. 52–3. Data on the capacity built comes from NFFO Fact Sheet 11, which includes all projects completed by December 2006: www.berr.gov.uk/files/file39336.pdf. Note that this fact sheet, like almost all UK official reports relating to NFFO and SRO, refers to capacity in 'MW dnc' and the declared net capacity (dnc) of a wind power project is defined by a Government statute (SI 1990 No 264) to be 43% of its sustained maximum power output, i.e. 43% of what is more normally referred to as the rated power output: www.opsi.gov.uk/si/si1990/Uksi_19900264_en_1.htm.

Projects smaller than about 3 MW competed against one another in a small wind band. (Though this limit was slightly different for each NFFO Order, and ranged from 1.8 MW in NFFO4 to 3.7 MW in NFFO3.) Prices in the small wind band in each Order were substantially higher than in the main wind band, reflecting the fact that small projects cost proportionately more to develop, build and maintain. Power purchase contract prices for these small wind projects averaged 5.3 p/kWh in NFFO3, reducing to 4.6 p/kWh in NFFO4 and to 4.2 p/kWh in NFFO5. Contracted capacities in the small wind bands were much smaller than in the main bands, totalling 46 MW in NFFO3, 24 MW in NFFO4 and 67 MW in NFFO5. The prices bid in Scotland for SRO contracts were generally a few tenths of a penny lower than NFFO bid prices, a consequence of the fact that wind speeds in Scotland are somewhat higher than in England and Wales.

69. See note 4 for the installed wind power capacities at the end of 2000; BWEA (2003) p. 73 lists the UK capacity year-by-year through the 1990s. The quoted total contracted wind power capacity of 2458 MW from the three later NFFO Orders and the three SRO Orders also includes 35 MW contracted in Northern Ireland in 1994 and 1996 under two NI-NFFO Orders. By the end of 2000 only about 260 MW of the contracted 2458 MW had been consented and built, in addition to about 150 MW from NFFO1 and NFFO2. Subsequently a further 200 MW of the later NFFO/SRO projects were consented and built to give a total of about 460 MW, i.e. only 19% of the contracted 2458 MW; see the NFFO Fact Sheet 11 referred to in note 68.

70. Visual impact, noise, birds, electromagnetic interference and shadow flicker are covered in EWEA (2004) pp. 179–86; see also SDC (2005) pp. 51–109, Hau (2006) pp. 537–61 and Gipe (1995) pp. 342–86. The Department for Business Enterprise and Regulatory Reform (BERR), which succeeded the Department of Trade and Industry (DTI) in June 2007, comments in its report *Wind Power: 10 Myths Explained* on a number of concerns that are sometimes also raised, including low-frequency sound and potential effects on tourism and house prices: www.berr.gov.uk/energy/sources/renewables/explained/wind/myths/page16060.html. See also the BWEA note, *Top myths about wind energy*: www.bwea.com/energy/myths.html.

71. See Chapter 5, note 44.

72. DTI (1996); a concise summary is provided by Legerton (1996). The issue of low-frequency noise was subsequently considered in some detail and a 2006 report 'concluded that there was no evidence of health effects arising from infrasound or low-frequency noise generated by wind farms'; see Hansard, 18 May 2007 column 1003W. The report has the BERR reference number URN 06/1412; www.berr.gov.uk/energy/sources/renewables/explained/wind/onshore-offshore/page31267.html. A later report commissioned by BERR and published as report URN 07/1235 investigated the overall level of wind farm noise complaints, as well as the particular issue of amplitude modulation (AM), and concluded that 'despite press articles to the contrary, the incidence of wind farm noise and AM in the UK is low'; www.berr.gov.uk/files/file40570.pdf. See also SDC (2005) pp. 75–81.

73. EWEA (2004) p. 194 reports that data from 42 surveys carried out in the UK between 1990 and 2002 showed, on average, that 77% of the public were in favour of wind energy with 9% against. It also noted that a summary on attitudes to wind power in the UK from 1990 to 1996 concluded that an 'overwhelming majority of residents in areas with a wind project are pro-wind, both in theory as a renewable energy source and in practice in their area, with an average of 8 out of 10 supporting their local wind farm'. The BERR report *Wind Power: 10 Myths Explained*, see note 70, notes similarly favourable attitudes in surveys that it commissioned. See also Gipe (1995) pp. 278–82; SDC (2005) pp. 82–5 and Freris (1998). A continuing high level of public support for wind power was confirmed by a May 2008 study undertaken for BERR, see BERR (2008c) p. 70.

74. Gipe (1995) p. 262. The quotation is from *Wind Energy Comes of Age* by P. Gipe, copyright © 1995 by John Wiley and Sons, Inc. and is reprinted with permission of John Wiley and Sons Inc.

75. Planning applications for any power-generating facility whose maximum sustained power output can exceed 50 MW are decided by a Government minister. No wind farm proposed in the 1990s was above this threshold.

76. Though Peter Edwards developed Britain's first wind farm on his own land at Delabole, in Cornwall, his achievement was exceptional; see Gipe (1995) pp. 278–9. Very few landowners were able to follow his example.

77. The required Environmental Impact Assessment (EIA) would typically include a full description of the proposed wind farm, showing the turbines to be used and where they would be sited, and accompanied by a comprehensive visual assessment. This would include computer-produced photo-montages from a wide range of locations around the wind farm. Background noise measurements made over an extended period at locations close to the nearest dwellings would need to be reported, together with a detailed assessment of what the noise levels would be when the wind turbines were operational; this would need to demonstrate compliance with the recommendations made in the Noise Working Group report (see note 72). The EIA would also show where access tracks would be built, where the cables between turbines would be buried and where the sub-station building would be located; it would also give details of how the wind farm would be built and demonstrate that there would be no significant adverse effect on the local hydrology. The results of comprehensive ecological assessments would also need to be included, covering all the existing fauna and flora on or close to the wind farm site with predictions from relevant experts as to the effect that the wind farm might have on them and – where appropriate – details of mitigating measures. The EIA would also need to demonstrate that there would be no significant adverse effect on television reception or on microwave communications in the area, and that there would be no significant adverse effect on the performance of military or civilian aviation radar systems. In

many cases there would also be some site-specific issues that would need to be addressed.

The 2005 report on 'Wind Power in the UK' produced by the Sustainable Development Commission covers in some detail the wide range of environmental issues that are usually considered in the Environmental Impact Assessment for a wind farm; SDC (2005) pp. 44–100 and 139–66.

78. For the purposes of the RO qualifying renewables include wind power (on land and offshore), hydro power (excluding projects built before 1990 and with a power output in excess of 20 MW, unless re-furbished), tidal power, wave power, power from landfill gas and power from the co-firing of biomass with fossil fuel; see *Energy Trends*: June 2008 pp. 17–21, BERR reference number URN 08/79b, at www.berr. gov.uk/files/file46668.pdf. See also DUKES (2007) pp. 170 and 192–3.

79. The specified percentage increased to 4.3% in the year commencing April 2003, then to 4.9% in 2004/5, to 5.5% in 2005/6, to 6.7% in 2006/7, to 7.9% in 2007/8, to 9.1% in 2008/9, to 9.7% in 2009/2010 and to 10.4% in April 2010. As originally legislated the obligation would remain at 10.4% until the end of March 2027, see the Government statute SI 2002 No 914, *The Renewables Obligation Order 2002*; www. opsi.gov.uk/si/si2002/uksi_20020914_en.pdf. The same percentages were specified by the corresponding statute in Scotland; see Scottish SI 2002 No 163, *The Renewables Obligation (Scotland) Order 2002*.

80. Most of the qualifying renewable electricity available in 2002/3 was provided by projects that had been built under NFFO, see DUKES (2007) pp. 173, 175 and 192–3. Changes made in 2001 to the arrangements for licensing electricity suppliers, together with the introduction of New Electricity Trading Arrangements (NETA), led to significant changes in NFFO contracts. From that year the Non Fossil Purchasing Agency (NFPA) became responsible for buying the output from all operating renewables projects that still had NFFO contracts, and they were required to pay generators the index-linked price specified in each NFFO contract for this output. The NFPA then sold the output from these NFFO projects to electricity suppliers through a continuing series of on-line auctions. See www.nfpa.co.uk/about.html and also www.nfpa.co.uk/auctionprices.html.

81. The system administrator is Ofgem (the Office of Gas and Electricity Markets), and suppliers must also provide evidence of their total electricity sales so that Ofgem can calculate the penalty payment that each supplier has to make. Ofgem is also responsible for issuing ROCs to renewable electricity generators, based on their qualifying monthly metered output.

82. ROCs corresponding to about 6 billion kWh (= 6 million MWh) were submitted at the end of 2002/3, leaving a shortfall of about 3 billion kWh which required payment of penalties totalling about £90 million. Sharing this penalty payment fund between the approximately 6 million ROCs submitted resulted in a repayment to suppliers of about £15 per ROC.

83. The NFPA auction website (see note 80) records that in 2001 the output from wind farms in the summer half-year was sold at an average price of 1.85 p/kWh; in the following winter half-year their output was sold at an average price of 2.84 p/kWh, reflecting the fact that in the UK the winter demand is substantially greater than in the summer. About two-thirds of the output from wind farms comes during these relatively windy winter months, so the weighted average selling price for wind in the period April 2001 through to March 2002 was 2.5 p/kWh. The weighted average selling price for the output from wind farms in the period April 2002 through to March 2003 (after the April 2002 commencement of the RO) was 6.5 p/kWh.

The Government introduced a Climate Change Levy in April 2001 which added 0.43 p/kWh to the cost of electricity generated using fossil fuels; electricity generated from renewable energy sources, such as wind power, was given exemption from this levy. The prices quoted for the output from wind farms both in 2001/2 and in 2002/3 include the approximately 0.4 p/kWh benefit of this levy exemption.

84. In the event ROC prices have remained high, and in the period 2002 through to 2008 they have ranged between 3.8 and 5.3 p/kWh; see www.e-roc.co.uk/trackrecord. htm. The year-by-year auction price paid for the output from wind farms (see note 80) has therefore also remained high; and helped by increased fossil fuel prices since 2003 the price has in 2008 exceeded 13 p/kWh (much to the benefit of the electricity supply companies that would only commit to long term power purchase contracts at prices in the range $4\frac{1}{2}$ to 5 p/kWh).

85. NFFO4 and NFFO5 prices averaged 3.5 p/kWh and 2.9 p/kWh respectively, see Table 8.3. SDC (2005) p. 29, Figure 10 (see also p. 128) shows that at these price levels the site average wind speed at hub height needs to exceed 8 m/s; the figure also shows that with long-term contract prices in the range 4.5 to 5 p/kWh the required hub height average wind speed reduces to about 6.5 m/s. Average wind speeds at 50 m height across much of Europe are shown – very approximately – in the map produced by the Risø National Laboratory, Denmark, and shown in SDC (2005) p. 14 and in EWEA (2004) p. 42; see also www.windpower.org/en/tour/wres/euromap.htm.

86. The end-of-year data plotted in Figure 8.10 come from the sources indicated in note 4. The installed UK wind power capacity at the end of 2004 was 900 MW but by the end of 2007 this had increased to 2425 MW to give an average of 510 MW per year over the three-year period.

It is worth noting that a feed-in tariff paying about 5 p/kWh would have achieved similar results, but at substantially lower cost to the electricity consumer. From a purely administrative perspective the RO had the advantage that its future cost could be accurately quantified, regardless of how much – or how little – renewable electricity generation it stimulated.

87. See Chapter 1 note 24; also Chapter 7.

88. BWEA (2003) p. 82; also Grainger and Den Rooijen (2001).

89. The Crown Estate is part of the hereditary possessions of the Sovereign, though it is managed by the Crown Estate Commissioners and its net income is paid to the UK Treasury, see www.thecrownestate.co.uk/about_us/our_history.htm. The Crown Estate is the landowner of the UK territorial seabed out to the 12 nautical miles (about 22 km) limit and the 2004 Energy Act gave it the right to issue leases for developments further offshore, out to the limit of the UK's Exclusive Economic Zone; as indicated in note 51 this can go out to 200 nautical miles though the similar rights of neighbouring countries usually reduce its extent. The total area of seabed controlled by the UK is about 870 000 km^2; see Carbon Trust (2008a) pp. 19–21.

90. *Windpower Monthly* vol. 17 no. 5, May 2001, pp. 44–6 lists the 18 sites and the developers responsible. Issues specific to consenting offshore wind projects are discussed by Trinick (2001).

91. The three areas were the Thames estuary, offshore from the Wash and offshore from the north-west coast of England, see DTI (2002) pp. 41–3. This same report also discusses Strategic Environmental Assessment and the offshore consenting process, see pp. 58–62 and pp. 64–70.

92. *Windpower Monthly* vol. 20 no. 1, January 2004, pp. 28–9 lists the 15 Round 2 sites and their proposed capacities, and names the developers who were initially responsible for the projects; there have been a number of subsequent changes of ownership.

93. Scroby Sands was completed in July 2004, just 2 km off the Norfolk coast from Great Yarmouth, and like North Hoyle used 30 Vestas 2 MW wind turbines. Kentish Flats was completed in July 2005, off the north Kent coast, and used thirty Vestas 3 MW turbines. Barrow, off the north-west coast, also used thirty Vestas 3 MW machines and was completed in March 2006. Burbo Bank, shown in Figure 8.12, was completed in July 2007 and used 25 Siemens 3.6 MW turbines. All these projects were in relatively shallow waters, typically 5 to 15 m deep, and used monopile foundations similar to those used at North Hoyle. The Beatrice project was also completed in July 2007, in waters more than 40 m deep and 22 km off the east coast of Scotland, to the north-east of Inverness and to the south of Wick. This used two 5 MW Repower wind turbines each with a diameter of 126 m to supply electricity to a nearby oil platform; and the greater depth of water required the use of a multi-piled jacket sub-structure, similar to those used for oil and gas platforms in what – to the oil and gas industry – is shallow water, see Carbon Trust (2008a) p. 45.

     These six projects have an overall capacity of 400 MW and with the addition of the 4 MW Blyth project (completed in 2000, see note 88) they gave the UK a total installed capacity of 404 MW at the end of 2007.
94. The combined Lynn and Inner Dowsing project was almost fully commissioned by the end of 2008, with 45 of the 54 Siemens 3.6 MW wind turbines operational. Robin Rigg (with 60 Vestas V90 3 MW turbines) is scheduled for completion in 2009, as are Gunfleet Sands (30 Siemens 3.6 MW) and Rhyl Flats (25 Siemens 3.6 MW). Offshore wind farms due for completion in 2010 include the first phase of Walney, a 450 MW Round 2 project whose 151 MW first phase will use 42 Siemens 3.6 MW machines. Another Round 2 project, the 504 MW Greater Gabbard, is scheduled for completion in 2011 using 140 Siemens 3.6 MW turbines. See *Windpower Monthly* vol. 24 no. 3, March 2008, pp. 104–5, *Windpower Monthly* vol. 24 no. 6, June 2008, pp. 52–4; see also the Crown Estate's interactive map at www.thecrownestate.co.uk/our_portfolio/interactive_maps/70_interactive_maps_marine.htm.
95. In 2004 wind farms on land typically cost about £700/kW to build while offshore wind farms cost about £1100/kW, though higher offshore wind speeds compensated in part for the extra cost; see *Windpower Monthly* vol. 21 no. 1, January 2005, p. 33 and also SDC (2005) pp. 29–30. By 2007 these costs had increased to about £900/kW on land and to about £1700/kW offshore; see *Windpower Monthly* vol. 24 no. 1, January 2008, p. 53. And in 2008 some offshore wind projects were costing as much as £2400/kW, see BERR (2008c) p. 88 and Carbon Trust (2008a) pp. 39 and 108, though the latter anticipates that there will be substantial cost reductions over the next decade.
96. The increase in five equal annual steps from 10.4% in the year commencing April 2010 to 15.4% in the year commencing April 2015 was announced in December 2003 and legislated in 2005 as Government statute SI 2005 No 926, *The Renewables Obligation Order 2005*; see www.opsi.gov.uk/si/si2005/uksi_20050926_en.pdf. The commitment to increase the RO to 20% on a 'guaranteed headroom' basis was made in the 2006 Energy Review report and re-stated in the May 2007 Energy White Paper, DTI (2007a) pp. 147–9. The 'guaranteed headroom' statement means that when renewable electricity generation in any year (up to 2027) exceeds 15.4% the Obligation will be increased (up to a maximum of 20%) just to the extent necessary to ensure that the required volume of renewable electricity generation is slightly greater; and the purpose of this arrangement is to ensure that there is no risk of ROC price collapse.

97. Between 2003 and 2007 wholesale electricity prices nearly doubled, from about 2.9 p/kWh to about 5.5 p/kWh; see Quarterly Energy Prices: June 2008 pp. 31 and 37, BERR reference number URN 08/276b: www.berr.gov.uk/files/file46669.pdf.
98. See *Windpower Monthly* vol. 23 no. 6, June 2007, pp. 29–30; see also DTI (2007a) pp. 147–53 and BERR (2008c) p. 92. The additional half a ROC for every MWh generated adds about 1.5 p/kWh to the long-term value of an offshore wind farm's output. Following the RO review wind farms on land will continue to receive 1 ROC per MWh.
99. *Windpower Monthly* vol. 23 no. 10, October 2007, pp. 57–62.
100. *Windpower Monthly* vol. 24 no. 7, July 2008, pp. 29–33, and BERR (2008c) pp. 57 & 86. See also the Crown Estate website; details of Round 3 are given at www.thecrownestate.co.uk/round3.
101. DUKES (2008) pp. 175–6.
102. *UK Renewable Energy Strategy*, BERR (2008c). The Secretary of State at the Department for Business, Enterprise and Regulatory Reform also asked the Renewables Advisory Board (RAB), an independent expert body sponsored by Government, to consider how the target of 15% energy from renewables by 2020 might be achieved. The RAB advised that electricity-generating renewables could provide 40% of UK electricity by 2020, with 70% of this provided by wind power. More specifically it proposed that the wind power capacity provided by wind farms on land should by 2020 be 13 000 MW, and that the installed offshore wind farm capacity should by then be 18 000 MW, see RAB (2008).
103. DUKES (2008) pp. 196–7 gives energy output data for wind and for other renewables. On a Renewables Directive basis the total electricity generation from renewables was 19.7 billion kWh in 2007, and was equal to 4.90% of UK electricity consumption; see also *Energy Trends*, June 2008 p. 19, www.berr.gov. uk/files/file46668.pdf. UK electricity consumption per home is about 4500 kWh/ year, see Chapter 1 note 4.
104. The Carbon Trust makes the case, in its October 2008 report on offshore wind power, that the UK should plan to install 29 000 MW of offshore wind power capacity by 2020, in addition to 11 000 MW on land. This would provide 31% of UK electricity needs at minimal (and potentially negative) additional cost to the electricity consumer; see Carbon Trust (2008a) pp. 2–3 and 27–8.
105. In the period between the end of 1990 and the end of 1997 wind projects with a total capacity of about 400 MW were actually built. However this figure was largely offset by the closure of about 300 MW of projects that had been built in the early 1980s; some of the turbines installed then proved too costly to maintain in operation when the initial 10-year premium price period of their power purchase contracts ended, and the contract price dropped to just a few cents per kWh.
106. These Standard Offer (SO) power purchase contracts were for 30 years, with the first 10 years at a specified price which price escalation provisions had increased to about 10 cents/kWh by 1990; see Chapter 5 note 72. Information on the Sky River project is given by Asmus (2001) pp. 185–6; see also *Windpower Monthly* vol. 7 no. 12, December 1991, p. 15.
107. One of these was the wind farm at Owenreagh in Northern Ireland, which used ten Z-40 wind turbines and was completed in January 1997.
108. Asmus (2001) pp. 172, 174–6 and 187–92; see also Righter (1995) p. 260 and *Windpower Monthly* vol. 14 no. 10, October 1998, p. 17.
109. Enron bought Zond in January 1997; see *Windpower Monthly* vol. 13 no. 6, June 1997, pp. 36–9. And in October 1997 Enron also bought Tacke, the German wind

turbine manufacturer that had collapsed financially two months earlier; see *Windpower Monthly* vol. 13 no. 11, November 1997, pp. 16–17. Asmus (2001) pp. 10 & 174–185 gives much additional information on the events leading up to Kenetech's bankruptcy in May 1996; see also *Windpower Monthly* vol. 12 no. 6, June 1996, pp. 20–21. Zond/Enron Wind purchased Kenetech's variable-speed drive patents after the latter's collapse, and were then able to use them to keep Enercon out of the US market; see Asmus (2001) pp. 158–9 and 185.

110.   The legislation that brought in the Production Tax Credit was the Energy Policy Act of 1992; see *Windpower Monthly* vol. 8 no. 11, November 1992, pp. 15–16 and Asmus (2001) p. 137. See also Wiser *et al.* (2007).

111.   States that were amongst the first to mandate specified levels of renewables development included Iowa, Minnesota and Texas. See *Wind Directions* vol. 18 no. 2, January 1999, pp. 16–19 and vol. 18 no. 5, July 1999, pp. 10–17; also *Windpower Monthly* vol. 15 no. 3, March 1999, pp. 34–40 and vol. 16 no. 3, March 2000, pp. 41–42.

112.   See *Windpower Monthly* vol. 16 no. 4, April 2000, pp. 42–7; also *Wind Directions* vol. 19 no. 4, May 2000, p. 13.

113.   *Windpower Monthly* vol. 18 no. 3, March 2002, pp. 20–1. GE had earlier been involved in the Federal wind programme in the 1970s and early 1980s, and had built the 2 MW rated Mod-1; see Chapter 5 and Figure 5.2.

114.   *Windpower Monthly* vol. 20 no. 3, March 2004, pp. 45–7; also *Wind Directions* vol. 23 no. 3, March 2004, pp. 18–19 and 38–41.

115.   GE Wind is profiled in *Windpower Monthly* vol. 21 no. 7, July 2005, pp. 38–9. Grid compliance issues are discussed in some detail in *Windpower Monthly* vol. 21 no. 7, July 2005, pp. 38–9. Figure 8.14 shows incidentally how steel towers are joined to the concrete foundations via a ring of bolts embedded in the foundation's cylindrical concrete upstand, see note 12.

116.   *Windpower Monthly* vol. 22 no. 3, March 2006, pp. 37–40.

117.   The PTC was extended in August 2005, for a period of two years from January 2006 through to the end of December 2007. Then in December 2006 it was extended for a further year, through to the end of December 2008; see Wiser *et al.* (2007) p. 2.

118.   *Windpower Monthly* vol. 24 no. 3, March 2008, pp. 40–42.

119.   *Windpower Monthly* vol. 23 no. 2, February 2007, p. 37; also *Windpower Monthly* vol. 24 no. 3, March 2008, p. 43 and no. 5, May 2008, pp. 40–2.

120.   Wiser and Bolinger (2008) indicate that wind power contributed 1.1% of total generation in the US in 2007.

121.   See Chapter 1 note 38.

122.   USDOE (2008) pp. 7–12, 17–18, 24 & 149. The 304 000 MW installed by 2030 would, it is estimated, give an annual electricity output of 1160 billion kWh, corresponding to an average capacity factor of 44%. The forecast total electricity consumption in 2030 is about 5800 billion kWh, 39% higher than in 2005.

123.   USDOE (2008) pp. 8–10, 48–54 and 124–6. See also *Windpower Monthly* vol. 23 no. 12, December 2007, pp. 42–3. Though several offshore wind projects are under development it seems unlikely that any will be built in the near future.

124.   USDOE (2008) pp. 11, 18–20 and 76–104.

125.   USDOE (2008) pp. 13–18 and 199–211. The required capacity of new coal-fired power stations would be reduced from 140 000 MW to 60 000 MW.

126.   Based on the data in the CIA World Factbook, see Chapter 6 note 26, the per capita electricity consumption in the US is 12 600 kWh/year. For European countries the corresponding figures are Denmark, 6200 kWh/year; Germany, 6630 kWh/year;

Spain, 6000 kWh/year; UK, 5730 kWh/year; and France, 7060 kWh/year. In China the per capita electricity consumption is 2150 kWh/year and in India it is just 430 kWh/year.

127. USDOE (2008) pp. 93–100.
128. *Windpower Monthly* vol. 24 no. 3, March 2008, pp. 43–6.
129. Ibid. pp. 105–6.
130. One project outside Japan and the US that used Mitsubishi wind turbines was the Llandinam wind farm, which was built in mid Wales in 1992. This used 103 Mitsubishi 300 kW machines, each with a 29 m diameter conventional three-bladed upwind rotor. See Hannah (1997) pp. 21–2; also *Wind Directions* vol. 12 no. 3, Winter 1992–3, pp. 1 and 14.
131. *Windpower Monthly* vol. 2 no. 7, July 1986, pp. 4–10; *Wind Directions* vol. 16 no. 3, April 1997, pp. 8–11 and *Windpower Monthly* vol. 21 no. 3, March 2005, pp. 65–6.
132. *Windpower Monthly* vol. 24 no. 5, May 2008, pp. 96–8. Also *Windpower Monthly* vol. 24 no. 3, March 2008, pp. 98–9 and Lewis (2007) pp. 5–9.
133. And as noted in Chapter 5 tubular towers – with their internal ladders – give weather protection to the maintenance engineers who have to climb up to access the machinery in the nacelle, an issue that is particularly important in regions (including northern Europe) where the most energetic winds occur in winter.
134. Parthan and Lemaire (2007).
135. *Windpower Monthly* vol. 9 no. 1, January 1993, pp. 24–6; see also Gipe (2004) pp. 13–14.
136. *Windpower Monthly* vol. 20 no. 3, March 2004, pp. 58–9; *Windpower Monthly* vol. 21 no. 12, December 2005, p. 28; *Windpower Monthly* vol. 22 no. 3, March 2006, pp. 45–6.
137. *Windpower Monthly* vol. 23 no. 3, March 2007, pp. 37–9; *Windpower Monthly* vol. 24 no. 4, April 2008, p. 82.
138. See Chapter 1 note 38.
139. *Windpower Monthly* vol. 22 no. 11, November 2006, pp. 27–8; *Windpower Monthly* vol. 24 no. 3, March 2008, pp. 97–8; *Windpower Monthly* vol. 24 no. 4, April 2008, pp. 34.
140. *Windpower Monthly* vol. 23 no. 12, December 2007, p. 56; also *Windpower Monthly* vol. 24 no. 4, April 2008, p. 34.
141. *Windpower Monthly* vol. 24 no. 7, July 2008, pp. 47–8; also *Windpower Monthly* vol. 23 no. 10, October 2007, pp. 49–55. £1 = CNY 12.9 was the average exchange rate in 2008.
142. *Windpower Monthly* vol. 24 no. 3, March 2008, p. 97; also *Windpower Monthly* vol. 24 no. 5, May 2008, p. 97 and *Windpower Monthly Special Report*, Opportunity and Risk in China, November 2008, p. 4.

Chapter 9 The future: from marginal to mainstream

1. See for example Chapter 5 note 1.
2. See the review of oil price changes in the 'costs compared with conventional generation' section of Chapter 1; also Chapter 1 note 33.
3. Exemplified in the UK by the 2008 Climate Change Act, which commits the UK to reduce all greenhouse gas emissions by 80% by 2050, see OPSI (2008); see also Weart (2003) and Stern (2007).
4. Ilex (2006) p. 1 and Smil (2003) p. 365.
5. See for example Socolow (2005), Evans (2007) pp. 75–80 and 173–4, DTI (2007a) pp. 172–9 and Stern (2007) pp. 249–53.

6. Socolow (2005) p. 40. Carbon dioxide emissions from coal-fired power stations are about 0.8 kg per kilowatt-hour of electricity generated, and are approximately twice as high as the emissions from gas-fired power stations. Carbon dioxide emissions associated with nuclear power and wind power (as a consequence of the energy and materials used in construction) are usually stated to be about 0.02 kg/kWh, see DTI (2006) p. 176; see also House of Commons (2006) pp. 37–9 which also reports other and somewhat higher data for nuclear power.
7. House of Commons (2008) pp. 4–5.
8. Stern (2007) p. 252 indicates that adding CCS will approximately double the cost of electricity generation from coal-fired power stations, but such cost estimates will remain very uncertain until a number of full-scale power stations with CCS have been built. And the Department of Trade and Industry noted, in the 2007 Energy White Paper, 'we cannot yet be sure that it will be technically feasible or economic to apply carbon capture and storage to electricity generation on a commercial scale'; DTI (2007a) p. 16.
9. House of Commons (2008) pp. 7–8.
10. See note 6; also SDC (2006) pp. 6–7.
11. As the House of Commons Environmental Audit Committee noted in its 2006 report 'no country has yet solved the problems of long-term disposal of high level waste', see House of Commons (2006) pp. 41–2; see also RCEP (2000) pp. 5 and 127, SDC (2006) pp. 7–8, 10 and 16–17 and Evans (2007) pp. 134–5.
12. Evans (2007) pp. 131–3; also House of Commons (2006) pp. 39–40.
13. House of Commons (2006) pp. 39–41; also SDC (2006) pp. 14–15.
14. SDC (2006) p. 17. Nuclear proliferation issues are further discussed in ibid. pp. 15–16 and 20, and in House of Commons (2006) pp. 39–41 and 71.
15. BERR (2008d) p. 61; see also DTI (2007b) pp. 66–70.
16. The chairman and chief executive of E.ON, Wulf Bernotat, gave the cost estimates in an interview with *The Times*, reported on 5 May 2008, and it is stated that the estimates were based on E.ON's experience as a partner in the construction of the 1600 MW Olkiluoto nuclear power plant in Finland, which when complete in about 2012 will be the first of the new 'third-generation' nuclear reactors. In Britain E.ON is the owner of Powergen. At the average exchange rate in 2008 of £1 = €1.26 the range of estimated costs corresponds to a nuclear power capital cost in the range £2500/kW to £3000/kW. And in the United States, in a hearing before the Florida Public Service Commission, Florida Power and Light indicated that the estimated cost of the two new nuclear power plants that it proposed to build at Turkey Point in south Florida would be in the range $5780/kW to $8070/kW, see for example Milborrow and Harrison (2009) p. 52; at the average exchange rate in 2008 of £1 = $1.86 this corresponds to a nuclear power capital cost in the range £3100/kW to £4300/kW.

In the closing months of 2008 the value of the pound declined substantially relative to both the dollar and the euro, and the average exchange rate in January 2009 was £1 = $1.45 = €1.09. If these exchange rates were to be sustained then the cost estimates for nuclear power plants given above would need to be further increased by about 20%. For comparison the nuclear power capital cost assumed in the UK Government's January 2008 'White Paper on Nuclear Power' (see note 15) was equivalent to £1750/kW; and the cost of energy from nuclear power stations is dominated by the capital cost.
17. See Chapter 1 note 14. In addition to the 87 000 billion kW flow of energy which the Earth receives from the Sun there is a relatively small flow of geothermal energy from beneath the surface; this is mostly the result of radioactive decay in the Earth's

crust and totals about 40 billion kW, i.e. an average of less than 0.1 W/m^2, see Smil (2003) p. 291.

18. Smil (2003) p. 293 indicates that on average tidal friction dissipates globally just under 3 billion kW. Though tides are primarily caused by the relative motion of the Earth and the Moon there is also a contribution from the relative motion of the Earth and the Sun.

19. The water reservoirs behind dams are significant sources of greenhouse gas emissions, and dam construction often results in the massive displacement of local people as well as severe adverse effects on aquatic biodiversity, both upstream and downstream. These and other environmental consequences are discussed in some detail by Smil (2003) pp. 246–259.

20. The first PV-powered satellite was Vanguard 1, launched in 1958; see Smil (2003) pp. 30–31.

21. See for example Evans (2007) pp. 81–114, Freris and Infield (2008) pp. 23–52 and Smil (2003) pp. 239–296. Smil notes, ibid. p. 274, that wind-powered electricity generation is seen as the most promising of all the new renewable technologies, and is far ahead of the others in terms of costs and operational reliability.

22. EC (2007) p. 19.

23. See for example the comments made in March 2008 by the UK Government's Chief Scientific Advisor. www.timesonline.co.uk/tol/news/environment/article3500954.ece

24. BERR (2008c) pp. 34–5 indicates that wind power is expected by 2020 to provide 68% of the 120 TWh of electricity from renewables; in 2006 wind power contributed just 4 TWh to the 19 TWh of electricity from renewables, see DUKES (2007) p. 193.

25. See Table 8.1; also Chapter 8 note 4 and Chapter 1 note 38.

26. See Hau (2006) pp. 232–244.

27. Wiser and Bolinger (2008) pp. 17 and 21; also Milborrow (2006) pp. 43–5 and Milborrow and Harrison (2009) pp. 52–3.

28. See Chapter 1 note 35.

29. See Chapter 1 note 36.

30. House of Commons (2006) pp. 45–6 indicates how the required rate of return can affect the calculated cost of energy; see also Milborrow (2006) and Milborrow and Harrison (2009).

31. Gas-fired power stations have low capital costs and the cost of the electricity they generate is dominated by the cost of their fuel; see Ilex (2006) p. 48.

32. In Germany the price paid in 2008 for wind-powered electricity generation was 8.03 cents/kWh, see Chapter 8 note 43; at the average exchange rate in 2008 of £1 = €1.26 this gives 6.4 p/kWh. The US figure of 3.3 p/kWh is for 2007, the latest available from Wiser and Bolinger (2008) p. 17; see also the US overview in Chapter 8. The price paid in China is for 2008; see the discussion in the China section in Chapter 8. The price paid in the UK for wind-powered electricity generation is the average for 2007 and 2008, and is calculated from the NFPA auction prices; see Chapter 8 note 80. The 1990 cost of wind-powered electricity generation was approximately 5.5 p/kWh, which at 2008 price levels becomes 9.4 p/kWh.

33. See for example the discussion of 'technology improvements on the horizon' in USDOE (2008) pp. 34–43; as is noted here wind technology 'is substantially less developed than fossil energy technology, which is still being improved after a century of generating electricity'.

34. Wiser and Bolinger (2008) p. 23 give the 2007 data for US wind farms; the German overview in Chapter 8 notes the 2007 capacity factor of 21%, see also Chapter 8 note 44.

35. DUKES (2008) p. 196.

36. One supposed advantage of the Renewables Obligation, at least from an administrative viewpoint, is that its cost can be accurately quantified for years in advance, regardless of how much – or how little – renewables capacity is actually built.

37. Though if the wind farms are spread over a large geographical area zero power would be very uncommon; Sinden (2005) p. 1 notes that low wind speed conditions affecting 90% or more of the UK occur in winter for an average of only about one hour every five years. The report also notes the good seasonal correlation in the UK between the availability of wind power and the demand for electricity; both are at their peak in the mid winter months.

38. Hau (2006) pp. 466–8 gives data covering a 31-year period (1967 to 1997) for a German coastal site showing that the year by year variations in a wind turbine's annual energy output would typically be less-than 10% of the long-term average energy output; variations exceeded 8% of the average for only 8 years out of 31, and the greatest variation in any year was 14%. And EWEA (2004) pp. 51–2 gives data on the average wind speed over a 20-year period (1979 to 1998) at an Irish coastal site which indicates that the standard deviation of the year by year variations was equal to about 5% of the long-term average wind speed; the standard deviation of the annual energy output of a wind farm at this location would consequently have been about 8% of the long-term average energy output.

39. See Chapter 1 note 22.

40. See the discussion in Chapter 6 in the section on 'shut-down in high wind speeds'; also Sinden (2005) p. 6.

41. UKERC (2006) pp. 41–2 reviews the fuel savings and carbon dioxide emission savings provided by wind power when it is used to meet up to 20% of the total electricity consumed, and notes that wind power's variability would reduce these savings by between zero and 7%, when compared with the savings that would be achieved if wind power was not variable. See also Chapter 1 note 27.

42. BERR (2008c) pp. 34–5. The Carbon Trust advocates a more ambitious renewables programme that would see wind power – mostly offshore – contributing 31% of UK electricity by 2020; see Carbon Trust (2008a) p. 27.

43. BP (2008) p. 44 indicates that 1 million tonnes of oil or oil equivalent produces about 4.4 billion kWh of electricity in a modern power station; and 1 tonne = 7.33 barrels.

44. The aggregate output from the many wind farms will lead to the temporary shut-down, in whole or part, of a number of fossil-fuelled power stations. If these were all gas-fired the carbon dioxide savings of about 0.4 kg/kWh (see note 6) would total about 33 million tonnes per year; if they were all coal-fired the carbon dioxide savings would be approximately double, i.e. about 66 million tonnes per year. The actual carbon dioxide savings will be somewhere in between these two limits. DTI (2007a) pp. 282 and 328 indicates that in 2020 the UK's total 'baseline' carbon dioxide emissions in 2020 will be about 550 million tonnes per year.

45. Carbon Trust (2008a) pp. 5 and 26–8.

46. USDOE (2008) p. 75.

47. Chapter 4 note 25.

48. House of Commons (2006) p. 44.

49. Ilex (2006) pp. 48–9. In the longer term demand management has the potential to significantly reduce the need for the supplementary power plants that are required to help maintain supply security; and these power plants could well be biomass-fuelled.

50. See for example UKERC (2006), Carbon Trust (2008a) and USDOE (2008); also Chapter 1 note 28.

51. BERR (2008c) p. 227.

52. Using the January 2009 exchange rate of £1 = €1.09, and with carbon dioxide emissions from a modern gas-fired power station of 0.4 kg/kWh; see note 6.
53. The Carbon Trust calculates a levelised cost in 2008 of about 10 p/kWh, see Carbon Trust (2008a) pp. 106–8; however in Denmark the contract to build the 200 MW Nysted/Rødsand offshore wind farm extension was won in April 2008 with a feed-in tariff that will pay 0.63 kroner/kWh, equal then to 6.5 p/kWh; see Chapter 8 note 32.
54. The vertical scale is in fact logarithmic and the dashed straight line corresponds to continuing exponential growth at the same rate as in the period 2001 to 2007, i.e. 25% per year. It would take growth at the reduced rate of 19% per year from the provisional capacity of 121 000 MW at the end of 2008 to reach 1 million MW by 2020.
55. EIA (2008) p. 61.
56. See for example USDOE (2008) pp. 61–4. The dominant material is steel, and wind turbines typically require about 100 tonnes per MW of installed capacity. The 2007 global production of just under 20 000 MW will therefore have used about 2 million tonnes of steel, which is approximately 0.2% of global annual steel production.
57. Hau (2006) pp. 708–9 shows this for wind turbines with diameters in the range from about 20 m to 90 m. The costs of the small grid-connected wind turbines that are sold for use on or close to individual dwellings are – relative to their rotor swept area – substantially higher. These small wind turbines are usually sited in urban or suburban areas where wind speeds are relatively low, and in many cases have an energy payback period that is well in excess of 10 years; see Carbon Trust (2008b), also BERR (2008c) pp. 136–40. Grid-connected small wind turbines generally represent a poor use of resources, but small wind-powered battery charging systems continue to be a cost-effective option in many locations that are not served by an electricity grid system.
58. Learning curves, and the effect of increased production volume on costs, are briefly discussed in Carbon Trust (2008a) pp. 48–9 and USDOE (2008) pp. 39–40.
59. See for example Hau (2006) pp. 741 and 748.
60. USDOE (2008); see the discussion in Chapter 8 under 'US overview'.
61. BERR (2008c) pp. 34–5, Carbon Trust (2008a) pp. 2 and 27.
62. See Milborrow and Harrison (2006), also *Windpower Monthly* vol. 23 no. 6, June 2007, pp. 33–4. The National Grid quotation comes from SDC (2005) p. 26.
63. The scenario proposed in USDOE (2008) for providing 20% of electricity from the wind by 2020 envisaged the installation of 304 000 MW of wind power capacity; 250 000 MW would be on land and the balance of 54 000 MW would be offshore. One option for delivering 50% by 2050 would be to double the capacity on land, with the balance of about 250 000 MW offshore. This would lead to a wind power density on land of 55 kW per square kilometre, somewhat less than in Germany and Denmark at the end of 2007.

   For the UK the scenario proposed in Carbon Trust (2008a) for providing 31% electricity from the wind by 2020 presumed the installation of 29 000 MW offshore and 11 000 MW on land. One option for providing 50% by 2050 would be to leave the installed capacity on land unchanged, and add an additional 22 000 MW offshore. The total wind power capacity in 2050 would then be 62 000 MW, with 51 000 MW offshore. At the 8 MW/km^2 that has been typical so far for offshore wind farms this wind power capacity would occupy a total area of about 6400 km^2, which represents less than 1% of the total UK sea area, and just 4% of all the UK sea area where the water depth is less than 60 m.

Appendix A: The power output from wind turbines

1. Sir Isaac Newton stated his three Laws of Motion in 1686; his third law states that for every Action (or Force, in today's terminology) there is an equal but opposite Reaction) See for example www.grc.nasa.gov/WWW/K-12/airplane/newton3.html.

2. Downstream from the turbine the natural turbulence in the wind gradually mixes the air that has been through the turbine (which has been slowed down) with the air that has gone round the turbine. This mixing process is effectively complete after about seven diameters downwind.

3. The speed of sound in air at normal sea-level temperatures is about 330 m/s; blade-tip speeds are typically in the range 60 m/s to 80 m/s, see EWEA (2004) p. 20. At the maximum normal operating wind speed for modern wind turbines (about 25 m/s) the air speed relative to the blade tip is then at most about 85 m/s. The Mach number is defined as the speed of the air flow divided by the speed of sound, and changes in the air density are negligibly small when the Mach number is less than about 0.3, i.e. when the air speed is less than about 100 m/s.

4. See for example Taylor (1983) pp. 19–21, Sharpe (1990) pp. 54–57, Burton *et al.* (2001) pp. 43–4, Manwell *et al.* (2002) pp. 85–6, Eggleston and Stoddard (1987) pp. 20–26.

5. Hau (2006) p. 29.

6. Both Sharpe (1990) and Wilson (1994) provide good introductions to the more detailed methods of calculating the performance of wind turbines; see also Eggleston and Stoddard (1987) pp. 15–65 and Manwell *et al.* (2002) pp. 83–119.

7. See for example Hau (2006) p. 143.

8. See Figure 5.9 for an example of a single-bladed modern wind turbine. There have been many recent two-bladed wind turbines – see for example Figures 5.3 and 7.8 – but relatively few with four blades, though some were deployed in California in the early 1980s; see Spera (1994) p. 617.

9. The curve for the modern wind turbine is representative of three-bladed megawatt-scale horizontal-axis wind turbines and the C_P values relate to the electrical power output, see for example Hau (2006) p. 491. For modern vertical-axis wind turbines (which are mostly two-bladed) the peak power coefficient will be of similar magnitude, but the optimum tip speed ratio will be somewhat lower, typically about 5; see Wilson (1994) p. 269. The performance of traditional windmills is reviewed in Appendix B, and the C_P values given in Figure A.2 are for the power to the millstones. The curve for American multi-bladed windmills is based on a combination of manufacturers' data and published independent test results, including Ball (1908) pp. 261–321, Wolff (1885) pp. 131–3 and Griffiths (1895). These multi-bladed windmills are mostly used for pumping water from boreholes, and the high solidity gives a relatively high torque at low wind speeds, so as to maximise the number of hours per year that some water can be pumped; C_P values are based on the useful pump power output, i.e. the volume of water delivered in a specified time and the height lifted, and so include pump losses.

10. See Chapter 2, note 16.

11. The power output is equal to the torque multiplied by the rotational speed; for a given power level if the rotational speed is low, the torque must be high, and vice versa.

12. Hoerner (1965) p. 3–15 indicates that the drag on a solid disc corresponds to a drag coefficient $C_D = 1.17$.

Appendix B: The performance of traditional windmills

1. Bennett and Elton (1899) p. 267.
2. Grist was the generic name for whatever foodstuff was to be ground, usually cereals such as wheat, barley and oats. However grist latterly came to refer mostly to animal feed. The grist ground at Little Dassett mill was a mixture of barley, wheat and beans; grist for animal feed was ground more coarsely, and so required less power. If, as at Little Dassett, the mill had two pairs of millstones the grist would usually be ground using one pair and wheat would be ground using the other. In Britain *corn* is the name given to the principal cereal in the region; in England usually wheat, in Scotland oats. In North America corn is the name given to maize, or Indian corn.
3. Bennett and Elton (1899) p. 268.
4. See Chapter 1, note 3.
5. Smeaton (1837) vol. 1, p. 169; see also Smith, D. (1981) p. 60. Smeaton's unit of volume measure was the Winchester bushel, which was equal to 35.2 litres and slightly smaller than the Imperial bushel (equal to 36.4 litres) that became the legal standard in the UK from 1826. The Winchester bushel has continued in widespread use in the United States.
6. Smeaton (1794) p. 29; see also Smith, N. (1981) p. 40.
7. This assumes, as seems reasonable, that the percentage power losses in the bearings and gearing of watermills are of similar magnitude to the losses in the bearings and gearing of windmills. Coulomb's measurements on oil-pressing windmills, discussed later in this appendix, indicate that the losses in the windshaft are about 10%; measurements made on Dutch drainage windmills and reported in Prinsenmolen Committee (1958) pp. 38 and 130, indicate that losses from the windshaft through to the scoop wheel shaft are an additional 15%.
8. Chapter 3 indicates that a bushel of wheat weighs about 28.6 kg; Smeaton's slightly smaller Winchester bushel would have a weight of about 27.7 kg.
9. Braudel (2002) p. 130 gives 3000 calories as the energy content of a kilogram of wheat; this is equivalent to 12.6 MJ (million joules), or to 3.5 kWh. The energy required to grind 1 kg of wheat between millstones is 0.014 kWh, which is just 0.4% of the grain's energy content.
10. See Gimpel (1979) p. 21, and Lawton (2004) p. 159.
11. A further cross-check is provided by the performance of the Albion flour mill built by Boulton and Watt in 1786 next to the Thames in London, see Chapter 3 note 54. This used two 50 hp steam engines to grind up to 160 bushels of wheat per hour, which corresponds to about 17 W to grind 1 kg/hr, though the precise power output of the steam engines is uncertain and the power provided was also used to sieve the output from the millstones.
12. Bennett and Elton (1899) p. 266.
13. Hassan and Sykes (1990) p. 13; Frost and Aspliden (1994) pp. 376–7.
14. Smeaton (1760). The papers Smeaton presented to the Royal Society in 1759 on the performance of waterwheels and windmill sails led them to award him the Copley medal, their highest award for original research. His tests on model waterwheels were significant in that they demonstrated conclusively that overshot wheels were much more efficient than undershot ones. The continuing interest in these papers led to them being published in book form, Smeaton (1794). See also Hills (1994) pp. 88–91, Smith, N. (1981) pp. 40–5, Prinsenmolen Committee (1958) pp. 22–7, and Golding (1955) pp. 13–15.

15. When a windmill is turning at its optimum speed (and for traditional windmills this corresponds to a sail tip speed which is about two and a half times the wind speed) the angle between the relative wind and the sail is small, see Figure 2.6(b). Aerodynamicists call this the angle of incidence and give it the symbol α (alpha). In aerodynamics terminology the *overall force* shown in the figure can be thought of as the sum of a *lift* force, which is perpendicular to the relative wind, and a *drag* force which is parallel to the relative wind, see Figure B.2; for efficient operation of the windmill the lift force needs to be large by comparison with the drag force. For structures that are effectively flat, such as the sails of traditional windmills (at least near the sail tips), the lift force is approximately constant once the angle α exceeds about 8°, but the drag force steadily increases, see Riegels (1961) p. 237; when α is 4° the drag force is only about one tenth of the lift force, at 8° it is one sixth, at 15° it is one quarter, at 20° it is one half and at 40° the drag force is about equal to the lift force.

　　As noted above at the sail tip the angle α is small, and the ratio of lift to drag is therefore high. However half-way between the sail tip and the central hub the circumferential speed is only half the tip speed; and as can be seen from Figure 2.6 (b) if the tip speed is replaced by a circumferential speed that is only half as large then the angle between the wind direction and the relative wind speed is substantially reduced, and the angle α between the relative wind and the sail is much increased. If the sail is untwisted this will result in a much increased drag force on the sails, which will reduce the power output. The drag can, however, be kept to a low level if the angle α between the relative wind and the sail is kept small by twisting the sail along its length, such that the angle between the sail and the plane of rotation (i.e. the 'weather') progressively increases towards the central hub. Brunnarius (1979) p. 14 notes that in Sussex the weather angle typically increased from about 4° at the tip to about 22° at the inner end.

16. Charles Coulomb (1736–1806) is best known for his work on the electrostatic forces between charged objects; the electrical unit of charge, the Coulomb, is named after him. His windmill measurements were reported first in 1781 to the Academie Royale des Sciences in Paris, see Coulomb (1785). They were subsequently published in book form with some of his other papers, Coulomb (1821) pp. 298–317. Coulomb's windmill measurements were often referenced in the nineteenth century, see for example Weisbach and Du Bois (1886) pp. 664–7 and Wolff (1885) pp. 125–8, but his pioneering work seems then to have been forgotten.

17. Hills (1994) pp. 173–8.

18. Stokhuyzen (1962) pp. 83–7.

19. The cup anemometer is the most frequently used instrument for measuring wind speed and consists of three or four hemispherical cups attached via short radial rods to a vertical central spindle, which is free to rotate. The drag force is greater when the wind blows into the open end of a cup than when it is blowing onto a cup's rounded surface. When exposed to the wind the cups therefore rotate around the vertical spindle, like a miniature drag-driven vertical-axis windmill, and the speed of rotation is proportional to the wind speed. For more details see Halliday (1990) pp. 34–5, Johnson (1985) p. 95 and Gipe (2004) pp. 49–52.

20. Prinsenmolen Committee (1958) p. 62 gives the 1936 test data, p. 67 gives the 1937 test data; p. 81 gives the corresponding air densities and the calculated values of the power coefficient.

21. Pearce (2002).

22. The weather angle on the Wicken windmill sails is 15° at the sail tips (increasing to 26° near the hub) and Pearce indicates that many windmills in Cambridgeshire and

East Anglia – where wide sails are favoured – have similar, high sail tip weather angles. Wide sails lead to a high solidity (such as Wicken's 29%) which in turn leads to a relatively low optimum tip speed ratio (≈ 2). And low tip speed ratios lead to an increase in the angle between the direction of the relative wind and the plane of rotation (see Figure 2.6). To avoid excessive drag the angle between the sails and the relative wind should be below about 15° (see note 15) and this can be achieved – for high-solidity rotors – by increasing the sail tip weather angle.

23. Prinsenmolen Committee (1958) pp. 68–70.

24. Close to the tip the Prinsenmolen's spar thickness is only 100 mm (0.7% of the sail length); at two thirds the distance from the hub to the tip, corresponding to the section shown in Figure B.2, the spar thickness is 200 mm; and at one third the distance from the hub to the tip the spar thickness is 310 mm. It should be noted that, like many other Dutch windmills built or rebuilt in the nineteenth century, the Prinsenmolen had iron spars. On English windmills spars continued to be made from wood and were consequently somewhat thicker, with a tip thickness typically about 1.5% of the sail length.

25. Prinsenmolen Committee pp. 98–106.

26. Consider two geometrically similar sails, one p times larger than the other, operating at the same tip speed ratio. (Smeaton's model windmills had an optimum tip speed of 2.6, virtually the same as the Prinsenmolen). The force on any object immersed in an airflow is proportional to its area and to the square of the air speed; the lift force on the sails is therefore proportional to their area multiplied by the square of the relative wind speed. At any given tip speed ratio the relative wind speed is proportional to the speed of the wind V approaching the windmill. The lift force on the sails is therefore proportional to their area and to the square of the wind speed, i.e. to $p^2 V^2$. The bending moment that the spar must be strong enough to withstand is proportional to this lift force multiplied by the length of the sail, and is therefore proportional to $p^3 V^2$. The stress that the bending moment produces in the spar (see for example Gordon (1978) pp. 377–9) is proportional to its thickness, and therefore to p; however it is also inversely proportional to the spar's second moment of area, and hence is also proportional to p^{-4}. Overall the stress in the spar is proportional to $(p^3 V^2)$ multiplied by p multiplied by p^{-4}; it is consequently proportional only to V^2. See also Gasch and Twele (2002) pp. 223–7.

27. It might be supposed that the relatively small size of Smeaton's model windmill sails could account, at least in part, for the fact that his careful measurements correspond to a power coefficient which is more that double what is measured on full size windmills. When comparing the flow around objects of similar shape but different size it is convenient to use a parameter called the Reynolds number, R, named after Osborne Reynolds (1842–1912) who was Professor of Engineering at Owens College, Manchester (a predecessor of the University of Manchester). His studies in the early 1880s revealed the significance of the Reynolds number, defined as $R = \rho V l / \mu$, where ρ (rho) is the fluid density, V is the fluid velocity, l is a length that indicates the size of the object and μ (mu) is the viscosity of the fluid. For air flows over aeroplane wings, or windmill sails, l is usually taken to be the distance from the leading edge to the trailing edge, which for windmills is the width of the sails; similarly the velocity V is usually taken to be the speed of the air relative to the sails, i.e. the *relative wind speed* shown in Figure B.2. The air density $\rho \approx 1.23$ kg/m³ (see Chapter 3, note 38) and the air viscosity $\mu \approx 18 \times 10^{-6}$ Ns/m². For a full size windmill operating in a typical wind speed of 8 m/s, and at a tip speed ratio of about 2.5, the *relative wind speed* is approximately 21 m/s; the width of the sails is typically in the range 2 to 3 m, and the Reynolds number for the flow

over the sails is therefore in the range 3 to 4 million. By contrast in Smeaton's experiments the whirling arm (see Figure B.1) moved the windmill models through the air at about 2 m/s. Smeaton found that his optimum tip speed ratio was about 2.6, which gives a *relative wind speed* – over the sails – of about 5.4 m/s. The width of the model sails was 0.14 m and the Reynolds number for the flow over them was therefore about 50 000.

The Reynolds number gives a measure of the relative importance of inertial forces and viscous forces in a fluid flow; and at low Reynolds numbers viscous forces are dominant. Though the Reynolds number for the flow over the sails of full size windmills is 60 to 80 times larger than for the flow over Smeaton's model windmill sails, in both cases the Reynolds number is high enough for inertial forces to be dominant and the flow patterns will therefore be similar. Riegels (1961) p. 237 shows how the lift and the drag on shapes that approximate to flat plates vary both with the angle of incidence (see note 15) and with the Reynolds number for Reynolds number values in the range 420 000 down to 42 000. As might be expected the lift and the important lift to drag ratio both reduce somewhat at the low end of this range, which would lead to the expectation that the performance of the model windmill sails would be somewhat *worse* than the performance of full size sails. As is discussed in this appendix the explanation for the superior performance of the model sails would seem to lie in the adverse effect that spar thickness has on the performance of full size sails.

28. Prinsenmolen Committee (1958) pp. 125–32. Adriaan Dekker had previously modified and streamlined the sails of a traditional Dutch windmill in a similar way, and tests in 1929 had given very positive results, indicative of a power coefficient $C_P \approx 0.19$; see Prinsenmolen Committee (1958) pp. 35–44.

29. Note that the empirical formulae for the power output from traditional windmills that were sometimes used in the nineteenth century, see for example Ball (1908) p. 254 and Encyclopaedia Britannica (1898) vol. 24 p. 109, incorrectly assumed that the power output was directly proportional to the total area of the sails. For solidities typical of traditional windmills, i.e. in the range between 15% and 30%, these formulae correspond to power coefficients in the range $C_P \approx 0.04$ to $C_P \approx 0.13$.

30. See Chapter 3, note 9.

31. See Chapter 3, note 39 and Appendix C.

32. This is the Rayleigh distribution of wind speeds, see Appendix C.

33. These stated average power outputs relate to a diameter, sail tip to sail tip, of 26 m. For other diameters the average power can be estimated using the fact that the power is proportional to the square of the diameter.

34. See Chapter 1, note 3.

35. Langdon (2004) pp. 142–3.

36. See Chapter 2, note 29.

37. See also Chapter 3, note 42.

38. Wulff (1966) p. 288.

39. Harverson (1991) p. 39.

40. See Table 3.1.

41. Harverson (1991) pp. 24–5.

42. Harverson (1991) pp. 4, 29 and 33.

Appendix C: Wind characteristics

1. See for example WMO (1981) pp. 5–19.

2. For a more detailed discussion of the variation of wind speed with height see Hassan and Sykes (1990) pp. 14–19, Gipe (2004) pp. 39–44 and Hau (2006) pp. 461–4.

3. See Hassan and Sykes (1990) pp. 19–21 and Gipe (2004) pp. 36–9. More general than the Rayleigh distribution, and applicable to a wider range of locations around the world, is the Weibull distribution. With this the probability p that the wind speed is greater than (or equal to) a specific value V_N is given by the expression $p = \exp[-(V_N/c)^k]$ where k and c are constants, see Johnson (1985) pp. 55–69, Manwell *et al.* (2002) pp. 56–59 and the Danish wind industry website www.windpower.org/en/tour/wres/weibull.htm. For most good wind regimes the so-called shape parameter k is within the range 1.5 to 3 and the scale parameter $c \approx 1.12V_A$, where V_A is the average wind speed; when $k = 2$ the Weibull distribution is identical to the Rayleigh distribution. At values of k above 2 the wind speed distribution is more peaked than that shown in the upper part of Figure C.1, and there is a reduction in the probability of very high wind speeds.
4. The availability is the number of hours in a period (such as a year) for which a wind turbine is – wind permitting – able to operate divided by the total number of hours in the period; it is usually expressed as a percentage.
5. The ratio between the average power in the wind and the power at the average wind speed is sometimes referred to as an energy pattern factor, or as a cube factor, see for example Golding (1955) pp. 28–35 and Gipe (2004) pp. 36–9. The value of 1.91 stated for the Rayleigh distribution is the limiting value that corresponds to a very large number of very narrow wind speed bands. For the more general Weibull distribution the energy pattern factor depends on the value of the shape parameter k; for example when $k = 2.5$ the energy pattern factor is 1.57, and when $k = 1.5$ the energy pattern factor is 2.72.

References

Amannsberger, K. and Hau, E. (1991). Status Report D: evaluation of the demand side of the market. In *Wind Energy in Europe*. Brussels: European Wind Energy Association.

Anderson, M. B., Gardner, P., Harris, A. and Tan, C. C. (1990). Comparison of measured and predicted results for a stall regulated 100 kW vertical axis wind turbine. In *Wind Energy Conversion 1990*, eds. T. Davies, J. Halliday and J. Palutikof. London: Mechanical Engineering Publications, pp. 335–46.

Argand, A. (1975). French investigations of large wind generators. In *Proceedings of the Second Washington Workshop on Wind Energy Conversion Systems*, ed. F. R. Eldridge. Report NSF-RA-N-75–050. Washington: National Science Foundation, pp. 173–81.

Armbrust, S. (1964). Regulating and control system of an experimental 100 kW wind electric plant operating parallel with an ac network. In *Proceedings of the Conference on New Sources of Energy, Rome, 1961, vol. 7, Wind Power*. New York: United Nations, pp. 201–5.

Armstrong, J. R. C., Greenwood, J., Lindley, D. and Smith, D. J. (1984). Installation, commissioning and operational aspects of the Orkney 20 m diameter WTG. In *Wind Energy Conversion 1984*, ed. P. J. Musgrove. Cambridge: Cambridge University Press, pp. 70–5.

Armytage, W. H. G. (1976). *A Social History of Engineering*. London: Faber and Faber.

Arnfred, J. T. (1964). Developments and potential improvements in wind power utilization. In *Proceedings of the Conference on New Sources of Energy, Rome, 1961, vol. 7, Wind Power*. New York: United Nations, pp. 376–81.

Asmus, P. (2001). *Reaping the Wind*. Washington DC: Island Press.

Baker, T. L. (1985). *A Field Guide to American Windmills*. University of Oklahoma Press.

Ball, R. S. (1908). *Natural Sources of Power*. London: Constable.

Bauters, P. (1982). The oldest references to windmills in Europe. In *Transactions of the Fifth Symposium of the International Molinological Society*, pp. 111–19.

Bedford, L. A. W. (1980). Review of the UK wind programme. In *Proceedings of the Second BWEA Wind Energy Workshop*, ed. P. J. Musgrove. London: Multi-Science Publishing, pp. 1–4.

Bedford, L. A. W. and Page, D. I. (1996). *The Development of Straight-bladed Vertical Axis Wind Turbine Generators in the United Kingdom 1980–1993*. Report ETSU-R-94. Harwell, Oxfordshire: Energy Technology Support Unit.

Beedell, S. (1975). *Windmills*. London: David & Charles.

Bennett, R. and Elton, J. (1899) *History of Corn Milling, vol. 2, Watermills and Windmills*. London: Simpkin, Marshall.

BERR (2008a). *Regional and Local Authority Electricity Consumption Statistics*. Report URN 08/487c. London: Department for Business Enterprise and Regulatory Reform. www.berr.gov.uk/whatwedo/energy/statistics/regional/regional-local-electricity/page36213.html.

BERR (2008b). *UK Energy in Brief July 2008*. Report URN 08/220. London: Department for Business Enterprise and Regulatory Reform. www.berr.gov.uk/whatwedo/energy/statistics/publications/in-brief/page17222.html.

BERR (2008c). *UK Renewable Energy Strategy: Consultation*. London: Department for Business Enterprise and Regulatory Reform. http://renewableconsultation.berr.gov.uk/consultation/consultation_summary.

BERR (2008d). *Meeting the Energy Challenge: a White Paper on Nuclear Power*. Report Cm 7296. London: TSO. www.berr.gov.uk/files/file43006.pdf.

Blom, L. H. (1999). *The Windmills of the Greek Islands*. Watford: The International Molinological Society.

Blyth, J. (1888). On the application of wind power to the generation and storage of electricity. *Proceedings of the Royal Philosophical Society of Glasgow*, **19**, 363–4.

Blyth, J. (1894). On the application of wind power to the production of electric currents. *Transactions of the Royal Scottish Society of Arts*, **13** (for 1892), 173–81.

BMU (2007). *EEG – The Renewable Energy Sources Act – the Success Story of Sustainable Policies for Germany*. Berlin: Federal Ministry for the Environment, Nature Conservation and Nuclear Safety. www.bmu.de/english/renewable_energy/downloads/doc/40066.php.

Böckler, G. A. (1661). *Theatrum Machinarum Novum*. Nürnberg: P. Fürsten, plate 49. http://kinematic.library.cornell.edu:8190/index.html#bockler1.

Bonnefille, R. (1976). French contribution to wind power development by EDF 1958–1966. In *Proceedings of the 1974 Stockholm Workshop on Advanced Wind Energy Systems*, vol. 1. Stockholm: STU/Vattenfall, pp. 1–17 to 1–22.

Bowers, B. (1982). *A History of Electric Light & Power*. London: Peter Peregrinus.

BP (2008). *Statistical Review of World Energy 2008*. London: BP. www.bp.com/statisticalreview.

Braasch, R. H. (1979). Darrieus vertical axis wind turbine program overview. In *Proceedings of the Fourth Biennial Wind Energy Conference and Workshop, Washington*. Report NTIS 791097. Springfield, Virginia: National Technical Information Service, pp. 39–58.

Braudel, F. (2002). *The Structures of Everyday Life*. London: Phoenix Press.

Breukers, S. (2006). *Changing Institutional Landscapes for Implementing Wind Power*. Amsterdam: Vossiuspers UvA – Amsterdam University Press.

Brown, A. (1984). Operating experience with the Howden HWP-300 aerogenerator. In *Wind Energy Conversion 1984*, ed. P. J. Musgrove. Cambridge University Press, pp. 59–69.

Brown, F. (1978). Power could come out of the Wash. *Electrical Review*, **202** no.7, 16–17, (17 February 1978).

Brunnarius, M. (1979). *The Windmills of Sussex*. London: Phillimore & Company.

Bullen, P. R., Mewburn-Crook, A. and Read, S. (1988). The engineering and economics of an augmentor for a vertical axis wind turbine. In *Wind Energy Conversion 1988*, ed. D. J. Milborrow. London: Mechanical Engineering Publications, pp. 221–6.

Burton, A. L. and Roberts, S. C. (1985). The outline design and costing of 100 m diameter wind turbines in an offshore area. In *Wind Energy Conversion 1985*, ed. A. D. Garrad. London: Mechanical Engineering Publications, pp. 269–77.

Burton, T., Sharpe, D., Jenkins, N. and Bossanyi E. (2001). *Wind Energy Handbook*. Chichester: John Wiley and Sons.

Butler, H. E. (ed.) (1949). *The Chronicle of Jocelin of Brakelond*. London: Nelson.

BWEA (1982). *Wind Energy for the Eighties*, eds. N. Lipman, P. Musgrove, and G. Pontin. Stevenage: Peter Peregrinus.

BWEA (2003). *BWEA 1978 – 2003 & Beyond*, eds. A. Hill, P. Musgrove, and G. Rajgor. London: British Wind Energy Association.

Cabinet Office Strategy Unit (2002). *The Energy Review*. London: The Cabinet Office. www.cabinetoffice.gov.uk/strategy/work_areas/energy.aspx.

Calvert, N. G. (1971). Windpower in Eastern Crete. *Transactions of the Newcomen Society*, **44**, 137–44 and Plate 32.

Cameron Brown, C. A. (1933). *Windmills for the Generation of Electricity*. Oxford: Oxford University Institute for Research in Agricultural Engineering.

Carbon Trust (2008a). *Offshore Wind Power: Big Challenge, Big Opportunity*. Report CTC743. London: The Carbon Trust. www.carbontrust.co.uk/Publications/publicationdetail.htm?productid=CTC743&metaNoCache=1.

Carbon Trust (2008b). *Small-scale Wind Energy*. Report CTC738. London: The Carbon Trust.

Carlin, P. W., Laxson, A. S. and Muljadi, E. B. (2002). *The History and State of the Art of Variable-Speed Wind Turbine Technology*. Report NREL/TP-500–28607. Golden, Colorado: National Renewable Energy Laboratory. www.nrel.gov/docs/fy01osti/28607.pdf.

Carter, J. M. (2007). North Hoyle offshore wind farm: design and build. *Proceedings of the Institution of Civil Engineers, Energy*, **160**, no. 1, pp. 21–9.

CEGB (1978). *Renewable Sources of Energy: the Prospects for Electricity*. Report G888. London: Central Electricity Generating Board.

CEGB (1980). *Costs of Producing Electricity*. Report G964. London: Central Electricity Generating Board.

Clare, R., Shaw, T. and Bossanyi, E. (1984). The economics and local taxation position of privately-owned wind turbines operated by consumers connected to the UK national electrical network. In *Wind Energy Conversion 1984*, ed. P. J. Musgrove. Cambridge: Cambridge University Press, pp. 426–44.

Claudi-Westh, H. (1975). A comparison of wind turbine generators. In *Proceedings of the Second Washington Workshop on Wind Energy Conversion Systems*, ed. F. R. Eldridge. Report NSF-RA-N-75–050. Washington: National Science Foundation, pp. 156–61.

Claudi-Westh, H. (1976). Some early Danish experiences – supplement. In *Proceedings of the 1974 Stockholm Workshop on Advanced Wind Energy Systems*, vol. 1. Stockholm: STU/Vattenfall, pp. 1–13 to 1–15.

Coulomb, C. A. (1785). Observations théoriques et expérimentales sur l'effet des mou-lins à vent et sur la figure de leurs ailes. *Memoires de l'Academie Royale des Sciences*, (for 1781) 65–81.

Coulomb, C. A. (1821). *Théorie des Machines Simples*. Paris: Bachelier.

Coutant, Y. (2001). *Windmill Technology in Flanders in the 14th and 15th centuries*. Watford, England: The International Molinological Society.

Cunliffe, B., Bartlett, R., Morrill, J., Briggs, A. and Bourke, J. (eds.) (2004).*The Penguin Illustrated History of Britain and Ireland*. London: Penguin Books.

Danish Energy Agency (1999). *Wind Power in Denmark Technology, Policies and Results*. Copenhagen. www.ens.dk/graphics/Publikationer/Forsyning_UK/Wind_Power99.pdf.

Danish Energy Authority (2002). *Wind Energy in Denmark, Status 2001*. Copenhagen.

Danish Energy Authority (2005). *Offshore Wind Power – Danish Experiences and Solutions*. Copenhagen. www.ens.dk/graphics/Publikationer/Havvindmoeller/uk_vindmoeller_okt05/index.htm.

Danish Energy Authority (2006). *Offshore Wind Farms and the Environment*. Copenhagen. www.ens.dk/graphics/Publikationer/Havvindmoeller/Offshore_wind_farms_nov06/index.htm.

Darby, H. C. (1977). *Domesday England*. Cambridge: Cambridge University Press.

Darrieus, G. (1926). *Turbine à Axe de Rotation Transversal à la Direction du Courant*. French patent no. 604,390.

Daudet, A. (Translation by F. Davies) (1978). *Letters from my Windmill*. London: Penguin Books.

DEFU (1981). *Large Scale, Electricity Producing Wind Energy Conversion Systems*. Report EEV 81–01E. Lyngby, Denmark: The Research Association of the Danish Electricity Supply Undertakings.

Delafond, F. H. (1964). Méthodes d'essais employées sur l'aérogénérateur 100 kW Enfield-Andreau de Grand Vent. In *Proceedings of the Conference on New Sources of Energy, Rome, 1961, vol. 7, Wind Power*. New York: United Nations, pp. 285–93.

Department of Energy (1976). *Energy Research and Development in the United Kingdom*, Energy Paper 11. London: HMSO.

Department of Energy (1977). *The Prospects for the Generation of Electricity from Wind Energy in the United Kingdom*, Energy Paper 21. London: HMSO.

Department of Energy (1979). *Energy Technologies for the United Kingdom*. Energy Paper 39. London: HMSO.

Department of Energy (1987). *Renewable Energy in Britain*. Harwell, Oxfordshire: Energy Technology Support Unit.

Department of Energy (1988). *Renewable Energy in the UK: the Way Forward*. Energy Paper 55. London: HMSO.

Divone, L. V. (1981). Wind energy development in North America. In *Proceedings of the International Colloquium on Wind Energy, Brighton, 1981*. Cranfield: BHRA Fluid Engineering, pp. 68–90.

Divone, L. V. (1984) Review of U.S. wind energy. In *Wind Energy Conversion 1984*, ed. P. J. Musgrove. Cambridge: Cambridge University Press, pp. 17–32.

Divone, L. V. (1994). Evolution of modern wind turbines. In *Wind Turbine Technology*, ed. D. A. Spera. New York: American Society of Mechanical Engineers, pp. 73–138.

Drees, J. (1976). Blade twist, droop snoot and forward spars. *Vertiflite*, **22** no. 5, 4–9.

DTI (1995). *The Prospects for Nuclear Power in the UK*. Report Cm 2860. London: HMSO.

DTI (1996). *The Assessment and Rating of Noise from Wind Farms, ETSU-R-97*. Report URN 96/1192. London: Department of Trade and Industry. www.berr. gov.uk/energy/sources/renewables/explained/wind/onshore-offshore/page21743. html.

DTI (2000). *Energy Projections for the UK*, Energy Paper 68. London: TSO.

DTI (2001). *Wind Power: Environmental and Safety Issues*. London: Department of Trade and Industry. www.berr.gov.uk/files/file17777.pdf.

DTI (2002). *Future Offshore*. Report URN 02/1327. London: Department of Trade and Industry. www.berr.gov.uk/files/file22791.pdf.

DTI (2003). *Our Energy Future – Creating a Low Carbon Economy*. Report Cm 5761. London: TSO.

DTI (2006). *The Energy Challenge*. Report Cm 6887. London: TSO.

DTI (2007a). *Meeting the Energy Challenge*. Report Cm 7124. London: TSO. www.berr. gov.uk/energy/whitepaper/page39534.html.

DTI (2007b). *The Future of Nuclear Power*. Report URN 07/970. London: Department of Trade and Industry. www.berr.gov.uk/files/file39197.pdf

DUKES (2007). *Digest of United Kingdom Energy Statistics, 2007*. London: TSO. www. berr.gov.uk/energy/statistics/publications/dukes/page39771.html.

DUKES (2008). *Digest of United Kingdom Energy Statistics, 2008*. London: TSO. www. berr.gov.uk/energy/statistics/publications/dukes/page45537.html.

Dyer, C. (1989). *Standards of Living in the Later Middle Ages*. Cambridge: Cambridge University Press.

EC (2007). *Renewable Energy Roadmap*. Report COM(2006) 848 final. Brussels: European Commission. http://ec.europa.eu/energy/energy_policy/doc/03_renewable_ energy_roadmap_en.pdf.

Edwards, P. (1993). The performance and problems of, and the public attitude to, the Delabole wind farm. In *Wind Energy Conversion 1993*, ed. K. F. Pitcher. London: Mechanical Engineering Publications, pp. 23–5.

Eggleston, D. M. and Stoddard, F. S. (1987). *Wind Turbine Engineering Design*. New York: Van Nostrand Reinhold.

EIA (2001). *Residential Energy Consumption Survey 2001*. Washington DC: Energy Information Administration. www.eia.doe.gov/emeu/recs/recs2001/ce_pdf/enduse/ ce1-2c_construction2001.pdf.

EIA (2008). *International Energy Outlook 2008*. Report DOE/EIA-0484. Washington DC: Energy Information Administration. www.eia.doe.gov/oiaf/ieo/index.html.

Eldridge, F. (1980). *Wind Machines*. New York: Van Nostrand Reinhold.

Elliott, D. E. (1975). Economic wind power. *Applied Energy*, **1**, 167–97.

Elliott, P. M., Gamble, C. R. and Lindley, D. (1990). WEG's California windfarm: a review of the first three years. In *Wind Energy Conversion 1990*, eds. T. Davies, J. Halliday and J. Palutikof. London: Mechanical Engineering Publications, pp. 15–17.

Encyclopaedia Britannica (1898). Ninth edition. Edinburgh: A. and C. Black.

Evans, R. L. (2007). *Fuelling Our Future*. Cambridge: Cambridge University Press.

EWEA (1991). *Wind Energy in Europe*. Brussels: European Wind Energy Association.

EWEA (1999). *Wind Energy The Facts*. Report DG XVII – 98/32. Luxembourg: European Commission. http://ec.europa.eu/energy/res/sectors/doc/wind_energy/ewea_the_facts.pdf.

EWEA (2004). *Wind Energy The Facts*. Brussels: European Wind Energy Association. www.ewea.org/index.php?id=91.

Flettner, A. (Translation by F. Willhofft) (1926). *The Story of the Rotor*. London: Crosby Lockwood.

Forbes, R. J. (1956). Power. In *A History of Technology*, vol. 2, ed. C. Singer. Oxford: Clarendon Press, pp. 589–622.

Forbes, R. J. (1958). Power to 1850. In *A History of Technology*, vol. 4, ed. C. Singer. Oxford: Clarendon Press, pp. 148–67.

Freris, L. and Infield, D. (2008). *Renewable Energy in Power Systems*. Chichester: John Wiley and Sons.

Freris, K. (1998). Love them or loathe them? Public opinion and windfarms. In *Wind Energy 1998*, ed. S. J. Powles. London: Professional Engineering Publishing, pp. 63–9.

Frost W. and Aspliden C. (1994). Characteristics of the wind. In *Wind Turbine Technology*, ed. D. A. Spera. New York: American Society of Mechanical Engineers, pp. 371–445.

Garrad, A. D. (1990). Forces and dynamics of horizontal axis wind turbines. In *Wind Energy Conversion Systems*, ed. L. L. Freris. London: Prentice Hall International, pp. 119–44.

Gasch, R. and Twele, J. (2002). *Wind Power Plants*. London: James and James.

Gimpel, J. (1979). *The Medieval Machine*. London: Futura Publications.

Gipe, P. (1995). *Wind Energy Comes Of Age*. New York: John Wiley and Sons.

Gipe, P. (2004). *Wind Power*. London: James & James.

Gleick, J. (2003). *Isaac Newton*. London: Harper Fourth Estate.

Golding, E. W. (1955). Reprinted 1976. *The Generation of Electricity by Wind Power*. London: E. & F. N. Spon.

Golding, E. W. (1961). *Windmills for Water Lifting and the Generation of Electricity on the Farm*. Informal Working Bulletin No. 17. Rome: Agricultural Engineering Branch, Food and Agriculture Organisation of the UN.

Golding, E. W. and Stodhart, A. H. (1949). *The Potentialities of Wind Power for Electricity Generation*. Technical Report W/T 16. London: The British Electrical and Allied Industries Research Association.

Gordon, J. E. (1978). *Structures, or Why Things Don't Fall Down*. London: Penguin Books.

Grainger, B. and Den Rooijen, H. (2001). Blyth offshore wind project. In *Wind Energy 2000*, ed. D. Still. London: Professional Engineering Publishing, pp. 75–86.

Gregory, R. (2005). *The Industrial Windmill in Britain*. Chichester: Phillimore.

Griffiths, J. A. (1895). Windmills for Raising Water, *Minutes of the Proceedings of the Institution of Civil Engineers*, **119**, 321–43.

Gross, R., Heptonstall, P., Leach, M., Anderson, D., Green, T. and Skea, J. (2007). Renewables and the grid: understanding intermittency. *Proceedings of the Institution of Civil Engineers, Energy*, **160**, no. 1, pp. 31–42.

Halliday, J. A. (1990). Wind resource – anemometry. In *Wind Energy Conversion Systems*, ed. L. L. Freris. London: Prentice Hall International, pp. 33–53.

Haldoupis, C. (1990). The wind regime in Lasithi Plateau. In *Wind Energy Conversion 1990*, eds. T. Davies, J. Halliday and J. Palutikof. London: Mechanical Engineering Publications, pp. 101–8.

Hannah, P. (1997). Windfarms of the UK – a 1997 update. In *Wind Energy Conversion 1997*, ed. R. Hunter. London: Mechanical Engineering Publications, pp. 17–22.

Harverson, M. (1991). *Persian Windmills*. Watford: The International Molinological Society.

Harverson, M. (2000). *Mills of the Muslim World*. The Fifth Rex Wailes Memorial Lecture. London: SPAB Mills Section.

Hassan, U. and Sykes, D. M. (1990). Wind structure and statistics. In *Wind Energy Conversion Systems*, ed. L. L. Freris. London: Prentice Hall International, pp. 11–32.

Hau, E. (2006). *Wind Turbines – Fundamentals, Technologies, Application, Economics*, 2nd edition. Berlin: Springer.

Hayes, G. (2003). *Beam Engines*. Princes Risborough: Shire Publications.

Hills, R. L. (1994). *Power from Wind: a History of Windmill Technology*. Cambridge: Cambridge University Press.

Hoerner, S. F. (1965). *Fluid-Dynamic Drag*. Brick Town, New Jersey: Hoerner Fluid Dynamics.

Holt, R. (1988). *The Mills of Medieval England*. Oxford: Blackwell.

House of Commons (1977). *The Development of Alternative Sources of Energy for the United Kingdom*. Select Committee on Science and Technology Third Report, Session 1976–77, HC534. London: HMSO.

House of Commons (1990). *The Cost of Nuclear Power*. Energy Committee Fourth Report, Session 1989–90, HC205. London: HMSO.

House of Commons (2006). *Keeping the Lights On: Nuclear, Renewables and Climate Change*. Environmental Audit Committee Sixth Report, Session 2005–06, vol. 1, HC 584–1, London: TSO. www.publications.parliament.uk/pa/cm200506/cmselect/cmenvaud/584/58402.htm.

House of Commons (2008). *Carbon Capture and Storage*. Environmental Audit Committee Ninth Report, Session 2007–86, HC 654, London: TSO. www.publications.parliament.uk/pa/cm200708/cmselect/cmenvaud/654/654.pdf.

Hütter, U. (1976). Review of development in West-Germany. In *Proceedings of the 1974 Stockholm Workshop on Advanced Wind Energy Systems*, vol. 1. Stockholm: STU/Vattenfall, pp. 1–51 to 1–72.

Ilex (2006). *The Balance of Power*. Godalming: WWF-UK. http://www.wwf.org.uk/filelibrary/pdf/ilex_report.pdf.

Inglis, D. R. (1978). *Wind Power and Other Energy Options*. Ann Arbor, Michigan: The University of Michigan Press.

Jacobs, M. L. (1964). Experience with Jacobs wind-driven electric generating plant, 1931–1957. In *Proceedings of the Conference on New Sources of Energy, Rome, 1961, vol. 7, Wind Power*. New York: United Nations, pp. 337–9.

Jamieson, P. and Hunter, C. (1985). Analysis of data from the Howden 300 kW wind turbine on Burgar Hill Orkney. In *Wind Energy Conversion 1985*, ed. A. D. Garrad. London: Mechanical Engineering Publications, pp. 253–8.

Jensen, M. L. and Bjerregaard, E. T. (1980). Tests performed on the 2 MW Tvind WECS. In *Proceedings of the Third International Symposium on Wind Energy Systems, Copenhagen, 1980*. Cranfield: BHRA Fluid Engineering, pp. 391–400.

Johnson, G. L. (1985). *Wind Energy Systems*. New Jersey: Prentice-Hall.

Jones, G. (2001). *The Millers, a Story of Technological Endeavour and Industrial Success, 1870–2001*. Lancaster: Carnegie Publishing.

Juul, J. (1956). Wind machines. In *Proceedings of the New Delhi Symposium on Wind and Solar Energy*. Paris: UNESCO, pp. 56–75.

Juul, J. (1964). Design of wind power plants in Denmark. In *Proceedings of the Conference on New Sources of Energy, Rome, 1961, vol. 7, Wind Power*. New York: United Nations, 229–40.

Kealey, E. (1987). *Harvesting the Air: Windmill Pioneers in Twelfth-century England*. University of California Press.

King, F. H. (1926). *Farmers of Forty Centuries*. London: Jonathan Cape.

Kovarik, T., Pipher, C. and Hurst, C. (1979). *Wind Energy*. Dorset: Prism Press.

Krohn, S. (1998). *The Wind Turbine Market in Denmark*. Copenhagen: Danish Wind Industry Association. www.windpower.org/media(487,1033)/the_wind_turbine_ market_in_denmark.pdf.

La Cour, P. (1901). Government experimental windmill at Askov. *Minutes of Proceedings of the Institute of Civil Engineers*, **145**, 387–9.

La Cour, P. (1905). *Die Windkraft und ihre Anwendung zum Antrieb von Elektrizitäts-Werken*. Leipzig: Nachfolger.

Landels, J. G. (1978). *Engineering in the Ancient World*. London: Chatto & Windus.

Langdon, J. (2004). *Mills in the Medieval Economy*. Oxford: Oxford University Press.

Lanoy, H. (1944). *Les Aéromoteurs Modernes*. Paris: Girardot.

Lawton, B. (2004). *Various and Ingenious Machines. Power and Transport, vol. 1*. Boston: Brill.

Leicester, R. J. (1989). The fair taxation of wind turbines, especially regarding local authority rating. In *Proceedings of the BWEA 1989 St. Andrews Workshop*, ed. J. Twidell. Glasgow: Energy Studies Unit, University of Strathclyde, pp. 1–5.

Legerton, M.L (1996). The recommendations of the noise working group. In *Wind Energy Conversion 1996*, ed. M. B. Anderson. London: Mechanical Engineering Publications, pp. 459–66.

Le Gouriérès, D. (1982). *Wind Power Plants*. Oxford: Pergamon Press.

Lewis, J. I. (2007). *A Comparison of Wind Power Industry Development Strategies in Spain, India and China*. www.resource-solutions.org/lib/librarypdfs/Lewis.Wind. Industry.Development.India.Spain.China.July.2007.pdf.

Lindley, D. and Musgrove, P. (1991). Status Report C: evaluation of the supply side of the market. In *Wind Energy in Europe*. Brussels: European Wind Energy Association.

Lindley, D., Musgrove, P., Armstrong, J. and Hitner, M. (1993). Early experience in UK wind farms. In *Proceedings of the 1993 European Community Wind Energy Conference, Travemunde*. Bedford: H.S. Stephens and Associates, pp. 759–62.

Lindley, D. and Stevenson, W. (1981). The horizontal axis wind turbine project on Orkney. In *Proceedings of the Third BWEA Wind Energy Conference*, ed. P. J. Musgrove. Cranfield: BHRA Fluid Engineering, pp. 16–32.

Lissaman, P. B. (1994). Wind turbine aerofoils and rotor wakes. In *Wind Turbine Technology*, ed. D. A. Spera. New York: American Society of Mechanical Engineers, pp. 283–321.

Lowe, J. E. and Wiesner, W. (1983). Status of Boeing wind-turbine systems. *IEE Proceedings Part A*, **130**, 531–36.

Lynette, R. (1988). Status of the US wind power industry. In *Wind Energy Conversion 1988*, ed. D. J. Milborrow. London: Mechanical Engineering Publications, pp. 15–24.

Lynette, R. and Gipe P. (1994). Commercial wind turbine systems and applications. In *Wind Turbine Technology*, ed. D. A. Spera. New York: American Society of Mechanical Engineers, pp. 139–214.

McGuigan, D. (1978). *Small Scale Wind Power*. Dorset: Prism Press.

MacKay, D. (2009). *Sustainable Energy – without the Hot Air*. Cambridge: UIT.

Manwell, J. F., McGowan, J. G. and Rogers, A. L. (2002). *Wind Energy Explained*. Chichester: John Wiley and Sons.

Marshall, Lord (1988). Conference opening speech to BWEA tenth annual conference. In *Wind Energy Conversion 1988*, ed. D. J. Milborrow. London: Mechanical Engineering Publications, p. xi.

Massy, J. (2006). Behind the British energy review. *Windpower Monthly*, **22** no. 10, 49–54, (October 2006).

Mathias, P. (1969). *The First Industrial Revolution*. London: Methuen.

Mays, I. D., Morgan, C. A., Anderson, M. B. and Powles, S. J. R. (1990). Experience with the VAWT 850 demonstration project. In *Proceedings of the 1990 European Wind Energy Conference, Madrid*. England: H. S. Stephens and Associates, pp. 482–7.

Mays, I. D., Musgrove, P. J., Morgan, C. A. and Hancock, G. (1988). The evolution of the straight-bladed vertical axis wind turbine. In *Wind Energy Conversion 1988*, ed. D. J. Milborrow. London: Mechanical Engineering Publications, pp. 187–94.

Meteorological Office (1976). *Maps of Hourly Mean Wind Speed over the United Kingdom 1965–73*. Climatological Memorandum 79. Bracknell: Meteorological Office.

Milborrow, D. (2006). Nuclear suddenly the competitor to beat. *Windpower Monthly*, **22** no. 1, 43–47, (January 2006).

Milborrow, D. (2007). Back to being a model of stability. *Windpower Monthly*, **23** no. 1, 47–50, (January 2007).

Milborrow, D. and Harrison, L. (2006). No limits to high wind penetration. *Windpower Monthly*, **22** no. 9, 51–6, (September 2006).

Milborrow, D. and Harrison, L. (2009). Wind rock solid as uncertainty reigns. *Windpower Monthly*, **25** no. 1, 51–5, (January 2009).

Morgan, C. A. and Mays, I. D. (1990). Progress with the VAWT 850 demonstration project. In *Wind Energy Conversion 1990*, eds. T. Davies, J. Halliday and J. Palutikof. London: Mechanical Engineering Publications, pp. 177–82.

Mosse, J. (1967). The Albion mills, 1784–1791. *Transactions of the Newcomen Society*, **40**, 47–60.

Musgrove, P. J. (1976a). Energy analysis of wave-power and wind-power systems. *Nature*, **262**, 206–7.

Musgrove, P. J. (1976b). Windmills change direction. *New Scientist*, **72**, 596–7.

Musgrove, P. J. (1977a). Wind energy systems and their potential in the UK. *Wind Engineering*, **1**, 235–40.

Musgrove, P. J. (1977b). The variable geometry vertical axis windmill. In *Proceedings of the International Symposium on Wind Energy Systems, Cambridge, 1976*. Cranfield: BHRA Fluid Engineering, pp. C7–87 to C7–100.

Musgrove, P. J. (1978a). Wind energy prospects in the UK. *Coal and Energy Quarterly*, **16**, 15–21.

Musgrove, P. J. (1978b). Offshore wind energy systems. *Physics Education*, **13**, 210–14.

Musgrove, P. J. (1978c). The prospects for wind energy in the UK. In *Proceedings of the July 1978 Symposium on Wind Power in the United Kingdom*. London: Multi-Science Publishing, pp. 80–90.

Musgrove, P. J. (1984). *Wind Energy Evaluation for the EC*. Report CDNA08996ENC. Luxembourg: Commission of the European Communities. http://bookshop.europa.eu/uri?target=EUB:NOTICE:CDNA08996:EN:HTML.

Musgrove, P. J. (1990a). Vertical axis WECS design. In *Wind Energy Conversion Systems*, ed. L. L. Freris. London: Prentice Hall International, pp. 295–306.

Musgrove, P. J. (1990b). Electricity privatisation and its implications for wind energy. In *Wind Energy Conversion 1990*, eds. T. Davies, J. Halliday and J. Palutikof. London: Mechanical Engineering Publications, pp. 1–4.

Musgrove, P. J. (1998). BWEA – Looking back and looking forward, 1978–1998–2018. In *Wind Energy 1998*, ed. S. J. Powles. London: Professional Engineering Publishing, pp. 19–24.

Needham, J. (1965). *Science and Civilisation in China, vol. 4, Physics and Physical Technology, Part 2, Mechanical Engineering*. Cambridge: Cambridge University Press.

Nickols, W. R. (1979). The design and construction of the Aldborough aerogenerator. In *Proceedings of the International Conference on Future Energy Concepts*. Conference Publication no. 171. London: Institution of Electrical Engineers, pp. 306–8.

Noakes, J., Oliver, A. and Morgan, C. (2000). The RES 52/1000. In *Wind Energy 1999*, ed. P. Hinson. London: Professional Engineering Publishing, pp. 199–205.

OPSI (2008). *Climate Change Act 2008*. London: Office of Public Sector Information. www.opsi.gov.uk/acts/acts2008/ukpga_20080027_en_1.

Oxera (2005). *What is the Potential for Commercially Viable Renewable Generation Technologies?* Report URN 05/1822. London: Department for Business, Enterprise and Regulatory Reform. www.berr.gov.uk/files/file21123.pdf.

Parthan, B. and Lemaire, X. (2007). *Wind Power in India: Behind the Success Story*. www.un.org/esa/sustdev/csd/csd15/lc/reep_windPower.pdf.

Pearce, D. L. (2002). Traditional windmill sails: design for power. *International Molinology*, **64**, 2–13.

Pedersen, B. M. and Nielsen, P. (1980). Description of the two Danish 630 kW wind turbines, Nibe-A and Nibe-B, and some preliminary test results. In *Proceedings of the Third International Symposium on Wind Energy Systems, Copenhagen, 1980*. Cranfield: BHRA Fluid Engineering, pp. 223–38.

Petersen, H. (1980). *The Test Plant for and a Survey of Small Danish Windmills*. Report Risø-M-2193. Roskilde, Denmark: Risø National Laboratory.

Powell, F. E. (1910). *Windmills and Wind Motors*. New York: Spon & Chamberlain. (Reprinted by Lindsay Publications, Illinois, 1985.)

Powles S. J. R., Anderson, M. B., Harris, A. and Groechel, K. (1988). Results from the UK 25 m vertical axis wind turbine programme. In *Wind Energy Conversion 1988*, ed. D. J. Milborrow. London: Mechanical Engineering Publications, pp. 47–56.

Price, T. J. (2005). James Blyth – Britain's first modern wind power pioneer. *Wind Engineering*, **29**, 191–200.

Prinsenmolen Committee (1958). *Research Inspired by Dutch Windmills*. Wageningen, Netherlands: H. Veenman en Zonen.

Putnam, P. C. (1948). *Power from the Wind*. New York: Van Nostrand Reinhold.

RAB (2008). *2020 VISION – How the UK can meet its target of 15% renewable energy*. London: Renewables Advisory Board. www.renewables-advisory-board.org.uk/vBulletin/showthread.php?t=136.

Ragwitz, M. and Huber, C. (2005). *Feed-in Systems in Germany and Spain, a Comparison*. Karlsruhe, Germany: Fraunhofer Institute. www.bmu.de/files/english/renewable_energy/downloads/application/pdf/langfassung_einspeisesysteme_en.pdf.

Rankine, W. J. M. (1866). *A Manual of the Steam Engine and Other Prime Movers*. London: Charles Griffin.

Rasmussen, B. and Øster, F. (1990). Power production from the wind. In *Wind Energy in Denmark 1990*. Copenhagen: Danish Energy Agency.

RCEP (2000). *Energy – The Changing Climate*. Royal Commission on Environmental Pollution, Twenty-second Report, Cm 4749. London: TSO. www.rcep.org.uk/newenergy.htm.

Rees, A. (1819). *Cyclopaedia; or Universal Dictionary of Arts, Sciences, and Literature*. London: Longman.

Reynolds, J. (1970). *Windmills and Watermills*. London: Hugh Evelyn.

Riegels, F. W. (1961). *Aerofoil Sections*. London: Butterworths.

Righter, R. W. (1995). *Wind Energy in America: a History*. Oklahoma: University of Oklahoma Press.

Risø (1989). *Catalogue of Danish Wind Turbines 1989*. Roskilde, Denmark: Risø National Laboratory.

Rockingham, A. P., Taylor R. H. and Walker, J. F. (1981). Offshore wind and wave power – a preliminary estimate of the resource. In *Proceedings of the Third BWEA Wind Energy Conference*, ed. P. J. Musgrove. Cranfield: BHRA Fluid Engineering, pp. 63–9.

Rogier, E. (1999). Les pionniers de l'électricité éolienne. *Systèmes Solaires*, **129**, 72–4.

Ryle, M. (1977). Economics of alternative energy sources. *Nature*, **267**, 111–17.

Savonius, S. J. (1931). S-rotor and its applications. *Mechanical Engineering*, **53**, 333–8.

Schmidt, W., ed. (1899). *Herons von Alexandria, vol. 1, Pneumatica et Automata*, Leipzig: Teubner.

SDC (2005). *Wind Power in the UK*. London: Sustainable Development Commission. www.sd-commission.org.uk/publications.php?id=234.

SDC (2006). *The Role of Nuclear Power in a Low Carbon Economy*. London: Sustainable Development Commission. www.sd-commission.org.uk/publications/downloads/SDC-NuclearPosition-2006.pdf.

Sektorov, V. R. (1934). Il primo impianto elettrico trifase aerodynamico a Balaklawa, *L'Elettrotecnica*, **23**, 538–42. Translated 1973 as NASA report TT F-14933. *The First Aerodynamic Three-phase Electric Power Plant in Balaklawa*. Washington, DC: National Aeronautics and Space Administration.

Sharpe, D. J. (1990). Wind turbine aerodynamics. In *Wind Energy Conversion Systems*, ed. L. L. Freris. London: Prentice Hall International, pp. 54–118.

Shearer, D. L. and Brown, A. (1986). The construction of a 26 MW wind park in the Altamont Pass region of California. In *Wind Energy Conversion 1986*, eds.

M. B. Anderson and S. J. Powles. London: Mechanical Engineering Publications, pp. 23–6.

Shepherd, D. G. (1994). Historical development of the windmill. In *Wind Turbine Technology*, ed. D. A. Spera. New York: American Society of Mechanical Engineers, pp. 1–46.

Simpson, P. B. and Lindley, D. (1980). Offshore siting of wind turbine generators in UK waters. In *Proceedings of the Second BWEA Wind Energy Workshop*, ed. P. J. Musgrove. London: Multi-Science Publishing, pp. 173–85.

Sinden, G. (2005). *Wind Power and the UK Wind Resource*. Oxford: Environmental Change Institute.

Skempton, A. W. (1981). *John Smeaton*. London: Thomas Telford.

Smeaton, J. (1760). An experimental enquiry concerning the natural powers of water and wind to turn mills and other machines, depending on a circular motion, part III, on the construction and effects of windmill sails. *Philosophical Transactions*, **51**, (for 1759), 138–74.

Smeaton, J. (1794). *Experimental Enquiry Concerning the Natural Powers of Wind and Water to Turn Mills and Other Machines Depending on a Circular Motion*. London: I and J. Taylor.

Smeaton, J. (1837). *Reports of the Late John Smeaton, F. R. S., Made on Various Occasions in the Course of his Employment as a Civil Engineer*. (First published 1797.) London: M. Taylor.

Smeaton, J. (1938). *Diary of His Journey to the Low Countries 1755*, ed. A. Titley from the original manuscript. London: Newcomen Society.

Smil, V. (1994). *Energy in World History*. Oxford: Westview Press.

Smil, V. (2003). *Energy at the Crossroads*. Cambridge, Massachusetts: MIT Press.

Smith, D. (1981). Mills and millwork. In *John Smeaton*, ed. A. W. Skempton. London: Thomas Telford, pp. 59–81.

Smith, N. (1981). Scientific work. In *John Smeaton*, ed. A. W. Skempton. London: Thomas Telford, pp. 35–57.

Socolow, R. H. (2005). Can we bury global warming? *Scientific American*, **293** no. 1, July 2005, 39–45.

Somerville, W. M. (2000). Fair Isle renewed. In *Wind Energy 1999*, ed. P. Hinson. London: Professional Engineering Publishing, pp. 263–80.

Spera, D. A. (1994). *Wind Turbine Technology*. New York: American Society of Mechanical Engineers.

Stern, N. (2007). *The Economics of Climate Change*. Cambridge: Cambridge University Press. www.sternreview.org.uk.

Stevenson, W. G. and Somerville, W. M. (1983). The Fair Isle wind power system. In *Wind Energy Conversion 1983*, ed. P. J. Musgrove. Cambridge: Cambridge University Press, pp. 171–84.

Stiesdal, H. (1999). Bonus' latest developments and commercial turbines. In *Proceedings of the 1999 European Wind Energy Conference, Nice*. London: James & James, pp. 465–8.

Stiesdal, H. and Kruse, H. (2000). The growth in turbine rating – a case study. In *Wind Energy 1999*, ed. P. Hinson. London: Professional Engineering Publishing, pp. 207–12.

Stoddard, W. (2002). The Life and Work of Bill Heronemus. *Wind Engineering*, **26**, 335–41.

Stodhart, A. H. (1976). Review of the UK wind power programme 1948–1960. In *Proceedings of the 1974 Stockholm Workshop on Advanced Wind Energy Systems*, vol. 1. Stockholm: STU/Vattenfall, pp. 1–23 to 1–34.

Stokhuyzen, F. (1962). *The Dutch Windmill*. Bussum, Holland: C.A.J. van Dishoeck. www. nt.ntnu.no/users/haugwarb/DropBox/The%20Dutch%20Windmill%20Stokhuyzen %201962.htm.

Stradanus, J. (1584). *Nova Reperta*. European cultural heritage online. http://libcoll. mpiwg-berlin.mpg.de/libview?mode=imagepath&url=/mpiwg/online/permanent/ einstein_ exhibition/sources/PZ39PA1P/pageimg.

Taylor, R. H. (1983). *Alternative Energy Sources*. Bristol: Adam Hilger.

Taylor, R. H. and Rockingham A. P. (1980) A comparison of studies of WECS economics for utility applications. In *Proceedings of the Second BWEA Wind Energy Workshop*, ed. P. J. Musgrove. London: Multi-Science Publishing, pp. 118–26.

Thompson, S. P. (1910). *The Life of William Thomson, Baron Kelvin of Largs*. London: MacMillan & Co.

Trinick, M. (2001). Consensus for offshore wind energy developments. In *Wind Energy 2000*, ed. D. Still. London: Professional Engineering Publishing, pp. 71–3.

Twidell, J., Gauld, R., Fern, D. and Burton, A. (1995). Recommissioning and operating the LS1 3 MW, Burgar Hill, Orkney wind turbine using novel crack repair. In *Wind Energy Conversion 1995*, ed. J. Halliday, London: Mechanical Engineering Publications, pp. 279–84.

UKERC (2006). *The Costs and Impacts of Intermittency*. London: UK Energy Research Centre. www.ukerc.ac.uk/ResearchProgrammes/TechnologyandPolicyAssessment/ TPAProjectIntermittency.aspx.

USDOE (2008). *20% Wind Energy by 2030*. Report DOE/GO-102008–2567. US Department of Energy. www.nrel.gov/docs/fy08osti/41869.pdf.

Vas, I. E. and South, P. (1980). A Review of the SERI Wind Energy Innovative Systems Program. In *Proceedings of the Third International Symposium on Wind Energy Systems, Copenhagen, 1980*. Cranfield: BHRA Fluid Engineering, pp. 61–74.

Vindmølleindustrien (1997). *Wind Power* Note no. 10. Copenhagen: Danish Wind Turbine Manufacturers Association.

Vowles, H. P. (1930). An inquiry into origins of the windmill. *Transactions of the Newcomen Society*, **11**, 1–14.

Wailes, R. (1954). *The English Windmill*. London: Routledge & Keegan.

Wailes, R. (1957). Windmills. In *A History of Technology*, vol. 3, ed. C. Singer, Oxford: Clarendon Press, pp. 89–109.

Walker, J. F. and Jenkins, N. (1997). *Wind Energy Technology*. Chichester: John Wiley and Sons.

Walton, J. (1974). *Water-mills Windmills and Horse-mills of South Africa*. Cape Town: C. Struik.

Warne, D. F and Calnan, P. G. (1977). Generation of electricity from the wind. *Proceedings of the Institution of Electrical Engineers*, **124** no. 11R, 963–85.

Watts, M. (1998). *Corn Milling*. Princes Risborough: Shire Publications.

Watts, M. (2004). Milling and millwrighting. In *Crafts in the English Countryside*, ed. E. J. Collins. London: Countryside Agency, pp. 188–210. www.craftsintheenglish-countryside.org.uk/full.html.

Watts, M. (2006). *Windmills*. Princes Risborough: Shire Publications.

Weart, S. R. (2003). *The Discovery of Global Warming.* Cambridge, Massachusetts: Harvard University Press. www.aip.org/history/climate.

Webb, W. P. (1931). *The Great Plains.* Boston: Ginn.

Weisbach, J. and Du Bois, A. (1886). *A Manual of the Mechanics of Engineering and of the Construction of Machines,* vol. 2. New York: John Wiley and Sons.

White, L. (1962). *Medieval Technology and Social Change.* Oxford: Clarendon Press.

Whitt, F. R. and Wilson, D. G. (1982). *Bicycling Science.* Cambridge, Massachusetts: MIT Press.

Wilson, R. E. (1994). Aerodynamic behaviour of wind turbines. In *Wind Turbine Technology,* ed. D. A. Spera. New York: American Society of Mechanical Engineers, pp. 215–82.

Wiser, R., Bolinger, M. and Barbose, G. (2007). *Using the Federal Production Tax Credit to Build a Durable Market for Wind Power in the United States.* Report LBNL-63583. Berkeley, California: Lawrence Berkeley National Laboratory. http://eetd.lbl.gov/EA/EMS/reports/63583.pdf.

Wiser, R. and Bolinger, M. (2008). *Annual Report on U.S. Wind Power Installation, Cost, and Performance Trends: 2007.* Report DOE/GO-102008–2890. U.S. Department of Energy. www1.eere.energy.gov/windandhydro/pdfs/43025.pdf.

WMO (1981). *Meteorological Aspects of the Utilisation of Wind as an Energy Source.* Technical Note 175. Geneva: World Meteorological Organisation.

Wood, E. S. (1995). *Historical Britain.* London: Harvill Press.

Wolff, A. R. (1885). *The Windmill as a Prime Mover.* New York: John Wiley and Sons.

Wollmerath, H. (1992). National wind energy programmes; Federal Republic of Germany. In *Wind Energy Conversion 1992,* ed. B. R. Clayton. London: Mechanical Engineering Publications, pp. 7–13.

Wulff, H. E. (1966). *The Traditional Crafts of Persia, their Development, Technology, and Influence on Eastern and Western Civilisation.* Cambridge, Massachusetts: MIT Press.

Yeomans, M. (2005). *Oil.* New York: The New Press.

Yergin, D. (1991). *The Prize.* London: Simon and Schuster.

Yorke, S. (2006). *Windmills and Watermills Explained.* Newbury: Countryside Books.

Young T. and McLeish, D. (1983). Construction, commissioning and operation of the 300 kW wind turbine at Carmarthen Bay. In *Wind Energy Conversion 1983,* ed. P. J. Musgrove. Cambridge: Cambridge University Press, pp. 296–302.

Index